Agriculture Waste Management and Bioresource

Agriculture Waste Management and Bioresource

The Circular Economy Perspective

Edited by

Suruchi Singh
Sunbeam College for Women, Varanasi, India

Pardeep Singh
Department of Environmental Studies, PGDAV College,
University of Delhi, New Delhi, India

Anu Sharma
Department of Environmental Sciences,
University of Jammu, India

Moharana Choudhury
Environmental Research and Management Division,
Voice of Environment (VoE) Guwahati,
Assam, India

WILEY

This edition first published 2023

Registered Offices
John Wiley & Sons, Inc., 111 River Street, Hoboken, NJ 07030, USA
John Wiley & Sons Ltd, The Atrium, Southern Gate, Chichester, West Sussex, PO19 8SQ, UK

Editorial Office
The Atrium, Southern Gate, Chichester, West Sussex, PO19 8SQ, UK

For details of our global editorial offices, customer services, and more information about Wiley products visit us at www.wiley.com.

Wiley also publishes its books in a variety of electronic formats and by print-on-demand. Some content that appears in standard print versions of this book may not be available in other formats.

Library of Congress Cataloging-in-Publication Data has been applied for

9781119808138 (Hardback): 9781119808336 (Adobe PDF): 9781119808350 (e-Pub)

Cover Design: Wiley
Cover Image: © nieriss/Shutterstock

Set in 9.5/12.5pt STIXTwoText by Straive, Pondicherry, India
Printed and bound by CPI Group (UK) Ltd, Croydon, CR0 4YY

C9781119808138_301122

Contents

List of Contributors

Sangita Agarwal
Department of Applied Science
RCC Institute of Information Technology
Kolkata, West Bengal, India

Md. Abdullah Aziz
Agricultural Statistics Division
Bangladesh Rice Research Institute (BRRI)
Gazipur, Bangladesh

Hangma Boro
Department of Botany
Rangapara College
Rangapara, Assam, India

Ratan Chowdhury
Department of Botany
Rangapara College
Rangapara, Assam, India

Papori Dahutia
Department of Physics
Dibrugarh University
Dibrugarh, Assam, India

Soumendra Darbar
Faculty Council of Science
Jadavpur University
Kolkata, West Bengal
India; Research and Development Division
Dey's Medical Stores (Mfg.) Ltd., Kolkata
West Bengal, India

Tathagata Deb
Department of Applied Science
RCC Institute of Information Technology
Kolkata, West Bengal, India

Sukhendu Dey
Department of Environmental Science
The University of Burdwan
Burdwan, West Bengal, India

Joystu Dutta
Department of Environmental Science
University Teaching Department
Sant Gahira Guru University
Ambikapur, Chhattisgarh, India

Joysurya Dutta
Symbiosis Institute of Operations
Management (SIOM), Nashik, Maharashtra

Apurba Ratan Ghosh
Department of Environmental Science
The University of Burdwan
Burdwan, West Bengal, India

Navamallika Gogoi
Department of Chemistry
Arunachal University of Studies
Namsai, Arunachal Pradesh, India

Nirmali Gogoi
Department of Environmental Science
Tezpur University
Tezpur, Assam, India

Anwesha Gohain
Department of Botany
Arunachal University of Studies
Namsai, Arunachal Pradesh, India

Avinash Pratap Gupta
Department of Environmental Science
Sant Gahira Guru University
Sarguja, Chhattisgarh, India

Ravindra Pratap Singh
Central Public Works Department
New Delhi, India

Sumi Handique
Department of Environmental Science
Tezpur University
Tezpur, Assam, India

R.Y. Hiranmai
School of Environment and Sustainable Development
Central University of Gujarat
Gandhinagar, Gujarat, India

Md. Shahadat Hossain
Training Division
Bangladesh Rice Research Institute (BRRI)
Gazipur, Bangladesh

Asha Humbal
School of Environment and Sustainable Development
Central University of Gujarat
Gandhinagar, Gujarat, India

Rani Jha
Department of Chemistry
Arunachal University of Studies
Namsai, Arunachal Pradesh, India

Sanjib Kalita
Department of Botany
Gauhati University
Guwahati, Assam, India

Mohammad Ashik Iqbal Khan
Plant Pathology Division
Bangladesh Rice Research Institute (BRRI)
Gazipur, Bangladesh

Vinod Kumar
Algal Research and Bioenergy Lab
Department of Chemistry
Uttaranchal University
Dehradun, Uttarakhand, India;
Department of Environmental Monitoring and Forecasting
Ecological Faculty
RUDN University
Moscow, Russia

A.K. Mall
Division of Crop Improvement
ICAR-Indian Institute of Sugarcane Research
Lucknow, Utter Pradesh, India

Varucha Misra
Division of Crop Improvement
ICAR-Indian Institute of Sugarcane Research
Lucknow, Utter Pradesh, India

Manisha Nanda
Department of Biotechnology
Dolphin (P.G.) Institute of Biomedical and Natural Sciences
Dehradun, Uttarakhand, India

Ajay Neearaj
School of Environment and Sustainable Development
Central University of Gujarat
Gandhinagar, Gujarat, India

Amrita Kumari Panda
Department of Biotechnology
Sant Gahira Guru Vishwavidyalaya
Sarguja, Ambikapur, Chhattisgarh, India

A.D. Pathak
Division of Crop Improvement
ICAR-Indian Institute of Sugarcane Research
Lucknow, Utter Pradesh, India

Bhawana Pathak
School of Environment and Sustainable Development
Central University of Gujarat
Gandhinagar, Gujarat, India

Md. Khairul Quais
Rice Farming Systems Division
Bangladesh Rice Research Institute (BRRI)
Gazipur, Bangladesh

Farhana Rahman
University of Technology Sydney (UTS)
Sydney, NSW, Australia

Vishal Rajput
Himalayan School of Bio-sciences
Swami Rama Himalayan University
Dehradun, Uttarakhand, India

Priyabrata Roy
Department of Molecular Biology and Biotechnology
University of Kalyani
Kalyani, West Bengal
India

Subham Roy
Department of Botany
Rangapara College
Rangapara, Assam, India

Ankita Saha
Department of Environmental Science
Tezpur University
Tezpur, Assam, India

Srimoyee Saha
Department of Life Science and Biotechnology
Jadavpur University
Kolkata, West Bengal, India

Kundil Kumar Saikia
Department of Environmental Science
Tezpur University
Tezpur, Assam, India

Palas Samanta
Department of Environmental Science
Sukanta Mahavidyalaya
University of North Bengal
Dhupguri, West Bengal, India

Santeshwari Srivastava
Division of Crop Improvement
ICAR-Indian Institute of Sugarcane Research
Lucknow, Utter Pradesh, India

Biman Kumar Sarma
Department of Zoology, Rangapara College
Rangapara, Assam, India

Payal Sen
Independent Researcher
Chembur East
Mumbai, Maharashtra, India

Tirthankar Sen
Department of Biosciences and Bioengineering
Indian Institute of Technology Guwahati
Guwahati, Assam, India

Tarakeshwar Senapati
Department of Environmental Science
Directorate of Distance Education
Vidyasagar University
Midnapore, West Bengal, India

Anu Sharma
Govt. SPMR College of Commerce
Constituent College of Cluster University of Jammu
Jammu, Jammu and Kashmir (UT), India

Pradeep Kumar Tiwari
Independent Researcher
New Housing Board Colony
Shahdol, Madhya Pradesh, India

Prachi Upadhyay
Independent Researcher
Forest Research Institute (Deemed-to-be) University
Dehradun, Uttarakhand, India

Monu Verma
Department of Environmental Engineering
The University of Seoul
Seoul, Republic of Korea

M.S. Vlaskin
Laboratory for Energy Accumulating Materials
Joint Institute for High Temperatures of the Russian Academy of Sciences
Moscow, Russia

1

Agricultural Waste as a Resource: The Lesser Travelled Road to Sustainability

Avinash Pratap Gupta[1], *Prachi Upadhyay*[2], *Tirthankar Sen*[3], *and Joystu Dutta*[1]

[1] *Department of Environmental Science, Sant Gahira Guru University, Sarguja, Chhattisgarh, India*
[2] *Forest Ecology and Environment Management Division, Forest Research Institute (Deemed-to-be) University, Dehradun, Uttarakhand, India*
[3] *Department of Biosciences and Bioengineering, Indian Institute of Technology Guwahati, Guwahati, Assam, India*

CHAPTER OUTLINE

1.1 Introduction

With the expansion of agricultural activities, the amount and type of wastes being produced from them have increased. As per the Organization for Economic Cooperation and Development and the Food and Agriculture Organization (OECD-FAO), 39.35 million tons of natural fibres are annually being obtained from plants by the farmers, which also contribute to waste production. China reportedly produces 56.2 million tons of

Agriculture Waste Management and Bioresource: The Circular Economy Perspective, First Edition.
Edited by Suruchi Singh, Pardeep Singh, Anu Sharma, and Moharana Choudhury.

agricultural wastes. The types of wastes and their amount are dependent on the country and their management (Nagalakshmaiah et al. 2016; Letcher and Vallero 2019; Kamel et al. 2020). Agricultural wastes are basically the outputs that are not utilized in products or residues produced from the harvest/rearing and processing of raw agricultural products. Although many agricultural wastes contain substances of human value or benefit, it is mostly the case that the economic value of these processed substances is lower than the cost of collection, transportation, and processing of the feedstock. Therefore, wastes are generated due to the economic infeasibility of processing or reusing discarded agricultural by-products.

Agricultural wastes may be in the many forms like solids, slurries, or liquids and can have both plant and animal origins. For example, agricultural wastes encompass

- animal wastes such as manure, carcasses and bones;
- food processing wastes such as corn stalks, fruit peels cores pits pulp stems and seeds, twigs, spoiled fruits, sugarcane bagasse, and molasses;
- crop discards such as corn stalks, fruit/vegetable drops, culls, and pruning;
- toxic agricultural wastes such as pesticides, herbicides, herbicides, and antifungals.

The composition of agricultural wastes can be different from location to location and is dependent on the particulars of the point of origin like the agricultural activities carried out and the regional specialization. Interestingly, the challenges in waste management as well as the opportunities faced by a region because of the primary residues generated are also related to the regional specialization. For example, in locations specifically used for animal husbandry, vast quantities of organic matter rich in nitrogen and nutrient-rich waste materials are generated alongside repugnant odours, microbial contamination, and greenhouse gas (methane) emissions. Manure production is therefore among the most feasible waste management routes for these regions in comparison with the regions devoted to crop production where nutrients and the organic matter deplete because of the nutritional demands exerted by the growing crops on the soil (Gontard et al. 2018).

During the past several decades, significant scientific and technological advancements have resulted in a gradual increase of agricultural yield and productivity throughout the world. This, however, has also resulted in a steady increase in the quantity of waste from livestock, agricultural crop residues, and by-products of agro-industries. Although highly accurate estimates of agricultural waste production are rare, it is believed that agricultural waste contributes significantly to the global total waste production. Over a decade ago, the world was estimated to produce nearly 1000 million tons of waste from agriculture (Agamuthu 2009). Organic wastes have been documented to account for up to 80% of the total solid waste being produced in a farm (Brown and Root Environmental Consultancy Group 1997). In 2012, organic wastes accounted for more than half of the fresh waste of the harvested crops representing a potential of 90 Million Tons Oil Equivalent (MTOE) which was far higher than the energy potential of other wastes such as round wood, municipal wastes, and tertiary forest residues (Elbersen et al. 2012). This scenario is likely to worsen if proper interventions are not put in place.

In this modern age of industrialization and globalization, economic progress is heavily intertwined with non-renewable resource utilization which adversely impacts environmental balance and biodiversity, jeopardizes global food security, and depletes our planet's non-renewable energy reserves (UNEP 2011). Agricultural waste can be transformed into economically valuable products, such as fertilizers, energy, and chemicals, by using specialized conversion processes. This is of critical importance so as to promote a sustainable bioeconomy and consequently decouple non-renewable resource utilization and economic progress from human well-being and environmental sustainability.

As per Indian Ministry of New and Renewable Energy (MNRE), on average, India generates half a billion metric tons (Mt) of crop residue every year. Although a significant fraction of this crop residue is used as fodder, fuel and for other domestic and industrial purposes, there still is an unutilized surplus of 140 Mt each year and 92 Mt of that are incinerated (NPMCR 2019). More agricultural wastes are generated in India than Bangladesh, Indonesia, and Myanmar combined. In fact, the agricultural waste volume that is incinerated in India is more than the net volume of agricultural waste produced by other countries in the region (Bhuvaneshwari et al. 2019).

In this review, we aim at constructing a picture of the state of agricultural waste and its generation. Agricultural wastes, their different types, and their characteristics have also been discussed. Apart from that, the key challenges and most promising opportunities in developing smart, sustainable, and efficient agricultural waste conversion mechanisms and residue management strategies are described.

1.2 Sources of Agricultural Waste Generation

Waste produced from agriculture can be broadly divided into two groups based on their composition namely:

- Lignocellulosic waste
- Proteinaceous waste

1.2.1 Lignocellulosic Waste

Lignocellulosic biomass is about 50% of the world's total biomass. Its annual production is around 10–50 billion tons (Mood et al. 2013; Kumar et al. 2018). It is the main constituent of the cell walls in plants and contains cellulose (40–60%), hemicellulose (20–40%), and lignin (10–25%) (Dionisi et al. 2015). The plant residues produced in agricultural field also come under this and are known as agricultural residues.

They can be grouped into two categories: field residues and process residues. Agricultural wastes which remain in the field after the process of crop harvesting has been concluded are referred to as field residues. Field residues function as feedstock for a number of processes designed to transform agricultural waste into economically valuable products such as animal feed, organic manure, raw materials for different manufacturing industries, and even various biochemicals of clinical and industrial significance. However, these processes also produce by-products which are known as process residues. Field residues and process residues can include a variety of waste products like molasses, husks, bagasse, seeds and seed pods, leaves, stems, straws, stalks, shells, and pulp. India being an agrarian nation produces nearly 686 million tons of lignocellulosic waste and 234 million tons of that are agricultural residues (Kapoor et al. 2020). But because of the current inefficiencies in plaguing agricultural waste management processes, a significant portion of the agricultural residues produced in India is underutilized. This also presents an opportunity for future initiatives aimed at processing agricultural waste and turning waste into wealth.

1.2.2 Proteinaceous Waste

Proteins are known part of our daily nutrition and biochemical function. They also are promising sources of raw material for polymers. The protein obtained from agricultural waste has a huge potential for various utilities. The waste produced from aquaculture and livestock are abundant in protein and can be utilized to derive novel ways to produce value-added products (Barone Justin and Schmidt Walter 2006). So, the proteinaceous waste can be from two major sources: (i) aquaculture and (ii) livestock.

1.2.2.1 Waste Produced from Aquaculture

The expansion of aquaculture has resulted in an increased feeding of feed for improved aquaculture productivity. In a proper aquaculture installation, approximately 30% of the feed fed is converted into solid waste. In aquaculture, the most prominent wastes generated are the metabolic wastes. It could be in soluble or suspended form. Since feeding rates are ambient temperature-dependent, rise in temperature leads to increased feeding. This consequently results in increase in volume of waste being generated. Optimized patterns of water flow in the production units are critical considerations for waste management as proper water flow minimizes fish faecal matter fragmentation and enables the rapid sedimentation of insoluble solid waste. Furthermore, if the non-fragmented waste like faeces is higher in amount, it becomes easier to capture them, and this greatly reduces the concentration of dissolved organic waste in the system (Mathieu and Timmons 1995; Obi et al. 2016).

1.2.2.2 Waste Produced from Livestock

Waste generated by livestock activities is produced from different sources. Slaughterhouses produce organic materials and solid wastes like manure, whereas animal's farms produce high amount of animal faeces and fodder waste. Urine, water produced during cage washing, water from bathing animals, and water from cleaning of slaughterhouses are part of the wastewater produced from livestock activities. Also, air pollutants like H_2S and CH_4 and odour are produced here. These wastes if not managed well contribute to greenhouse gas production and affect the soil and water quality. These also act as a source of bad odour and site for parasites and germs making them a threat for human health and environment (Obi et al. 2016). Poultry industry is also expanding with increase in consumption of poultry products. The feather left out from the birds are contributing significantly to the waste. This waste is mainly composed of protein (90%), specifically keratin which is insoluble and is hard to degrade naturally. Hence, better management is needed to utilize these wastes (Jana et al. 2020).

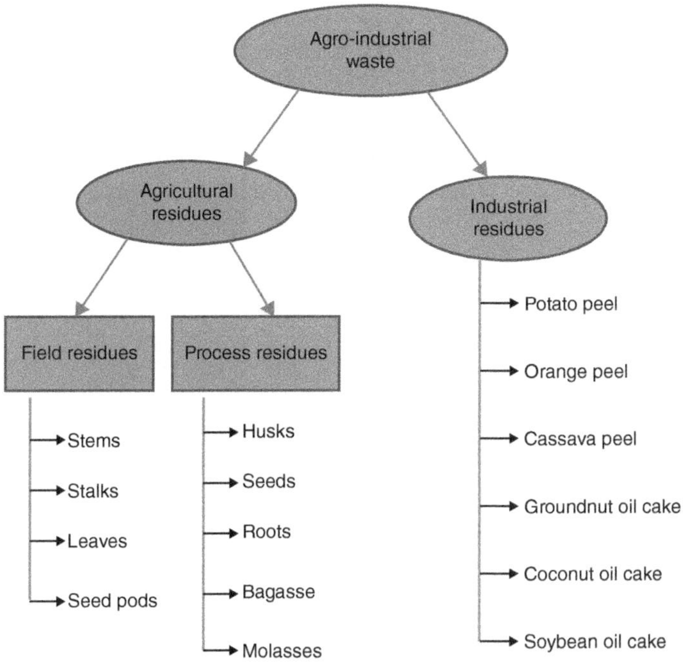

Figure 1.1 Agro-industrial waste materials. *Source:* Sadh et al., 2018 / Public Domain / CC BY 4.0.

Figure 1.1 shows the different types of Agro-industrial waste of their potential sources.

Table 1.1 Chemical composition of residues.

Name	Carbon (%)	Nitrogen (%)	Hydrogen (%)
Wheat bran	49.81	0.7	6.11
Rice bran	48.39	0.89	5.43
Oat bran	40.62	0.31	4.31
Corn cob	53.61	1.91	8.97
Sugar cane bagasse	48.32	0.2	7.84
Redgram husk	42.23	ND	5.52
Green gram husk	43.17	1.86	ND
Bengal gram husk	42.45	0.75	5.82
Black gram husk	39.61	2.34	5.87
Ground nut oil cake	40.34	2.31	4.21
Spent coffee	34.51	0.36	3.95
Spent tea	32.61	0.31	ND
Pineapple waste	45.68	0.61	3.97
Palmoil fiber	41.4	1.67	5.44
Coconut oil cake	48.16	1.69	5.15

ND not detectable.
Source: Thadikamala and Reddy Shetty (2013) / Springer Nature.

1.3 Agricultural Waste Characterization

Knowing the characteristics of agricultural waste is important for formulating an effective waste management system. Wastes produced from agriculture vary in both their quality and quantity. Wastes produced by the processing of crops and from livestock need both solid and liquid waste management. Knowing the characteristics of waste helps in reusing the wastc, recovery of useful components, production of useful products like biofuels, biofertilizers, and overall management (Loehr 2012). Table 1.1 shows the chemical composition of waste having different resources. The wastes can be categorized on the basis of their sources as given in the following sections:

1.3.1 Food Processing

The variation in wastes produced during food processing is due to the type of product, methods used to harvest the raw materials, and the processing methods used. For example, wastes generated from fruit and vegetables are often acidic and contain good amount of organic matter. Also, wastes from fruit and vegetable processing from operations like peeling and blanching have high percentage of total dissolved organic solids. The analysis of edible part of vegetables has shown the presence of water (74–94% of total weight), carbohydrate (3.2–19.1% of total weight), protein (1.1–6.7% of total weight), and fat (0.1–0.5%). Also processes like washing contribute to production of wastewater (Loehr 2012). Apart from this, crop residues left after harvest are also the major waste produced and are of high concern, especially in countries like India. They comprise cellulose, hemicellulose, lignin, nitrogenous material, chlorophyll, and other inorganic material (Bhuvaneshwari et al. 2019). Food wastes also consist of natural as well as anthropogenic toxins such as solanine and dioxin which needs to be excluded to make the quality control systems more efficient. Also, the less use of potentially hazardous products in food processing will lead to the generation of less toxic waste making its management easy (Pandey and Dwivedi 2020).

1.3.2 Meat and Poultry Processing Industry

The main waste produced in this is from killing, dehairing, removal of hide, processing, and other clean-up operations. It contains blood, inorganic and organic solids, salts, and chemical used during processing. Usually at a beef slaughterhouse, 50-lb blood, 50-lb paunch manure, and 40-lb animal manure per animal can be produced. At poultry processing farms, waste is generated by killing, scalding, defeaturing, evisceration, washing, chilling, and clean-up activities. Quantity and quality of waste generated totally depend on how processes are being handled. With the expansion of poultry industry, water usage, biological oxygen demand (BOD), and solid waste produced have also increased. Now, water used for each bird is about 12–15 gal, BOD produced is 60 lb/1000 birds, and solid waste produced is about 47 lb/1000 birds. These industries are not much affected by seasonal variations, and with expansion of these industries, the waste produced is also of high concern (Loehr 2012).

1.3.3 Livestock Waste

The characteristics of animal waste may be defined on the basis of changes in environment and overall productivity of the animals. The volatile matter present in the waste is found by summing up the percent of protein, fat, fibre, and non-fibre found in feed after digestion.

The fibre and non-fibre extracts are measured by analysing the lignin, cellulose, and hemicellulose contents. Wastes generated from grass-fed animals are found to be less biodegradable compared to concentrate-fed animals. The characteristics of the waste depend on the digestibility of feed and feed composition. The faeces produced mainly have undigested food which was not broken down by the bacteria. Undigested proteins are the part of faeces, and undigested excess nitrogen is removed through urine. Livestock waste can be solid, liquid, or semi-solid. So, depending on the waste type, the process to handle waste should be formulated with the traditionally used methods. It is necessary to device newer methods and proper training of individuals for waste handling (Loehr 2012).

1.4 Waste to Wealth: Agricultural Waste Management Approaches

Among the existing renewable resources, biomass is seen to be found abundantly. It can be used as an energy source or can be transformed into other usable forms. In the present global scenario, impending energy shortages are an area of immense concern. Using biomass as a source of energy is a probable solution for the energy shortage. Furthermore, biomass utility is not only limited for energy generation, but it can also be utilized in making a variety of commercial products. A lot of work has been carried out to make the utilization of biomass efficient, and the major ones are described in the further sections and subsections.

1.4.1 Natural Fibres

With increasing plastic pollution and poor recycling rate (only 9% of plastic waste is recycled), the need to develop bio-based polymers has aroused. Two main strategies have been taken up for this: first by deriving biodegradable polymers from bio-based products like agriculture waste and second by using non-biodegradable polymers and modifying their properties accordingly. Both the strategies are effective, if the products can be successfully marketed and commercialized (Maraveas 2020).

In another study by Kuan et al. (2011), the physiochemical properties of the residues from alkali-treated raw cereal materials like wheat straw and corn cob were studied. They produce good amount of fibrous wastes and contain high dietary fibres (49.87–68.65%) as well. In contrast to the soluble fibrous waste, the cereal materials and the left insoluble fibrous residues have essential minerals, the ability to scavenge 2,2-diphenyl-1-picrylhydrazyl radical, good emulsification and the stability of emulsion, mineral binding capacity, and good water and oil holding capacity. Also, the production of nanofibres using silk fibroin (SF) by the process of electrospinning is successfully done. Zhang et al. (2012) studied various properties of fresh coir fibre and reported absorption coefficient to be 0.8 at 20 mm thickness and frequency more than 1360 Hz. They also found that an increase in thickness helps in improved low-frequency sound absorption.

There have been a lot of studies analysing the scope of different agricultural wastes like grape and tomato pomace, green tea extracts, coconut shells, vegetable wastes, rice husks, fruit peels, grapefruit seed extract, and waste vegetables in the production of biopolymers. But the key issue faced is collecting the waste as the mechanism of collection and sorting of waste is not effective enough to acquire the waste completely. Also, the variability of the waste globally demands variation in the standard production protocols. But, bio-based products from waste need to be made commercially for the overall global sustainability (Maraveas 2020).

1.4.2 Agriculture

In 2016, the agricultural waste was analysed as a stabilizer of soil, and it was seen that these wastes were not disposed off properly in India. So, ashes produced from these materials at different concentrations (3, 6, 9, 12, and 15%) were taken, and California bearing ratio (CBR) and standard proctor tests were carried out. The parameters studied were specific gravity (2.662), liquid limit (66), plastic limit (26.62), plasticity index (39.99), free well index (23.08), optimum moister content (26.11), and maximum dry density (1.445). It is seen that during milling paddy, 78% of the total weight is because of broken rice grains and bran and rest is contributed by husk. This husk is used as a source to generate steam in parboiling process in the mills. The husk is composed of organic volatile matter (75%) and rest contributes to ash production during firing process which is also known as rice husk (RHA). RHA has about 85–90% of amorphous silica (Maiz et al. 1997; Harshwardhan and Upadhyay 2017).

In another study by Jansirani et al. (2012), production of vermicast and its utilization using organic waste were analysed. Vermicasts produced by the decomposition of tea waste, coir of coconut, and sugarcane bagasse were the main parameters to be studied. So, there is scope for using the agricultural waste for betterment of agricultural produce only. It could be used for energy generation, as compost, as feed for livestock, and for many things (Duque-Acevedo et al. 2020).

1.4.3 Bioenergy Production

Biofuels are fuels that are produced from biomass instead of being purified from the planet's underground reserves of fossil fuels such as crude oil. Bioethanol is produced via the alcoholic fermentation of glucose derived from sustainable and renewable sources like agricultural wastes and food discards. They have a comparatively lesser negative impact on the environment than conventional routes of industrial-scale ethanol production. Bioethanol is chemically undisguisable from ethanol except that they have different substrates and carbon footprints. Bioethanol is a type of biofuel. By the year 2023, ethanol is expected to be responsible for around 67% of the global growth in conventional biofuels, while the remaining 33% is expected to be attributed to biofuels such as biodiesel and hydrotreated vegetable (OECD-FAO 2018). In 2011, bioethanol production increased by 85 billion litres all across the world (Avci et al. 2013; Saini et al. 2014). By the end of 2024, the global bioethanol production and consumption will increase to about 134.5 billion litres (Bušić et al. 2018). Several studies have explored agro-industrial residues like rice straw, potato, sweet potato and beet discards, corn stalks, and sugarcane bagasse (Duhan et al. 2013; Kumar et al. 2014; Kumar et al. 2016) for the production of biofuel. Faraco and Hadar (2011) in their study found that residues obtained from cereal crops, olive tree, tomato, and grape were rich in waste with lignin and cellulose. They also have the potential to be utilized as raw material for producing ethanol at large scale.

Apart from ethanol, butanol is also produced from lignocellulosic biomass. Butanol is not only more energy rich than ethanol, but it can also be mixed with gasoline and can even be used instead of gasoline in car engines with minimal modifications. Lignocellulosic biomass obtained from agricultural wastes like corn stovers, fibres, and straw is generally subjected to complex pre-treatment procedures that yield simple sugars which are subsequently treated with butanol-producing microbial cultures (Ezeji and Blaschek 2010). Over the years, a variety of studies have reported biofuels production from agricultural residues like rice straw, potato and sugar beet discards, sawdust, corn stalks, and sugarcane bagasse (Duhan et al. 2013; Kumar et al. 2014; Kumar et al. 2016). They are supposed to present a viable substitute to fossil fuel-based fuels. Moreover, biofuels made from agricultural wastes or other comparable feedstocks recycle carbon sequestered from the atmosphere thereby helping slow down the process of global warming. Since agricultural residues are cheaper, are easier to source, and have fast harvest times, they are a superior choice over woody forest biomass (Limayema and Ricke 2012). Therefore, biodiesel production using agricultural wastes disincentivizes deforestation and forest ecosystem destruction. Furthermore, agricultural waste-based biofuel production also reduces the need to allocate precious farmland to dedicatedly grow suitable crops such as non-edible oilseed crops (jatropha oil, castor oil, and tall oil) exclusively for biodiesel production (Koçar and Civaş 2013). Modern science and technological and even social progress are critically dependent on an easily available, stable, consistent, and low-priced energy source resistant to natural, social, and economical perturbations. Agricultural residue-based biofuels can be used to meet that demand. India, being an agrarian nation, holds the calibre to be among the global leaders in renewable, sustainable biofuel production from agricultural waste.

Methane gas generation by digesting agricultural wastes anaerobically is another promising technology that achieves that bioconversion of agricultural wastes high in organic material (Timbers and Downing 1977). During the digestion, methane is not formed in pure form, but as a mixture of gas which is called as biogas. This is a renewable source of energy with numerous advantages. Along with the gas, it produces hugely fertile manure which can be utilized in fields (Kapoor et al. 2020). This microbial fermentation-based digestion of agricultural waste in the absence of oxygen helps to produce methane that involves two steps. Acid-forming bacteria first digest volatile solids to produce organic acids which act as the substrate for a consortium of methane-producing micro-organisms and then methane is produced. Biogas constitutes methane (50–70%), carbon dioxide (25–45%), nitrogen (0.5–3%), and hydrogen gas (1–10%) with traces of H_2S. The heating value of the gas ranges between 18 and 25 MJ/m^3 (Timbers and Downing 1977). This gas can help in electricity generation and also for cooking purpose (Kapoor et al. 2020).

Over the past 50 years, the world's meat production has more than tripled. To keep up with the world's insatiable hunger for animal proteins, animal husbandry and livestock rearing activities have increased significantly in most countries including India (Ritchie and Max 2017). However, this has also significantly increased the waste being generated from livestock activities. Malodorous gases such as hydrogen sulphide, ammonia, and greenhouse gases like methane are also released into the atmosphere as a direct consequence of rearing livestock, particularly ruminant livestock. Anaerobic digestion simplifies the treatment and disposal of animal waste to a considerable extent by stabilizing the sludge and producing relatively odour-free digestion sludge; hence, it offers an alternative energy source.

1.4.4 Biofertilizer Production

The tropical climate of India not only favours the growth of crops but also promotes the growth of pathogenic microorganisms, pests, and weeds. Therefore, farmers are highly reliant on chemical pesticides and herbicides to avoid the destruction of their valuable produce. Although the use of pesticides and herbicides in moderation can help increase agricultural productivity, farmers often end up overusing and even abusing these chemicals. Furthermore, the packaging and containers using which these chemicals are distributed at retail outlets are often discarded indiscriminately by farmers in open fields and waterbodies. The Plant Protection Department (PPD) estimates that standard packaging tends to retain approximately 1.8% of the chemicals in it (Dien and Vong 2006). Most farmers tend to vastly underestimate the environmentally detrimental effects concentrated doses of chemical pesticides and herbicides can have on the environment and its active ecosystems. Improper storage of unused pesticides and herbicides also can lead to serious environmental risks as the chemicals may leak or leach out of their primary containment thereby contaminating the surrounding environment. Chemical fertilizers play a pivotal role in modern agriculture. They are primarily composed of nitrogen, phosphorus, and potassium. Like pesticides and herbicides, fertilizers too are often abused by farmers in hopes of increasing their agricultural output by a substantial margin (Hai and Tuyet 2010). The absorption rate of such fertilizer compounds depends on a variety of parameters such as the physiochemical properties of the land, plant types, and fertilization technique used (Thao 2003). A fraction of the excess fertilizer is retained in the soil, whereas most of the remainder ends up in various waterbodies such as ponds, lakes, streams, and rivers via surface run-off and irrigation canals. A small part of the excess fertilizer also evaporates or gets denitrated resulting in air pollution. Therefore, given the current situation, natural fertilizers derived from agricultural wastes of both plant and animal origin are advantageous over conventional chemical fertilizers. Organic fertilizers can be applied alongside chemical fertilizers or as a complete replacement for chemical fertilizers. The nutrients provided by chemical fertilizers can easily be supplemented by manure. Poultry manure has been reported to be rich in phosphorous and is said to positively impact the agricultural productivity (Mokwunye 2000). Adding manure in soil improves its fertility as it increases the nutrient retention capacity or cation exchange capacity, water holding capacity, and overall physical condition and stability.

One of the areas of focus in agricultural research is boosting the natural rate of soil carbon sequestration using soil amendment techniques, as it can enhance agricultural productivity. These techniques have the capability to mitigate Green house gas (GHG) emissions as well as to decrease the volume of agricultural waste being generated. This is because sequestration of carbon requires substantial residence time in the soil as well as chemical protection from oxidation into carbon dioxide or reduction to methane (Srinivasarao et al. 2013). Over the years, partially burnt products such as black carbon and biochar have been receiving increasing interest for soil amendment applications because of their ability to be a long-term carbon sink with a very low rate of chemical transformation (Izaurralde et al. 2001; McHenry 2009). Biochar is a fine-grained product, which is rich in carbon and porous in nature. It consists of varying proportions of carbon (C), hydrogen (H), oxygen (O), nitrogen (N), sulphur (S), and ash, derived from the pyrolysis of biomass (Amonette and Joseph 2009; Masek 2009). In pyrolysis systems, biomass such as agricultural waste is heated and the temperature ranges between 400 and 600 °C, in anaerobic conditions. This prevents oxidation of the material thereby avoiding release of carbon dioxide. Instead, a part of the biomass is vaporized, and a carbon-rich char is left behind, which is referred to as biochar.

Biochar being porous in nature helps to improve soil water retention capacity and surface area, when amended to soil. It also actively influences and interacts with the soil matrix, soil microbiome, and plant roots, thus promoting retention of nutrients and triggering a number of biogeochemical reactions which help to enrich the soil (Lehmann and Joseph 2009). In fact, there are several reports describing soil quality improvements such as improved soil pH stability, increased earthworm population, and decreased fertilizer requirement, on amendment with biochar (Tryon 1948; Gaunt and Cowie 2009). The applications of biochar are not limited to agriculture. Biochar finds varied uses in a number of domains, including water treatment, construction, food technology, cosmetic manufacturing, metallurgy, chemical engineering, and many more.

Natural fertilizers made from organic remains such as agricultural waste are known as compost. They are formed by the process of composting (Tuomela et al. 2000; Misra et al. 2003). Composting is considered as a safe method for managing waste. It is an aerobic process where microbes act on complex degradable matter and turn them into organic and inorganic by-products (Toledo et al. 2018; Ayilara et al. 2020). These transformed by-products can be utilized as biofertilizers and for enhancing soil quality. It is also known to boost the overall agricultural yield and productivity (Ayilara et al. 2020). They are also known to be more resistant to drought, disease, and toxicity (Misra et al. 2003; Lei et al. 2010). Furthermore, compost promotes the nutrient uptake, cycling, and microbial activity of soil and hence is a viable substitute to harmful chemical fertilizers and pesticides. Large-scale compost using agricultural waste is an ambitious undertaking which, if implemented appropriately, will be of significant social and economic benefit. However, the production of organic fertilizers from agricultural waste at scale is a challenging undertaking requiring significant upfront technical, economical, and time investment. There are reports discussing the technical and economical common challenges plaguing organic waste composting projects (Hettiarachchi et al. 2018). One of the prominent issues is the lack of a steady, stable, and secure market for the product. This is, however, not an issue for small-scale, on-site fertilizer production plants as farmers can make compost from their own agricultural waste and feed in back into their own land, thus creating a self-sufficient circular supply chain. Small-scale production of biofertilizers in a domestic set-up, i.e. composting, is also a viable alternative which has attractive economic incentives especially for economically downtrodden farmers who have easy access to agricultural waste. Furthermore, crop residue, animal manure, and food discards are rich in organic content which make them ideal raw materials for composting purposes.

A bio-composting study carried out by Pratap Singh and Prabha (2017) in Uttar Pradesh, India, reported notable increases in the agronomic properties of rice and wheat when the soil was supplemented with bio-compost (Pratap Singh and Prabha 2017). The compost preparation was mostly composed of agricultural waste materials such as wheat straw, rice straw, and leaves. Wheat and rice straw, vegetable crop, leaves contributed to the biomass used, and compost was prepared. It constituted of total solids (45%), organic matter (26.7%), carbon (15.3%), and total nitrogen (1.36%). The compost produced was rich in organic matter and carbon and provided sufficient nitrogen and phosphorus to the crops (Brady and Weil 1966). This also helped to enhance the soil health by supporting microorganism and necessary native flora and fauna (Nielsen and Angelidaki 2008; Wall et al. 2015). This study also showed the contribution by compost layer of 1 in. thickness. The total nitrogen added to the soil was about 1 ton/ha, carbon was 13.3 ton/ha, and organic carbon was 24 ton/ha. And also, it helped in the availability of other important nutrients like phosphorus, potassium, magnesium, and iron. (Pratap Singh and Prabha 2017).

Composting has shown to be advantageous in terms of enhancing productivity, providing disease resistance, and many other things. Its use should be spread across by utilizing the biodegradable waste which is combusted (Ayilara et al. 2020).

1.4.5 Value-Added Products

Agricultural residues such as wheat and rice bran, corn cobs, gram husks, coffee bean discards, oil cakes, and inedible fruit wastes are rich in nutrients and can therefore be utilized by a variety of different industries as value-added products.

In a study by Igor and Ruta (2015), the possibility to use potato peel which is produced in the food production unit was studied. Also, the use of potato peel powder was seen as a partial flour replacement, and it did not affect the fermentation and other properties.

Another study done in 2016 showed the potential usage of coffee waste to produce bricks. The study analysed two main parameters, namely CW (coffee waste) ratio and temperature. Also, compressive strength and shrinking density were the properties considered in the study. A control brick and three bricks made of coffee waste with different percentages were exposed to high temperature (1050 °C). Apart from that, the manufactured brick was tested for any possible heavy metal leaching. It was seen that with the addition of CW there was a linear increase in shrinkage, but it was complying with the minimum standard of 8% and overall, a good quality brick was produced. Therefore, it can be concluded that CW could be used to produce fire clay bricks, and different proportions of CW can be used. Also, it also becomes an alternative way of coffee waste disposal and a low-cost waste additive for brick production (Loeppert and Suarez 1966).

The leftover parts of different fruits and vegetables are commonly thrown away as wastes and are considered to be of no use. But several research works have shown their utility (Parashar et al. 2014). For example, pomegranate peels are documented to have a number of bioactive compounds like phenolic acids, flavonoids, and hydrolysable tannins which have promising applications in food preservation using natural compounds (Chia-Hui et al. 2020). In a recent report by Kumar and colleagues, the authors explored a variety of novel applications of fruit and vegetable peels, ranging from the production of microbiological media and fortified probiotics to reducing agents in metal nanoparticle synthesis and biosorbents (Kumar et al. 2020).

Solid state fermentation (SSF) can help to increase the antioxidant activity of various substrates with the help of substrates. Pineapple waste can also be used as a substrate in

SSF, and it also produced anticancer agents along with antioxidants. The skin and the central part of the fruit were used, and it was concluded that using fermented pineapple waste increased protein content, fibre content, phenols, and antioxidants. Hence, it can be used as a source for antioxidants (Rashad et al. 2015).

1.4.6 As Adsorbents for the Removal of Heavy Metals

Industrialization and urbanizations have led to numerous environmental issues including release of heavy metals in excess. The main problem associated with heavy metals is that unlike organic pollutants which degrade to non-toxic components, heavy metals do not break into harmless form (Gupta et al. 2001). It is also a matter of concern as heavy metals pose a threat to different life forms. Different studies have been carried out to resolve this matter, and it has been determined that adsorption is a highly effective technique for heavy metal removal in effluent treatment plants. Activated carbon has also been found effective (Chand et al. 1994). But lately, agricultural waste has aroused as an effective, low-cost alternative to treat effluents with heavy metal using adsorption. The low-cost agricultural waste can be many by-products like sugarcane bagasse (Mohan and Singh 2002), rice husk (Ayub et al. 2002), sawdust (Ajmal et al. 1996), coconut husk (Tan et al. 1993), and oil palm shell (Khan et al. 2003), and many of them are researched by scientists for heavy metal removal.

In 2014, a study was done in Chennai where agricultural waste was used as an adsorbent to remove iron present in drinking water as its concentration was 0.3 mg/l. Sugarcane bagasse and coconut coir were chosen as the solid-phase extra coir for removing iron and calculating the amount removed, and they were found to exhibit excellent iron removal ability. Also, parameters like pH, dose of adsorbent, initial iron concentration, and contact time were analysed, and it was reported that coconut coir is much more effective than sugarcane bagasse (Misra 1968; Harshwardhan and Upadhyay 2017).

Apart from this, there have been a number of studies done to assess the potential held by agricultural wastes as adsorbents to remove heavy metals from water (Bhatnagar and Sillanpää 2010; Farooq et al. 2010; Bilal et al. 2013). They can be used as low-cost ion exchange material as well as adsorbents to remove chromium, lead, cadmium, nickel, etc. This will provide a cost-effective source for water purification and also help in the management of agricultural waste (Sabir et al. 2021).

1.4.7 Animal Feed

A variety of agricultural by-products and wastes are produced globally, and they hold the potential to replace the conventional materials used as feed for animals (Ajila et al. 2012). While raising animals, approximately 50–60% of the cost is covered by feed and related things. Hence, developing a cost-effective nutritious feed is required for supporting the growing livestock (Uyeh et al. 2021). Getting good quality animal feed is still an issue in developing countries, despite the efforts being put to find alternatives (Leng et al. 1992). The residues left post-harvesting are a good source of fibre and have a low amount of protein, starch, and fat. The traditionally used method to feed and increase livestock with forage and pasture along with grain and protein concentrate is no longer effective to meet the increased demand of livestock in future. Also using grains and protein adds on to the competition with human food. Hence, using agricultural residues as a food source for livestock is a good alternative (Hussein and Sawan 2010).

In a study by Uyeh et al. (2021), several by-products from agriculture like cottonseed meal, mushroom medium, distillers' wet grain, citrus peel, pineapple peel, rice bran, and soy sauce cake were tested. They then proposed an integrated model for animal feed formulation and production. It was also tested for the nutrient's composition so that the nutrition requirement can also be taken care of.

1.5 Challenges in Agricultural Waste Management

The very first challenge faced is related with the environmental and economic components of agricultural waste. There is a need of early prediction tools which would be useful to guide the users and policy-makers. For quantifying impact of a product on environmental components, life cycle assessment (LCA) is a popular and reliable methodology to be used. But the issue faced in this is the lack of inventory data related to agricultural residue chain (e.g. Avadi et al. 2016). This is because either the data is available for a bench or pilot scale project or the particular agriculture residue chain is not assessed at all.

The second challenge faced are the various limitations faced by the current technologies employed for the conversion of agricultural wastes into useful products. For example, although anaerobic digestion is an effective approach for biogas and fertilizer production which utilizes a variety of different agricultural wastes, its effectiveness is limited in a number of scenarios. Many times, agricultural residues are replaced by energy crops as primary feedstock in anaerobic digesters, which indirectly also affect the land use pattern (Njakou Djomo et al. 2015).

The third challenge is to overcome the issues of the development of molecules and materials from agricultural wastes. Till date, only a handful of chemicals and polymers are bio-based. As per Fiorentino et al. (2017), it is 3% of chemicals and according to Aeschelmann et al. (2017), it is 2% of polymers. But with increasing knowledge about effects of petro-derived chemicals, the demand and need for its alternative have increased. It is important to find alternatives not only because of ill effects of petro-based chemicals but also because mineral oil which is being used as raw material is becoming scarce and other issues, such as increases in plastic-based waste. The main issue faced in the conversion of agricultural waste into biomaterial is associated with the energy to be consumed, degradation process involved, the chemical nature and complexity of the product, contaminants present, and overall social perception. However, if we are successful in producing organic acids, enzymes, biodegradable plastics, and other products from agricultural waste, the economic values added in generating electricity, producing animals feed stock, and other applications will be twice (Kiran et al. 2015).

The fourth challenge is related to integrating agricultural waste and related business to economy. Previously, the management focused on single resource and its conversion to a single final product valorization; therefore, chain is modified accordingly in terms of their functionality, energy consumption, etc. Cross-chain valorization of agricultural waste is quite challenging because of heterogeneity of resources. By-products streams are often bulky and add cost. Clustering of different production chains into one is a way to make valorizations feasible. Still the economic values of the main products influence the major part of decision-making in business. Also, awareness regarding valorization and associated opportunities in different sectors and acceptability of agriculture residue-based products by consumers is low.

The fifth and final challenge is associated with the management of the material and related information to resolve the issues associated with unbalanced distribution of nutrients, contamination, accumulation, and residue conversion-related issues. In past, the agricultural developments in EU led to rise of many environmental, socio-economic, and technical issues. While the production of exported food and feed nutrient is often depleted, it remains in excess at regions of livestock breeding. In areas with agricultural waste conversion technologies like biogas plant, often the primary feedstock gets replaced by specific crops rather than agricultural residues for economic reasons often due to agriculture and energy policy measures. Agricultural residues conversion depends on different sectors, and hence it is important to increase awareness and intersectoral interaction (Gontard et al. 2018).

1.6 Conclusion

Agriculture is the primary activity carried out in most of the countries across the world making it significant. It forms the basis for many industries and supports livelihood of many, and because of this, it becomes important to make this as self-sustainable process. There still has been no significant work to make agricultural waste as a source of regular income. Although agriculture has significantly expanded, technologies to utilize the leftover resources have not been developed much. Hence, it is important to work beyond the traditional boundaries of agriculture and come up with innovative solutions. There have been the development of technologies utilizing certain agricultural wastes, but still there is a lot of scope to come up with more self-sustaining technologies which could provide economic benefits as well as add-on benefits to the farmers. There is a need of sector wise focus on type of waste, nature of waste and then come up with solutions utilizing the traditionally established methods along with deriving new technologies. That way, the agriculture sector can become profitable in terms of produce as well as leftovers.

References

Aeschelmann, F., Carus, M., Baltus, W. et al. (2017). Bio-based building blocks and polymers: global capacities and trends 2016–2021. Nova Institute GmbH and Europeans Bioplastics Association. http://biobased.eu/downloads/bio-based-building-blocks-and-polymers-global-capacities-andtrends-2016-2021-2/ (accessed 14 December 2021).

Agamuthu, P. (2009). Challenges and opportunities in Agro-waste management: an Asian perspective. Inaugural meeting of First Regional 3R Forum in Asia 11–12 November, Tokyo, Japan, 2009.

Ajila, C.M., Brar, S.K., Verma, M. et al. (2012). Bio-processing of agro-byproducts to animal feed. *Critical Reviews in Biotechnology* 32 (4): 382–400.

Ajmal, M., Rao, R.A.K., and Siddiqui, B.A. (1996). Studies on Removal and Recovery of Cr (VI) from electroplating wastes. *Water Research* 30 (6): 1478–1482.

Amonette, J. and Joseph, S. (2009). Characteristics of biochar: micro-chemical properties. In: *Biochar for Environmental Management: Science and Technology* (ed. J. Lehmann and S. Joseph), 33–52. London, UK: Earth Scan.

Avadi, A., Nitschelm, L., Corson, M., and Vertes, F. (2016). Data strategy for environmental assessment of agricultural regions via LCA: case study of a French catchment. *The International Journal of Life Cycle Assessment* 21 (4): 476–491.

Avci, A., Saha, B.C., Dien, B.S. et al. (2013). Response surface optimization of corn stover pretreatment using dilute phosphoric acid for enzymatic hydrolysis and ethanol production. *Bioresource Technology* 130: 603–612.

Ayilara, M.S., Olanrewaju, O.S., Babalola, O.O., and Odeyemi, O. (2020). Waste management through composting: challenges and potentials. *Sustainability* 12 (11): 4456.

Ayub, S., Ali, S.I., and Khan, N.A. (2002). Adsorption studies on the low-cost adsorbent for the removal of Cr (VI) from electroplating wastewater. *Environment Pollution Control Journal* 5 (6): 10–20.

Barone Justin, R. and Schmidt Walter, F. (2006). Nonfood applications of proteinaceous renewable materials. *Journal of Chemical Education* 83 (7): 1003.

Bhatnagar, A. and Sillanpää, M. (2010). Utilization of agro-industrial and municipal waste materials as potential adsorbents for water treatment – a review. *Chemical Engineering Journal* 157 (2–3): 277–296.

Bhuvaneshwari, S., Hettiarachchi, H., and Meegoda, J.N. (2019). Crop residue burning in India: policy challenges and potential solutions. *International Journal of Environmental Research and Public Health 16*: 832.

Bilal, M., Shah, J.A., Ashfaq, T. et al. (2013). Waste biomass adsorbents for copper removal from industrial wastewater – a review. *Journal of Hazardous Materials* 263: 322–333.

Brady, N.C. and Weil, R.R. (1966). *The Nature and Properties of Soils*, 14. Upper Saddle River, NJ, USA: Prentice Hall.

Brown and Root Environmental Consultancy Group (1997). Environmental review of national solid waste management plan. Interim report submitted to the Government of Mauritius.

Bušić, A., Marđetko, N., Kundas, S. et al. (2018). Bioethanol production from renewable raw materials and its separation and purification: a review. *Food Technology and Biotechnology* 56 (3): 289–311.

Chand, S., Aggarwal, V.K., and Kumar, P. (1994). Removal of hexavalent chromium from the wastewater by adsorption. *Indian Journal of Environmental Health* 36 (3): 151–158.

Chia-Hui, C., Ho, S.N., Hu, P. et al. (2020). Food preservative sorbic acid deregulates hepatic fatty acid metabolism. *Journal of Food and Drug Analysis* 28: 2.

Dien, B.V. and Vong, V.D. (2006). Analysis of pesticide compound residues in some water sources in the province of Gia Lai and DakLak. Vietnam Food Administrator.

Dionisi, D., Anderson, J.A., Aulenta, F. et al. (2015). The potential of microbial processes for lignocellulosic biomass conversion to ethanol: a review. *Journal of Chemical Technology and Biotechnology* 90: 366–383.

Duhan, J.S., Kumar, A., and Tanwar, S.K. (2013). Bioethanol production from starchy part of tuberous plant (potato) using Saccharomyces cerevisiae MTCC-170. *African Journal of Microbiology Research* 7: 5253–5260.

Duque-Acevedo, M., Belmonte-Ureña, L.J., Plaza-Úbeda, J.A., and Camacho-Ferre, F. (2020). The management of agricultural waste biomass in the framework of circular economy and bioeconomy: an opportunity for greenhouse agriculture in Southeast Spain. *Agronomy* 10 (4): 489.

Elbersen, B., Startisky, I., Hengeveld, G. et al. (2012). Atlas of EU biomass potentials. *Biomass Futures Project Deliverable* 3: 3.

Ezeji, T.C. and Blaschek, H.P. (2010). Butanol production from lignocellulosic biomass. In: *Biofuels from Agricultural Wastes and Byproducts* (ed. H.P. Blaschek, T.C. Ezeji and J. Scheffran), 19–37. https://doi.org/10.1002/9780813822716.ch3.

Faraco, V. and Hadar, Y. (2011). The potential of lignocellulosic ethanol production in the Mediterranean Basin. *Renewable and Sustainable Energy Reviews* 15 (1): 252–266.

Farooq, U., Kozinski, J.A., Khan, M.A., and Athar, M. (2010). Biosorption of heavy metal ions using wheat based biosorbents – a review of the recent literature. *Bioresource Technology* 101 (14): 5043–5053.

Fiorentino, G., Ripa, M., and Ulgiati, S. (2017). Chemicals from biomass: technological versus environmental feasibility: a review. *Biofuels, Bioproducts and Biorefining* 11 (1): 195–214.

Gaunt, J. and Cowie, A. (2009). Biochar greenhouse gas accounting and emission trading. In: *Biochar for Environmental Management: Science and Technology* (ed. J. Lehmann and S. Joseph), 317–340. London, UK: Earthscan.

Gontard, N., Sonesson, U., Birkved, M. et al. (2018). A research challenge vision regarding management of agricultural waste in a circular bio-based economy. *Critical Reviews in Environmental Science and Technology* 48 (6): 614–654.

Gupta, V.K., Gupta, M., and Sharma, S. (2001). Process development for the removal of lead and chromium from aqueous solution using red mud – an aluminum industry waste. *Water Research* 35 (5): 1125–1134.

Hai, H.T. and Tuyet, N.T.A. (2010). Benefits of the 3R approach for agricultural waste management (AWM) in Vietnam. Under the Framework of joint Project on Asia Resource Circulation Policy Research Working Paper Series. Institute for Global Environmental Strategies supported by the Ministry of Environment, Japan.

Harshwardhan, K. and Upadhyay, K. (2017). Effective utilization of agricultural waste: review. *Journal of Fundamentals of Renewable Energy and Applications* 7: 5.

Hettiarachchi, H., Meegoda, J.N., and Ryu, S. (2018). Organic waste buyback as a viable method to enhance sustainable municipal solid waste management in developing countries. *International Journal of Environmental Research and Public Health* 15: 2483.

Hussein, S.D.A. and Sawan, O.M. (2010). The utilization of agricultural waste as one of the environmental issues in Egypt (A Case Study). *Journal of Applied Sciences Research* 6 (8): 1116–1124.

Igor, S. and Ruta G. (2015). Latvia University of Agriculture, Research for rural development.

Izaurralde, R.C., Rosenberg, N.J., and Lal, R. (2001). Mitigation of climate change by soil carbon sequestration: issues of science, monitoring, and degraded lands. *Advances in Agronomy* 70: 1–75.

Jana, A., Halder, S.K., Dasgupta, D. et al. (2020). Keratinase biosynthesis from waste poultry feathers for proteinaceous stain removal. *ACS Sustainable Chemistry & Engineering 8* (48): 17651–17663.

Jansirani, D., Nivethitha, S., and Singh, M.V.P. (2012). Production and utilization of vermicast using organic wastes and its impact on Trogonella foenum and Phaselous aureus. *International J. Research in Biological Science* 2 (4): 287–289.

Kamel, R., El-Wakilb, N.A., Dufresne, A. et al. (2020). Nanocellulose: from an agricultural waste to a valuable pharmaceutical ingredient. *International Journal of Biological Macromolecules* 163: 1579–1590.

Kapoor, R., Ghosh, P., Kumar, M. et al. (2020). Valorization of agricultural waste for biogas based circular economy in India: a research outlook. *Bioresource Technology* 304: 123036.

Khan, N.A., Shaaban, M.G., and Hassan, M.H.A. (2003). Removal of heavy metal using an inexpensive adsorbent. *Proceedings of UM Research Seminar 2003 Organized by Institute of Research Management and Consultancy (IPPP)*, University of Malaya, Kuala Lumpur.

Kiran, E.U., Trzcinski, A.P., and Liu, Y. (2015). Platform chemical production from food wastes using a biorefinery concept. *Journal of Chemical Technology & Biotechnology* 90 (8): 1364–1379, 2017.

Koçar, G. and Civaş, N. (2013). An overview of biofuels from energy crops: current status and future prospects. *Renewable and Sustainable Energy Reviews* 28: 900–916.

Kuan, C.Y., Yuen, K.H., Bhat, R., and Liong, M.T. (2011). physicochemical characterization of alkali treated fractions from corncob and wheat straw and the production of nanofibres. *Food Research International* 44 (9): 2822–2829.

Kumar, A., Duhan, J.S., and Gahlawat, S.K. (2014). Production of ethanol from tuberous plant (sweet potato) using Saccharomyces cerevisiae MTCC-170. *African Journal of Biotechnology* 13 (28): 2874–2883.

Kumar, A., Sadh, P.K., and Surekha, D.J.S. (2016). Bio-ethanol production from sweet potato using co-culture of saccharolytic molds (*Aspergillus* spp.) and Saccharomyces cerevisiae MTCC-170. *Journal of Advanced Biotechnology* 6 (1): 822–827.

Kumar, A., Ogita Shinjiro, Y., and Yuan, Y. (2018). Biofuels: greenhouse gas mitigation and global warming, agro-industrial lignocellulosic waste: an alternative to unravel the future. *Bioenergy* 16: 291–305.

Kumar, A., Singh, P., Gupta, V., and Prakash, B. (2020). Application of nanotechnology to boost the functional and preservative properties of essential oils. In: *Functional and Preservative Properties of Phytochemicals* (ed. B. Prakash), 241–267. Academic Press.

Lehmann, J. and Joseph, S. (2009). Biochar systems. In: *Biochar for Environmental Management: Science and Technology* (ed. J. Lehmann and S. Joseph), 147–168. London, UK: Earthscan.

Lei, Z., Chen, J., Zhang, Z., and Sugiura, N. (2010). Methane production from rice straw with acclimated anaerobic sludge: effect of phosphate supplementation. *Journal of Bioresource Technology* 101: 4343–4348.

Leng, R.A., Choo, B.S., and Arreaze, C. (1992). Practical technologies to optimize feed utilization by ruminants. In: *Legume Trees and Other Fodder Trees as Protein Sources for Livestock* (ed. A. Speedy and P.L. Pugliese), 145–120. Rome, Italy: FAO.

Letcher, T.M. and Vallero, D.A. (2019). *Waste: A Handbook for Management*. Academic Press.

Limayema, A. and Ricke, S.C. (2012). Lignocellulosic biomass for bioethanol production: current perspectives, potential issues and future prospects. *Progress in Energy and Combustion Science* 38: 449–467.

Loehr, R. (2012). *Agricultural Waste Management: Problems, Processes, and Approaches*. Elsevier.

Loeppert, R.H. and Suarez, D.L. (1966). *Carbonate and Gypsum*. American Society of Agronomy.

Maiz, I., Esnaola, M.V., and Millan, E. (1997). Evaluation of heavy metal availability in contaminated soils by a short sequential extraction procedure. *The Science of the Total Environment* 206: 107–115.

Maraveas, C. (2020). Production of sustainable and biodegradable polymers from agricultural waste. *Polymers* 12 (5): 1127.

Masek, O. (2009). Biochar production technologies. http://www.geos.ed.ac.uk/sccs/biochar/documents/BiocharLaunch-OMasek.pdf (accessed 6 December 2021).

Mathieu, F. and Timmons, M.B. (1995). *Techniques for Modern Aquaculture* (ed. J.K. Wang). St. Joseph, MI: American Society of Agricultural Engineers.

McHenry, M.P. (2009). Agricultural biochar production, renewable energy generation and farm carbon sequestration in Western Australia, Certainty, uncertainty and risk. *Agriculture, Ecosystems and Environment* 129: 1–7.

Misra, R. (1968). *Ecology Workbook*. Oxford and IBH Publ. Co., Ecology.

Misra, R.V., Roy, R.N., and Hiraoka, H. (2003). *On Farm Composting Methods*. Food and Agricultural Organization of the United Nations: Rome, Italy.

Mohan, D. and Singh, K.P. (2002). Multi-component adsorption of cadmium and zinc using activated carbon derived from bagasse – an agricultural waste. *Water Research* 36: 2304–2318.

Mokwunye, U. (2000). Meeting the phosphorus needs of the soils and crops of West Africa: the role of indigenous phosphate rocks. Paper presented on Balanced Nutrition Management systems for the Moist Savanna and Humid Forest Zones of Africa at a symposium organized by IITA at Ku Leuva at Cotonun, Benin Republic, October 9–12.

Mood, S.H., Golfeshan, A.H., Tabatabaei, M. et al. (2013). Lignocellulosic biomass to bioethanol, a comprehensive review with a focus on pretreatment. *Renewable and Sustainable Energy Reviews* 27: 77–93.

Nagalakshmaiah, M., Mortha, G., and Dufresne, A. (2016). Structural investigation of cellulose nanocrystals extracted from chili leftover and their reinforcement in cariflex -IR rubber latex. *Carbohydrate Polymers* 136: 945–954.

Nielsen, H.B. and Angelidaki, I. (2008). Codigestion of manure and industrial organic waste at centralized biogas plants: process imbalances and limitations. *Water Science and Technology* 58: 1521–1528.

Njakou Djomo, S., Witters, N., Van Dael, M. et al. (2015). Impact of feedstock, land use change, and soil organic carbon on energy and greenhouse gas performance of biomass cogeneration technologies. *Applied Energy* 154: 122–130.

NPMCR (2019). http://agricoop.nic.in/sites/default/files/NPMCR_1.pdf (accessed 18 December 2021).

Obi, F.O., Ugwuishiwu, B.O., and Nwakaire, J.N. (2016). Agricultural waste concept, generation, utilization and management. *Nigerian Journal of Technology* 35 (4): 957–964.

OECD-FAO (2018). *Agricultural Outlook*. OECD-FAO https://doi.org/10.1787/19991142.

Pandey, S. and Dwivedi, N. (2020). Utilisation and management of agriculture and food processing waste. In: *Innovations in Food Technology* (ed. P. Mishra, R.R. Mishra and C.O. Adetunji). Singapore: Springer.

Parashar, S., Sharma, H., and Garg, M. (2014). Antimicrobial and antioxidant activities of fruits and vegetable peels: a review. *Journal of Pharmcognosy and Phytochemistry* 3: 160–164.

Pratap Singh, D. and Prabha, R. (2017). Bioconversion of agricultural wastes into high value biocompost: a route to livelihood generation for farmers. *Advances in Recycling and Waste Management* 2: 137.

Rashad, M.M., Mahmoud, A.E., Ali, M.M. et al. (2015). Antioxidant and anticancer agents produced from pineapple waste by solid state fermentation. *International Journal of Toxicological and Pharmacological Research* 7 (6): 287–296.

Ritchie, H. and Max, R. (2017). Meat and dairy production. Our World in Data.

Sabir, A., Altaf, F., Batool, R. et al. (2021). Agricultural Waste Absorbents for Heavy Metal Removal. In: *Green Adsorbents to Remove Metals, Dyes and Boron from Polluted Water*, Environmental Chemistry for a Sustainable World, vol. 49 (ed. Inamuddin, M. Ahamed, E. Lichtfouse and A. Asiri). Cham.: Springer.

Sadh, P.K., Duhan, S., and Duhan, J.S. (2018). Agro-industrial wastes and their utilization using solid state fermentation: a review. *Bioresources and Bioprocessing* 5: 1. https://doi.org/10.1186/s40643-017-0187-z.

Saini, J.K., Saini, R., and Tewari, L. (2014). Lignocellulosic agriculture wastes as biomass feedstocks for second-generation bioethanol production: concepts and recent developments. *Biotech*, https://doi.org/10.1007/s13205-014-0246-5.

Srinivasarao, C.H., Venkateswarlu, B., Lal, R. et al. (2013). Sustainable management of soils of dryland ecosystems for enhancing agronomic productivity and sequestering carbon. *Advances in Agronomy* 121: 253–329.

Tan, W.T., Ooi, S.T., and Lee, C.K. (1993). Removal of chromium (VI) from solution by coconut husk and palm pressed fibre. *Environmental Technology* 14: 277–282.

Thadikamala, S., Reddy Shetty, P., Intensification of Fructosyltransferases and FructoOligosaccharides production in solid state fermentation by *Aspergillus awamori* GHRTS, 2013 *Indian J. Microbiol.* 53, 337–342. https://doi.org/10.1007/s12088-013-0380-5.

Thao, L.T.H. (2003). Nitrogen and phosphorus in the environment. *The Journal of Surgical Research* 15 (3): 56–62.

Timbers, G.E. and Downing, C.G.E. (1977). Agricultural biomass wastes: utilization routes. *Canadian Agricultural Engineering* 19 (2): 84–87.

Toledo, M., Siles, J., Gutiérrez, M., and Martín, M. (2018). Monitoring of the composting process of different agroindustrial waste: influence of the operational variables on the odorous impact. *Waste Management* 76: 266–274.

Tryon, E.H. (1948). Effect of charcoal on certain physical, chemical, and biological properties of forest soils. *Ecological Monographs* 18: 81–115.

Tuomela, M., Vikman, M., Hatakka, A., and Itavaara, M. (2000). Biodegradation of lignin in a compost environment: a review. *Bioresource Technology* 72: 169.

UNEP (2011). Towards a green economy: pathways to sustainable development and poverty eradication. https://sustainabledevelopment.Un.Org/Index.Php?Page=view&type=400&nr=126&menu=35. Retrieved December 18, 2021, from https://sustainabledevelopment.un.org/index.php?page=view&type=400&nr=126&menu=35

Uyeh, D.D., Ha, Y., and Park, T. (2021). Animal feed formulation: rapid and non-destructive measurement of components from waste by-products. *Animal Feed Science and Technology* 274: 114848.

Wall, D.H., Nielsen, U.N., and Six, J. (2015). Soil biodiversity and human health. *Nature* 528: 69–76.

Zhang, Z., Gonzalez, A.M., Davies, E.G.R., and Yang, L. (2012). Agricultural wastes. *Water Environment Research* 84 (10): 1386–1406.

2

Sustainable Physical Methods Used for the Management of Agricultural Waste Biomass

Farhana Rahman[1], *Md. Abdullah Aziz*[2], *Md. Khairul Quais*[3], *Md. Shahadat Hossain*[4], *and Mohammad Ashik Iqbal Khan*[5]

[1] *University of Technology Sydney (UTS), Sydney, NSW, Australia*
[2] *Agricultural Statistics Division, Bangladesh Rice Research Institute (BRRI), Gazipur, Bangladesh*
[3] *Rice Farming Systems Division, Bangladesh Rice Research Institute (BRRI), Gazipur, Bangladesh*
[4] *Training Division, Bangladesh Rice Research Institute (BRRI), Gazipur, Bangladesh*
[5] *Plant Pathology Division, Bangladesh Rice Research Institute (BRRI), Gazipur, Bangladesh*

CHAPTER OUTLINE

Agriculture Waste Management and Bioresource: The Circular Economy Perspective, First Edition.
Edited by Suruchi Singh, Pardeep Singh, Anu Sharma, and Moharana Choudhury.

Source: Gabriel / Adobe Stock.

2.1 Introduction

Efficient resource utilization is indispensable to ensure a circular economy where almost no unrecoverable waste is generated. There are various types of wastes that are generated from different sectors (Figure 2.1). In this chapter, we will be focusing on agricultural waste biomass.

Agricultural waste biomass plays a vital role in production and energy utilization within a circular economy. In a circular bioeconomic system, the use of waste biomass has to be adopted from the producer (raw materials) to consumers (final product) throughout the value chain process. Therefore, it is also important to consider the product design and generated waste management systems.

The amount of waste generation is increasing day by day. It is predicted that global waste generation will nearly double by 2050. Most of the growth will come from the developing countries, and the Asian countries are also responsible for such a higher amount of waste generation (Figures 2.2 and 2.3). Agriculture is the largest contributor of any resource sector in the field of economy as well as a large generator of waste materials. Agriculture sector includes cultivation of plants, fungi, animals, and other life forms for fibre, food, drugs, biofuel, and other products which are used to sustain and improve human life (Sharma et al. 2020a). Waste management helps to maintain a healthy environment in a proper manner for the farm animals. Additionally, this could reduce the requirement of the fertilizers and other nutrients needed for the crop production. The effective management of agricultural waste biomass ensures sustainable agriculture through nutrient cycling in

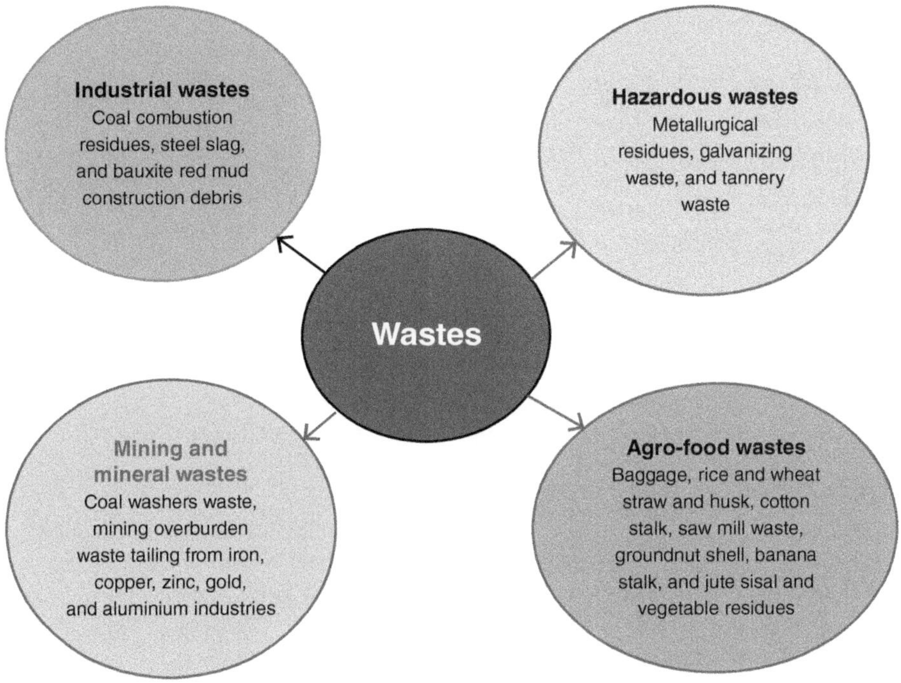

Figure 2.1 Different types of wastes and their generation from various sectors. *Source:* Sharma et al. (2020a).

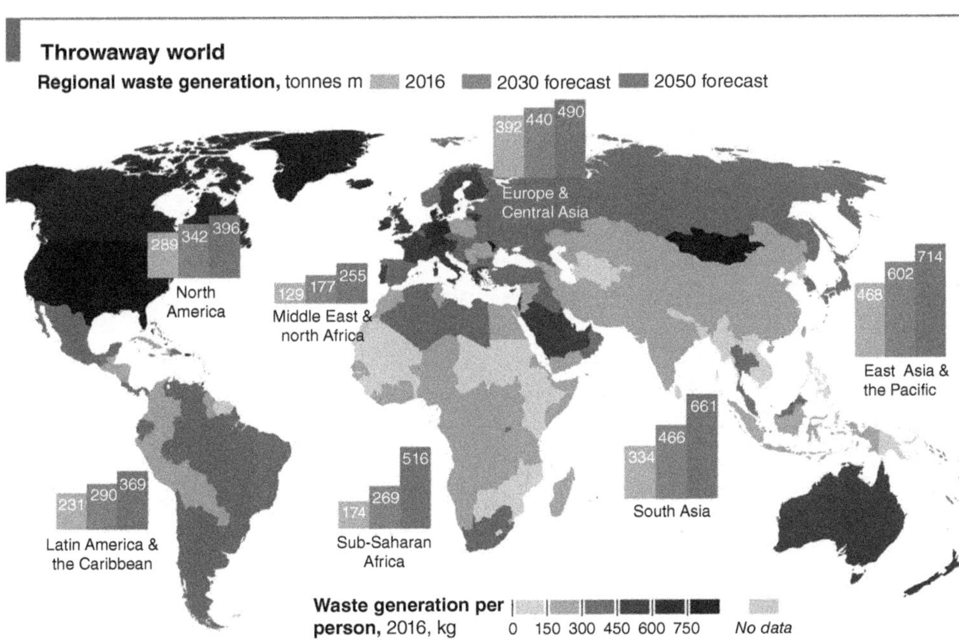

Figure 2.2 Global waste generation will nearly double by 2050. Most of the growth will come from the developing countries. *Source:* Adapted from World Bank, The Economist 2018.

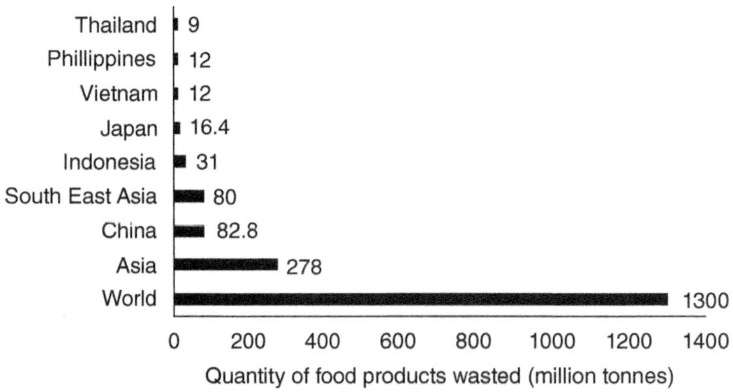

Figure 2.3 Amount of food waste generated at various Asian countries. *Source:* Kiran et al. (2014).

local, regional, and global scale (Principato 2018; Campos et al. 2020). Agricultural waste is mainly composed of organic wastes which can be categorized as field waste (husk, weeds, cattle wastes, and straw), animal waste (animal dung, waste feeds, and urine), agro-industrial waste (sugarcane, peels, and molasses), plant waste (carbon dioxide and water vapour), and natural waste (food waste and green waste) (Khan et al. 2009; Karthikeyan et al. 2018; Kibler et al. 2018; Lee 2018).

2.2 Major Crops for Agricultural Waste Generation

The major crops produced all over the world, particularly in Europe, the United States, Canada, Brazil, Argentina, China, and India, are rice, wheat, oats, maize, soybean, barley, rapeseed, sugarcanes, corns, sugar beets, and others.

All the major crops produced huge amount of agriculture residues globally of about 3.3 Gt annually (Table 2.1).

Table 2.1 Cumulative generation potential of agricultural residues in selected countries.

Country of origin	Amount of residue (Mt fresh weight/year)
China	716
The United States	682
India	605
Europe	580
Brazil	451
Argentina	148
Canada	105
Total	**3287**

Source: Bentsen et al. (2014).

The cereal crops were the major contributor of agricultural waste among the other agricultural crops (Nakamura 2009; Centore et al. 2014). Cereal crops waste includes stem, leaf, and sheath materials that comprised 66% of agricultural waste biomass, and more that 60% of this waste produced in the developing and least-developing countries of the world (Smil 1999). In the European Union countries, about 23 Mt/per annum agricultural dry waste were derived from cereals (IEA 2015); besides, almost 368 Mt/per annum wastes were generated in a form of straw in India (Hiloidhari et al. 2014). While, about 649 Mt cereal wastes were generated annually in China (Jiang et al. 2012). The second agriculture waste generating crop is sugarcane, while other agriculture biomass comes from oil crops, roots, tubers, nuts, fruits, and vegetables (Otles and Kartal 2018).

2.2.1 Cereal Crops

The unutilized residue generated from cereal crops globally was about 517 Mt/per year (Yevich and Logan 2003). According to the projection of FAO, there will be an increase in agriculture land use of about 13% which is equivalent to 120 M ha particularly in the developing countries by 2030. Besides, the cereal production has to be increased of about 0.9% all over the world from 2005/2007 to 2050 (Figure 2.4) (FAO 2012; Slade et al. 2012). Therefore, there is a possibility of growing more crops in each unit of land that may create more agricultural waste (Bentsen et al. 2014; Principato 2018).

2.2.2 Oil Crops

Like cereals, there was an increasing trend in the oil crops cultivation that was raised by 2.5% from 1999 to 2015 (FAO 2003; FAO 2015). The major countries with growing economy like China and India have increased their oil crop production to fulfil their growing needs (FAO 2015). In European regions, rapeseed cultivation was intensified in a faster way to produce biofuel energy (Koerbitz 1999; Raneses et al. 1999; EPRS 2015).

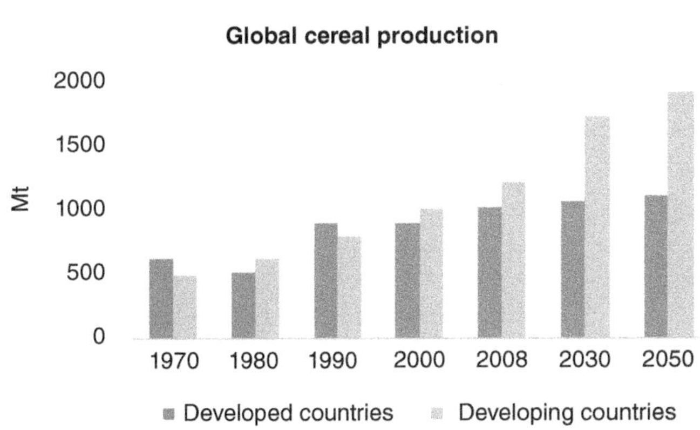

Figure 2.4 Projected global demand for cereal crops. *Source:* Tripathi et al., 2019 / Public Domain / CC BY 4.0.

2.3 Predominant Agriculture Residue of Agricultural Crops

2.3.1 Cereal Straw

The cereal straw is the leftover part when the product or its seed has been collected during harvesting and processing of the crop. This dry stalk can be used as fodder, bedding for domestic animals, compost for the crop production, and other purposes. In some cases, this straw is burned in the field. However, some countries have banned this practice because of air pollution (CO_2 emission). The cost generated from collection of straw through raking, baling, and others are almost $25 per tonne. It has been seen that coal and petroleum have 10–20 times higher volumetric energy density compared to straw. Besides, a high cost is generated within the compaction technologies that required to be adopted for producing straw pallets and wafers that develop high mass density (Demirbas 2011; Principato 2018).

2.3.2 Rice Husk

The most common agricultural residue found all over the world is the rice husk (Tripathi et al. 2019). During the processing stage of rice, rice husk is usually separated from the grain which is uniform in nature having valuable flow features compared to other types of biomass. For achieving these qualities, the biofuel industries adopted some technologies like gasification by utilizing the rice husk to be a unique fuel. There are some limitations on using rice husk as it contains higher amount of silica that is responsible for ash and slag issues in the boiler during the combustion stage (Guo et al. 2018; Havrysh et al. 2021).

2.3.3 Corn Stovers

The corn stover is the upper portion of the corn and consists of stalk, tassel, leaves, cob, husk, and silk. Almost 5% of stover is used for the animal food and bedding, and rest of the part is incorporated in the field. Corn stover may also be used as a fuel and converted into biofuels, as well as a fibre for pulp manufacturing in the paper industries and a board for a variety of other industries. (Demirbas 2011; Guo et al. 2018).

2.3.4 Sugarcane Bagasse

Sugarcane produces comparatively higher biomass because of its higher photosynthesis efficiencies than other cereals. Every tonne of sugarcane produces 135 and 130 kg sugar and dry bagasse, respectively. Dry bagasse has a heating value of 19.2 MJ/kg dry basis, widely used in many sugar mills as fuel to cook the cane and to evaporate the water from the cane syrup. However, generally, a huge amount of bagasse is disposed through the traditional ineffective incineration process (Demirbas 2011; Ong et al. 2018).

All the agricultural and food waste can be utilized for various purposes, e.g. as animal feed, construction of earthen houses, and most importantly for energy production in some industries and businesses such as sugar mills, rice mills, and others (Figure 2.5).

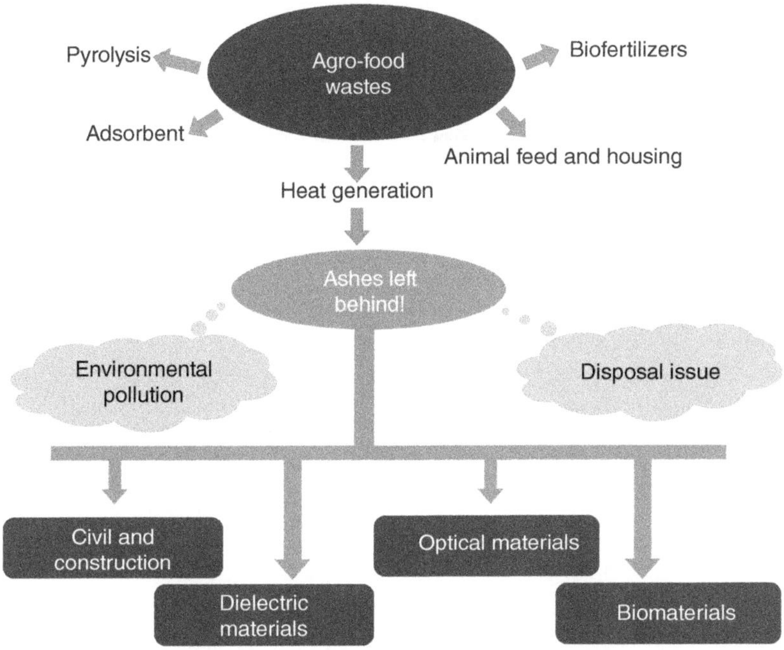

Figure 2.5 Applications of waste derived from agro-food and the generated ashes from the waste. *Source:* Singh and Bahadur (1999); Goel et al. (2012); Kaur et al. (2014); Danewalia and Singh (2016); Jha et al. (2018); Sharma et al. (2020a).

2.4 Forest Biomass

Forest biomass could be divided into the primary forest products which include industrial round wood, stem wood, the primary forest residues counting logging residues, and secondary forest residues including bark, black liquor, sawdust along with post-consumer wood, construction, and demolition wood. Table 2.2 represents the top forest residues producer countries of the world, and Table 2.3 shows the potential of energy generation from biomass residues in the field, in the state of Punjab (north-west Punjab state), India.

Table 2.2 Major global producers of forest residues.

Country of origin	Potential forest stock residues (46% of the total stock) (metric tonne)
Russian Federation	5718
Indonesia	2221
The United States	2078
Brazil	1613
China	807

(Continued)

Table 2.2 (Continued)

Country of origin	Potential forest stock residues (46% of the total stock) (metric tonne)
Sweden	316
France	308
Finland	246
India	232
Philippines	162
Poland	132
Norway	81
Austria	80
Republic of Korea	65
South Africa	52
Canada	50
Thailand	40
The United Kingdom	15
Japan	0
Germany	0
Australia	0
Subtotal of 21 top selected countries	14 218

Source: IEA CCC, IEA Clean Coal Centre, London, UK (2014).

2.5 The Application of Biomass Waste

The wastes generated from different sources are responsible for various environmental impacts on cultivable lands, nearby surface, and ground water. This adverse effect can be minimized by efficient utilization of biomass waste residues. In the power plant, the biomass waste residues usually processed through incineration for energy production; however, the ashes and the point source CO_2 can be combined in the production of carbonated materials which is mainly used for construction. By effectively using biomass ash in these circular management systems, landfill areas can be maintained, construction material resources can be reinforced, and CO_2 emissions and related adverse impacts may be reduced (Paul et al. 2018; Tripathi et al. 2019).

It is important to keep in mind that not all the types of biomass ashes have the potential to go through the direct carbonation process, many studies suggested that some of them are readily carbonated because of the presence of minerals. To resolve this problem, the non-reactive biomass ashes to CO_2 can be utilized in their 'raw' texture together with reactive biomass ashes to generate composite materials. So, the generation of carbonated-cemented composite materials used mainly for construction can be manufactured with outmost care through combining the biomass ashes with the raw wastes. In case of developing countries, particularly countries based on predominant agricultural activities, the enormous amount of waste was generated (Tripathi et al. 2017), and as there was ongoing rapid urbanization that required more

Table 2.3 Potential of energy generation from biomass residues in the field, in the state of Punjab, India (2015–2016).

Crops	Area, 000 (ha)	Production, (mt)	Residue production ratio	Heating values of crop residue, (MJ/kg)	Gross biomass generation, (mt/Yr)	Surplus biomass (mt/Yr)	Power generation potential from surplus biomass, (MW/Yr)
Wheat	3506.0	16.068	1.80	15.54	28.92	7.23	890.8
Rice	3046.0	11.803	1.70	17.15	20.07	5.02	682.0
Cotton, American	335.0	0.065	3.80	17.40	0.25	0.06	8.5
Barley	170.0	0.315	1.30	18.16	0.41	0.10	14.7
Maize	127.0	0.468	2.30	17.39	1.08	0.27	37.1
Sugarcane	92.0	0.638	0.33	20.00	0.21	0.05	8.3
Potato	91.6	2.366	0.81	16.00	1.92	0.48	60.8
Rapeseed and mustard	31.6	0.037	1.80	17.00	0.07	0.02	2.2
Barley	12.0	0.044	1.30	18.16	0.06	0.01	2.1
Moong	10.5	0.004	1.25	15.50	0.01	0.00	0.2
Sesamum	8.2	0.002	1.20	14.35	0.00	0.00	0.1
Sunflower	5.5	0.010	3.80	17.53	0.04	0.01	1.3
Cotton Desi	4.0	0.001	3.80	17.53	0.00	0.00	0.1
Arhar	2.8	0.003	2.50	18.58	0.01	0.00	0.2
Gram	1.9	0.002	1.30	16.02	0.00	0.00	0.1
Groundnut	1.7	0.003	2.30	15.56	0.01	0.00	0.2
Total	7445.8	31.82			53.0	13.3	1708.6
References	Statistical abstract of Punjab (2018)	Statistical abstract of Punjab (2018)	Hiloidhari and Baruah (2011a); Hiloidhari and Baruah (2011b); Hiloidhari et al. (2014); Singh et al. (2008);	Hiloidhari and Baruah (2011a); Hiloidhari and Baruah (2011b); Hiloidhari et al. 2014; Singh et al. 2008;	Jat et al. (2020)	Jat et al. (2020)	Jat et al. (2020)

Source: Singh et al. (2008); Hiloidhari and Baruah (2011a, b); Hiloidhari et al. (2014); Statistical abstract of Punjab (2018); Jat et al. (2020)).

construction materials, the new bio-based products can be introduced by replacing traditional virgin materials that can ensure significant benefit to the environment, society, and the economy of a country (National Environmental Agency 2018; Tripathi et al. 2019).

2.5.1 Biomaterial Production

The increasing trend of world population creates pressure on food production that is the biggest factor of enormous waste generation, particularly the agro-food waste. In the Section 2.3, all the predominant agriculture residues of agricultural crops were discussed. Most of the generated agro-food waste remains untreated that is not utilized for another purposes (Sharma et al. 2020b).

The untreated agro-food waste can be applied for various purposes in an environmentally friendly way. These wastes can be used as a biofertilizer, used as a construction element for building earthen houses, roads, etc., can be applied as animal feed, and also can be used for energy production as biofuel and generation of heat in the small-scale industries. These wastes can also be applied as a resource material for updating modern tools with advanced functions by reducing the use of minerals. The generated ashes produced in the industries after generating heat can be utilized for various purposes, particularly for the removal of toxic metals from contaminated wetlands, synthesis of glass materials, and in bioengineering, visual, and dielectric tools. All these techniques used for the production of cost-effective value-added tools that can help to minimize the impact of environmental pollution and also help to address waste management issues by creating a sustainable alternative path (Sharma et al. 2020b).

In Figure 2.6, the various bio-glassed and bioceramics are being produced. These products are mainly used for biomedical purposes such as orthopaedics and dentistry.

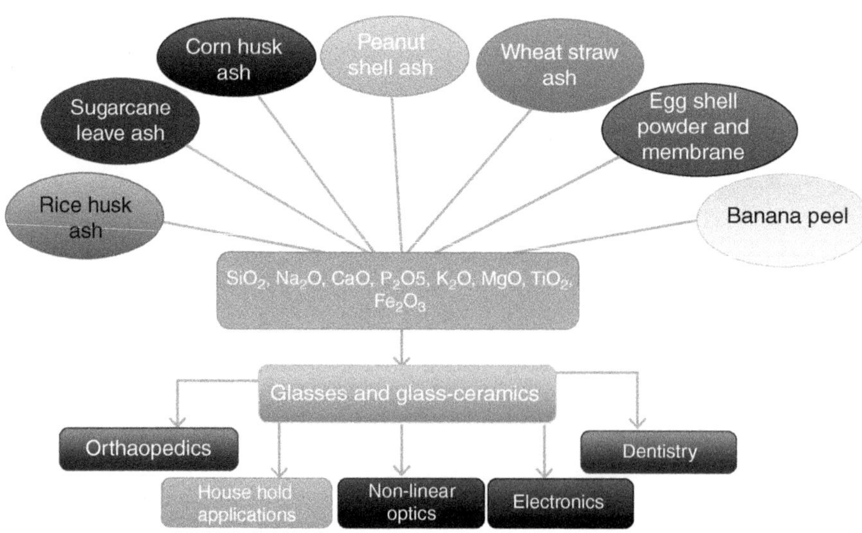

Figure 2.6 In the ashes of agro-food waste, various oxides are found. These are applied efficiently to synthesize glasses and glass ceramics that can be utilized in various purposes. *Source:* Singh and Bahadur (1999); Goel et al. (2012); Kaur et al. (2014); Danewalia and Singh (2016); Jha et al. (2018); Sharma et al. (2020b).

2.6 Challenges in Biomass Residues Utilization

The agricultural biomass wastes or residues are comprised crop stalks, leaves, roots, fruit peels, seeds, nut shells, and others that usually disposed of or incinerated as a part of supply of feedstock materials. There are challenges associated with biomass waste residue utilization that occurs during the several stages of crop processing, viz. production, harvesting, and processing along with retail or consumer loss (FAO 2011). There are several stages of waste streams of secondary and tertiary phases of sugarcane processing along with the waste generated during primary biomass processing (Schieber et al. 2001). So, it was apparent that the amount of agriculture waste biomass seemed to be almost the same as by-product generated from crop cultivation, harvesting, and processing. In the developing countries, biomass waste was poorly treated and remained open on the cultivated land to be decomposed in a natural way. As a result, most of the biomass waste remained unused which creates adverse impact on the surrounding environment and causes negative impact on living bodies (Tripathi et al. 2019).

2.7 The Waste Management Concept

The waste management process is comprised several stages such as collection, transportation, processing, recycling or disposal, and finally monitoring of waste materials. In general, the whole waste management system follows the procedure of collection, transportation, pre-treatment, processing, and final abatement of residues. The main aim of waste management procedure is to ensure sanitary living standard by decreasing the number of unwanted elements appears or leaves the community as well as by creating awareness on reuse of elements within the community (Principato 2018).

The following goals need to be accomplished through the waste management concept:

i) Decreasing the total number of generated waste by waste recycling and refuse;
ii) Recycling and re-entry of specific groups of substances into production cycles as secondary raw material or energy producer;
iii) Re-entry of waste generated from biological sources into the natural cycle;
iv) Following the best-suited options of the amount of waste reduction that can be disposed as 'appropriate' landfills;
v) There would be flexible concept on waste quantity fluctuation and the elements of domestic wastes.

There is a variation in the waste management scenario between the developing and industrialized countries. The transfer of proven technology between these groups of countries often seems inadequate (Campbell 1993). Thus, it is so important to consider the local factors that include the characteristics and components of the waste and seasonal variation in climate, the social structures and norms, the culture and behaviour for handling and disposal of solid waste, their political structures, and the existing awareness on the most predicted limited resource that is obvious to occur.

2.8 Waste Management Systems

Conventional parameters are not enough to differentiate the generated waste from various sources (Kan 2009). There are some innovative ways to categorize the waste based on classical characteristics. In the following sections, there will be a discussion of different types of waste management procedure.

2.8.1 Disposal

The disposal method is known as cohesive approach that is required to manage huge quantity of waste originated from different sources during the production process. The solid wastes usually have been disposed in open land over many years. This sort of landfill disposal has gained popularity for being the most cost-effective way of waste management (Demirbas 2011).

2.8.2 Landfill

Landfill is another method of waste management that required storing waste in a tightly sealed container. Normally, the storage containers have been designed in such a way that can prevent the degradation of various wastes during landfill procedure. It helps to preserve the quality of environment by inhibiting the toxic contamination associated with the various types of waste. The sealed containers aid in faster decomposition rate of the normal waste, as well as the organic waste such as papers, leaves, stems, and grasses due to the absence of water and air (Kormi et al. 2018). The severe phases of decomposition in landfills are discussed as follows:

i) The initial phase of decomposition normally takes less than seven days where the oxygen has been removed from the waste with the help of aerobic bacteria (Augenstein and Pacey 1991).
ii) The second phase is known as anaerobic acid phase where a variety of hydrolytic and fermentative bacterial species have been used to hydorlize the polymers into soluble sugars, amino acids, long-chain carboxylic acids, and glycerol (Micales and Skog 1997). There are certain types of polymers found in the diverse group of wastes, such as cellulose, hemicellulose, proteins, and lipids. The use of methane seems to be useful after a year from the beginning of the landfill procedure if it has been exploited economically (Figure 2.7).

Finally, the principal elements of landfill gas generated from the decomposed organic material arise in the formation of household waste. It usually takes place with the presence of naturally occurring bacteria in the anaerobic condition.

2.8.3 Incineration

There is another popular method of waste disposal known as incineration. The process of incineration usually involved combustion of generated waste (Kan 2009). Incineration, along with other waste disposal methods that require high temperature for processing, is

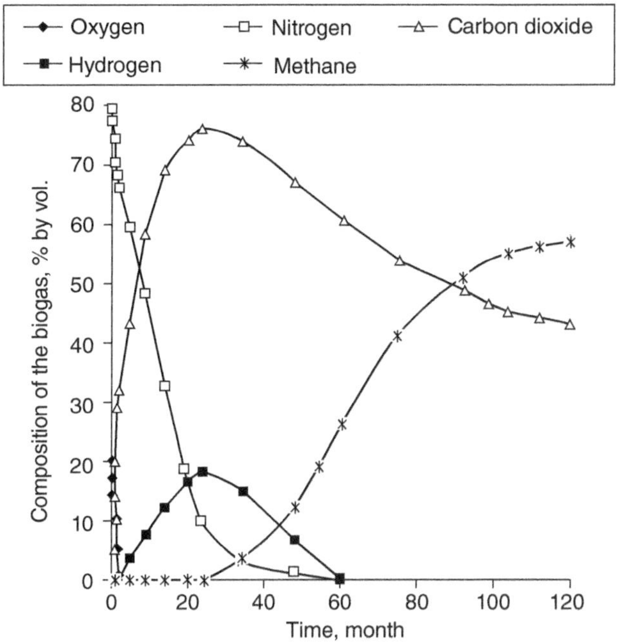

Figure 2.7 Production of biogas components with time in landfill decomposition. *Source:* Demirbas (2011) / With Permission of Elsevier.

also known as 'Thermal treatment'. The solid waste incineration system is designed, constructed, and operated following some certain criteria (Yang et al. 2007). Usually, collected wastes are processed by the incinerator for the treatment of input wastes that generated energy as a form of heat during the combustion stage (Principato 2018).

There are so many controversies raised for using the combustion stage of waste disposal as it causes air pollution by emitting greenhouse gases (GHGs) into the air. Moreover, solid residues are developed after the incineration of solid wastes. Although incineration generates huge amount of inorganic compounds, its residues usually comprised high carbon compound due to partial incineration during this method of waste disposal. Besides, there is lesser extent of carbon dioxide emission during the combustion of fossil carbon (Awasthi et al. 2019).

Figure 2.8 shows the incineration of agriculture residue from paddy crop field taken from Jat et al. (2020).

2.8.4 Biogas

The transformation of organic component to a gas is known as biogas, through a process called anaerobic digestion (AD). Biogas consists of methane and carbon dioxide with a less amount of hydrogen sulphide. Methane has wider applications for household activities such as cooking, heating, and others as a principle element of biogas (Mortier et al. 2016). The chemical composition of biogas is almost similar to the composition of natural gas. An anaerobic condition was created in physical sealed structure known as the biogas plant

Figure 2.8 Burning of residue generated from paddy crop field. *Source:* Jat, et al., (2020), ELSEVIER.

(biodigester). The anaerobic condition in the biogas plant stimulates different chemical and microbiological interactions among various elements; the biogas is produced as a final output generated from that reaction through the decomposition of raw input. The major content of biogas is methane (Balat 2008; Kormi et al. 2018). Biogas can be utilized as a fuel for engine, gas turbines, fuel cells, boilers, industrial heaters, and manufacturing processes of certain chemicals along with other applications. The waste treatment procedure of biodegradable materials can be processed together with AD, followed by the aerobic composting; it used to take place before the process of landfilling (Anyaoku and Baroutian 2018).

At present, the similar species of anaerobic bacteria that generate natural gas also produce methane. This type of bacteria is the primitive form of life in this planet. They originated before the photosynthetic processes started by green plant species; huge amount of oxygen was released to the atmosphere from that photosynthesis mechanism. However, the primitive anaerobic bacteria separated organic elements or decomposed without the present of oxygen and generated biogas as a by-product (Guo et al. 2018). There are four major stages of the AD process:

i) Hydrolysis;
ii) Acidogenesis;
iii) Acetogenesis; and
iv) Methanogenesis.

The AD plays a significant role in the decomposition of food and agriculture waste and other biomass in landfills. It is a biochemical process for transforming biogenic solid waste into a steady output such as humus (Liu et al. 2016; Lee 2018). However, aerobic transformation requires oxygen for the metabolic process of the species like aerobic microorganisms responsible for substrate degradation. The transformation takes place through aerobic process is composting and generating from agriculture waste. The composting seems helpful for further agricultural production activities as it provides and develops applicable materials such as mulch, soil additives, and amendments. It can also be used as organic fertilizer (Cerda et al. 2018).

Methane gas is one of the major by-product of biogas generation through an anaerobic system. The amount of methane production depends on the type and nature of biological elements; the amount varies from waste to waste (biodegradable). In case of pure cellulose, the ratio of methane and carbon dioxide is 50–50 for biogas production. However, the methane concentration differs for mixed waste feedstocks and that was about 40–60% by volume. The amount of methane raised higher for fats and oils compared to other waste types, and it generates approximately 70% from the production of biogas (Jutidamrongphan 2018).

2.8.5 Biomass and Biowaste

The leftover part of agriculture process that cannot be considered as food is known as biomass. There are different types of biomass resources generated from various sources such as woody and herbaceous species, wood dust and waste, residues come from agriculture and industrial practices, waste generated from paper production, municipal solid waste, waste generated from food processing industry, animal husbandry waste, waste generated from aquatic plants and animal, and others sources (Joshi and Visvanathan 2019). The important organic components that comprise the biomass are cellulose, hemicelluloses, and lignin.

Biomass energy can be utilized by following four principal steps, such as:

i) Direct combustion;
ii) Gasification;
iii) Pyrolysis; and
iv) Anaerobic digestion.

In the developing countries, the burning of dry agricultural residues was found for cooking and heating, mainly at the suburban and rural areas. Dry biomass was also found to go through the combustion process to generate electricity, which can generate methane, hydrogen, and carbon monoxide; sometimes, it was converted to liquid fuel. The fermentation process was applied mainly for the wet forms of biomass such as the wet waste from sewage sludge, cattle manure, and liquid food waste from food industry, and so on, that can generate fuel as well as organic fertilizers (Chen et al. 2016). It is easier to convert the biomass waste directly into the liquid fuel that can meet the transportation needs of the land, water, and air route vehicles. In the process of biomass conversation, solar and wind power may play a significant role for their applications (De Clercq et al. 2017).

2.8.6 Vermicomposting

Vermicomposting is a process that takes place in the mesophilic condition, often known as natural decomposition by micro-organisms. Different types of wastes go through the vermicomposting process with the presence of various earthworm species. The waste substrate was fragmented in the earthworms' intestine that generates the improved physicochemical qualities of the waste by increasing organic matter decomposition. The biochemical degradation of the waste takes place with the presence of micro-organisms in the earthworms' gut. Finally, the waste transformed into fine, homogenized manure comprised plenty of nutrients, abundant microbial population, humic substances, growth hormones, and enzymes that enhance effective growth of various types of plants. In addition, the transformation of waste into useful products results a circular bioeconomy that is essential for ensuring a sustainable development (Sharma and Garg 2017).

2.9 The Choice of Technology

Many technologies depending on the type and nature of waste, quality, source of origin, and local condition have been used for waste management (Demirbas 2011). The desired technology can be selected based on the following criteria:

i) Cost-effectiveness of the technology;
ii) Less land requirement for management;
iii) Very few adverse effects on the air, water, and land quality;
iv) Generation of more energy with low amount of waste; and
v) Reduction of mass volume to a lesser extent (Kumar 2000).

2.10 Sustainable Physical Methods for Agricultural Biomass Management

There are many management processes of agricultural biomass, viz. composting, incineration, and recycling.

2.10.1 Composting

Composting is a method in which organic matter present in the agriculture waste decomposed by anaerobically and aerobically through the chemical process and finally converted into humus. The three stages of composting are as follows:

i) **Preparation of agricultural waste:** the waste should be reasonably free of soil, dirt, and visible surface. They should be arranged properly; thus, they would burn with a minimum of smoke.
ii) **Decomposition:** waste is usually decomposed by three ways: made system, anaerobically decomposition, and vermiculture decomposition.
iii) **Management of recycling agro-wastes.**

2.10.2 Application of Captured CO_2 as an Alternative Measure

There is a worldwide demand for the conservation of energy, the reduction of CO_2 emissions, and the reuse of generated waste through a carbon-efficient management solution. The method of capturing and storing carbon is popular that also reflects within the

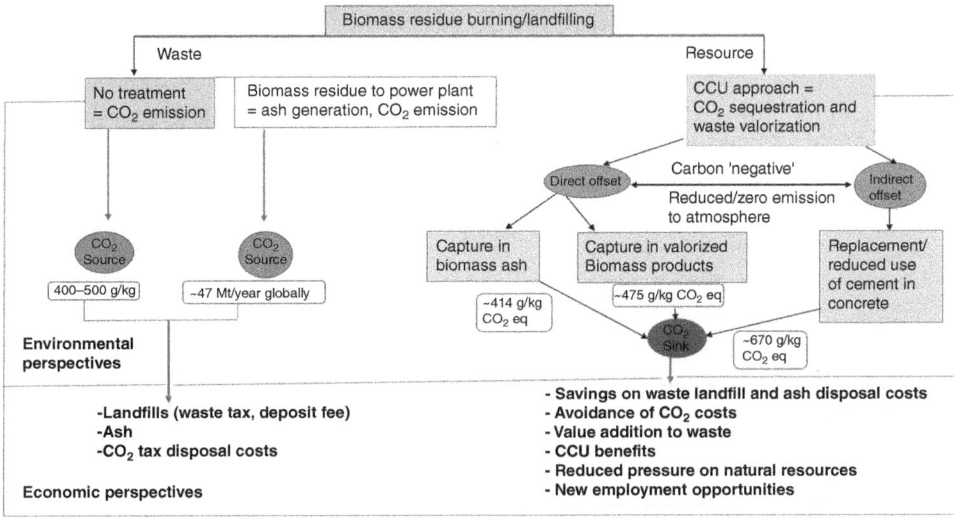

Figure 2.9 Application of agricultural waste biomass in terms of environmental and agricultural perspective by utilizing carbon capture and utilization (CCU) technique. *Source:* Tripathi et al., 2019 / Public Domain / CC BY 4.0.

commitment made by industrialized nations for decreasing the level of CO_2 emission. Carbon dioxide capture and utilization (CCU) has been developed by advanced technology, where carbon dioxide has been used as a feedstock. Several elements have been generated through this conversion by using CO_2 that comprised construction materials, plastics, and fossil fuels (Armstrong and Styring 2015). On the other hand, the efficient use of CCU technology will be partially depending on the straight use of point emissions of carbon monoxide and carbon dioxide by applying cost-effective sorbents that can be generated from waste as well (Kaithwas et al. 2012). It is important to mention that CO_2 is utilized intentionally to condition cementitious elements such as calcium silicate-based materials (Berger et al. 1972) and concrete articles (Maries and Hills 1986).

In order to solidify cement-based waste forms, the carbonation process seems useful (Lange et al. 1996 and Lange et al. 1997), and it also helps in the stabilization of the soil that is contaminated with heavy metal pollution from zinc, copper, and lead (Whitehead et al. 2003). The managed carbonation of biomass-derived wastes can play a significant role by treating the waste through carbonation to reduce the risk and to generate engineered elements (Figure 2.9). Also, the carbonation of biomass ash is important in terms of its traditional aspect (Garcia-Garcia et al. 2017).

2.11 Biomass Waste Management: Adverse Effect of Biomass Waste

Human health will be under risk if there is an increasing trend of unregulated disposal of agricultural biomass that can contaminate the surface and ground waters. The high amount of agricultural biomass is also responsible for eutrophication and emission of GHGs, NO, and N_2O resulting in higher rate of global warming compared to CO_2 (Koch 2010; Chun 2015).

Approximately less than two giga tonnes of crop residues are going for the process of incineration (Devi et al. 2017) that results in 18% emission of CO_2, including enormous particles and black carbon around the world (FAO 2014; Jain et al. 2014). The extent and type of pollution from biomass rely on the components of biomass and the way they are used for energy production. For example, low-efficient cookers used in open space particularly in developing countries are responsible for the degradation of indoor air quality (Chum et al. 2011). The environmental impact arising from biomass seems to be complex having positive and negative impact. The biomass renewability and carbon neutrality are defined as having positive impact, whereas the change in land use, loss of carbon in soil along with nutrient reserve, depletion of biodiversity, and lack of water are accounted for negative environmental impact. Therefore, the use of biomass for energy production activities needs to be monitored.

2.12 Sustainable Use of Biomass Resources

From many studies on energy consumption, it was apparent that there was a need to decrease the use of coal and other natural resources that were identified within the energy policy in the developing countries; these non-renewable resources can be replaced by renewable energy, and application of agricultural biomass in a sustainable way make a great difference in a positive way to reduce the consumption of natural resources (IEA 2002–2003). The United Kingdom would be one of the examples for this, where coal was not used for a while during the summer of 2019, and they met their needs from alternative sources of energy (Schieber et al. 2001). In the past 25 years, a significant development was observed in reusing the biomass residues. The alternative fuels and biochar production was observed as a successful initiative for generating electricity (Ergüdenler and Işiğigür 1994; Paula et al. 2011; Miles and Miles 1992). Some of the agricultural biomasses helped in renewable bioenergy production. For example, maize, sweet sorghum, miscanthus, switchgrass, Kenaf (herbaceous species), and eucalyptus from woody plants help in the production of crop-based biohydrogen (Saratale et al. 2019). However, the use of microbial-based biomass has been growing (Saratale et al. 2019), whereas the plant-based residues continued to be of the major priority for fuel and feedstocks (Sumner et al. 1985; Kwapinski 2010 and Speratti et al. 2018).

2.13 A Case Study of Southeast Spain: Managing Agriculture Waste Biomass

The economic development throughout the world was based on fossil fuel for many decades. There was a highest increase in the extraction of these non-renewable resources that was up to 113% since the end of the last century, and a massive amount of wastes were generated due to over-exploitation of resources. All these circumstances led to think about a sustainable development model that will make sure to produce more crops and generate energy with a least use of natural resources. This model will minimize GHGs emission and reduce the level of pollution and solid waste generation.

The Mediterranean region, particularly the province of Almeria, Spain, has been taken as a case study, focusing on the significance of intensive horticulture. Duque-Acevedo et al. (2020) conducted a study of the main agriculture food products of this region; they found that waste generated from agricultural field can be efficiently used for many purposes, especially for energy production.

There was a greenhouse land area of about 31 614 ha in Almeria along with a cultivable land of around 45 680 ha, which has two crop production cycles annually. The greenhouse land area has a physically increase of about 16%, and crop cultivable lands were expanded of about 14% in the last 18 years. This high amount of land areas makes this province (Almeria) as the highest concentration of greenhouse-protected crop land areas not only in Spain but also in the whole world. The greenhouse crop land areas in this province were comprised 6 districts and 37 towns. The largest greenhouse crop areas were found at the Campo De Dalias district of about 21 545 ha that is equivalent to about 68% of areas; among this, around 12 756 ha of areas ha are situated in El Ejido town that is equivalent to 40% of the total area. Alto Andarax is another district of this province, where lesser extent of greenhouse crop-protected areas were found (66 ha). As per the high to low cultivation and crop production rate, the agricultural areas were categorized as tomato (high), pepper, courgette, watermelon, cucumber, aubergine, melon, and green beans (low). Among these crops, tomato, pepper, courgette, and watermelon are the long-cycle agricultural crops that comprise lager cultivation land areas with high production rate (García-Raya et al. 2019).

A massive amount of agricultural waste biomass were generated annually from these high-productive crop cultivation lands that comprise both long-cycle and short-cycle agricultural crops. The waste biomass residues were generated as a form of plant waste that includes roots, stems, leaves, and fruits from the processing stage (pruning, maintenance, and final elimination of all the crop plants at the finishing stages of production cycle). In the province, Almeria, the amount of agricultural waste biomass generated from horticulture was about 1 370 743 tonnes per year; among these wastes, 515 901 were generated from tomato and 271 460 tonnes were from pepper annually. These two crops, tomato and pepper, generated more than 50% of total agricultural waste residues annually. The high amount of agricultural biomass created enormous problems to the producers particularly during the productive months of summer in every year. During the summer time, the long-cycle crop production time has ended up in 75% of waste generation. The generated agricultural waste biomass was shifted to the private waste management authorities and approved recycling plants for the treatment of generated waste and its recovery. Other parts of this waste were applied as a raw product for the food of animals. A few cases were seen when the growers can manage their waste on their own at their own land. A comparatively small portion of these was applied for compost generation inside the greenhouse. Campo de Nijar and Bajo Andarax were other districts of this province, where agriculture crop production activities were taken place; the management of generated waste from plants was done by the producers, and the applications of biofertilizers were applied to those greenhouses who are involved in organic farming.

There was a lack of integrated management of greenhouse plant residues in the province of Almeria; however, there were enormous opportunities that would be promoted for the recovery and reuse of the agricultural waste to reduce its adverse impact on the environment. Several initiatives were taken for implementing the sustainable waste management at the province of Almeria in Spain. Based on many research and specific studies on generated agricultural waste biomass from the greenhouse crop lands, it was observed that there was a great potential of effective utilization of these agricultural waste biomasses to produce energy, biofuels, and bioproducts. The plant waste from greenhouse crop lands in the process of gasification and heating, as well as the recyclability and reusing in compost making, was considered as an organic amendment in the greenhouse crop land in that province. In addition, from the national to international level, a number of strategic and regulatory frameworks have been reinforced to introduce and implement the recovery of waste generated from crop land, an important achievement to improve the utilization of waste for different purposes. The wastes from two dominated crops, tomato and pepper, provide an opportunity to transform the wastes into green fertilizers that make the system and the economy more circular and help to get rid of linearity of traditional agricultural method.

2.14 Conclusion

Agriculture and animal farming leave out wastes after they are harvested. Since agriculture is on a large scale, the agricultural wastes generated cannot be ignored and should be managed in an efficient way through the agricultural waste management programme. Improper dispose of agricultural wastes can be harmful to the environment and human beings as they contain various chemicals, minute metal particles, pharmaceuticals, and pathogens. Attention should be focused on minimizing wastage by creating a more efficient sustainable agricultural supply chain. Recycling of agricultural wastes into useful products could generate other sets of wastes, which may serve as raw materials for another useful product, thereby necessitating the continuous recycling of agricultural wastes until every potential waste is converted into wealth.

References

Anyaoku, C.C. and Baroutian, S. (2018). Decentralized anaerobic digestion systems for increased utilization of biogas from municipal solid waste. *Renew. Sust. Energ. Rev.* 90: 982–991. https://doi.org/10.1016/j.rser.2018.03.009.

Armstrong, K. and Styring, P. (2015). Assessing the potential of utilization and storage strategies for post-combustion CO_2 emissions reduction. *Front. Energy Res.* https://doi.org/10.3389/fenrg.2015.00008.

Augenstein, D. and Pacey, J. (1991). Landfill methane models. In: Proceedings from the technical sessions of SWANA's 29th annual international solid waste exposition, SWANA, Silver Spring, MD.

Awasthi, M.K., Chen, H., Awasthi, S.K. et al. (2019). Greenhouse gas emissions through biological processing of solid waste and their global warming potential. In: *Biological Processing of Solid Waste* (ed. S. Kumar, Z. Zhang, M.K. Awasthi and R. Li), 111–127. CRC Press.

Balat, M. (2008). Progress in biogas production processes. *Energy Educ. Sci. Technol.* 22: 15–36.

Bentsen, N.S., Felby, C., and Thorsen, B.J. (2014). Agricultural residue production and potentials for energy and materials services. *Prog. Energy Combust. Sci.* 40: 59–73.

Berger, R.L., Young, J.F., and Leung, K. (1972). Acceleration of hydration of calcium silicates by carbon dioxide treatment. *Nat. Phys. Sci.* 240: 16–18.

Campbell, D.J.V. (1993). Waste management needs in developing countries. In: proceedings of the Sardinia conference 1993, Environmental Safety Center, Building 7.12, AEA Technology, Harwell, Oxfordshire, UK.

Campos, P., Miller, A.Z., Knicker, H. et al. (2020). Chemical, physical and morphological properties of biochars produced from agricultural residues: implications for their use as soil amendment. *Waste Manag. (Elmsford) 105*: 256–267. https://doi.org/10.1016/j.wasman.2020.02.013.

Centore, M., Hochman, G., and Zilberman, D. (2014). Worldwide survey of biodegradable feedstocks, waste-to-energy technologies, and adoption technologies. In: *Modelling, Dynamics, Optimization and Bioeconomics I*, Springer Proceedings in Mathematics and Statistics 73, vol. 73 (ed. A.A. Pinto and D. Zilberman). Cham: Springer https://doi.org/10.1007/978-3-319-04849-9_11.

Cerda, A., Artola, A., Font, X. et al. (2018). Composting of food wastes: status and challenges. *Bioresour. Technol.* https://doi.org/10.1016/j.biortech.2017.06.133.

Chen, G., Liu, G., Yan, B. et al. (2016). Experimental study of co-digestion of food waste and tall fescue for bio-gas production. *Renew. Energy* 88: 273–279. https://doi.org/10.1016/j.renene.2015.11.035.

Chum, H., Faaij, A.P.C., Moreira, J.R. et al. (2011). Bioenergy. In: *IPCC Special Report on Renewable Energy Sources and Climate Change Mitigation* (ed. O. Edenhofer et al.). Cambridge, United Kingdom and New York, NY, USA: Cambridge University Press. 10.1017/CBO9781139151153.006.

Chun, A.M.S. (2015). *Ground Rules for Humanitarian Design* (ed. A.M.S. Chun and E. Brisson). London: Wiley Publishers ISBN 978-1-118-36159-7.

Danewalia, S.S. and Singh, K. (2016). Magnetic and bioactive properties of MnO_2/Fe_2O_3 modified Na_2O-CaO-P_2O_5-SiO_2 glasses and nanocrystalline glass-ceramics. *Ceram. Int.* 42: 11858–11865.

De Clercq, D., Wen, Z., Gottfried, O. et al. (2017). A review of global strategies promoting the conversion of food waste to bioenergy via anaerobic digestion. *Renew. Sust. Energ. Rev.* 79: 204–221. https://doi.org/10.1016/j.rser.2017.05.047.

Demirbas, A. (2011). Waste management, waste resource facilities and waste conversion processes. *Energy Convers. Manag. 52* (2): 1280–1287. https://doi.org/10.1016/j.enconman.2010.09.025.

Devi, S., Gupta, C., Jat, S.L., and Parmar, M.S. (2017). Crop residue recycling for economic and environmental sustainability: the case of India. *Open Agric.* 2: 486–494.

Duque-Acevedo, M., Belmonte-Ureña, L.J., Plaza-Úbeda, J.A., and Camacho-Ferre, F. (2020). The management of agricultural waste biomass in the framework of circular economy and bioeconomy: an opportunity for greenhouse agriculture in Southeast Spain. *Agronomy (Basel) 10* (4): 489. https://doi.org/10.3390/agronomy10040489.

EPRS (2015). European Parliamentary Research Service. Members' Research Service PE 568.329.

Ergüdenler, A. and Işiğigür, A. (1994). Agricultural residues as a potential resource for environmentally sustainable electric power generation in Turkey. *Renew. Energy* 5: 786–790.

FAO (2003). World agriculture: towards 2015/2030 An FAO Perspective. (Earthscan, London). http://www.fao.org/3/a-y4252e.pdf (accessed 19 March 2021).

FAO (2011). Global forest products facts and figures. http://www.fao.org/fileadmin/user_upload/newsroom/docs/2011%20GFP%20Facts%20and%20Figures.pdf (accessed 19 March 2021).

FAO (2012). World Agriculture Towards 2030/2050: The 2012 revision ESA E Working Paper No. 12-03. http://www.fao.org/fileadmin/user_upload/esag/docs/AT2050_revision_summary.pdf (accessed 19 March 2021).

FAO, (2014). Global forest resources assessment. FAO Forestry Paper 163. Food and Agricultural Organization of the United Nations, Rome, Italy; 378 p.

FAO (2015). World agriculture: towards 2015/2030. Summary Report ISBN 92-5-104761.

Garcia-Garcia, G., Woolley, E., Rahimifard, S. et al. (2017). A methodology for sustainable management of food waste. *Waste Biomass Valoriz.* 8: 2209–2227. https://doi.org/10.1007/s12649-016-9720-0.

García-Raya, P., Ruiz-Olmos, C., Marín-Guirao, J. et al. (2019). Greenhouse soil biosolarization with tomato plant debris as a unique fertilizer for tomato crops. *Int. J. Environ. Res. Public Health* 2019 (16): 279.

Goel, A., Kapoor, S., Rajagopal, R.R. et al. (2012). Alkali-free bioactive glasses for bone tissue engineering: a preliminary investigation. *Acta Biomater.* 8: 361–372.

Guo, W., Zhou, Y., Zhu, N. et al. (2018). On site composting of food waste: a pilot scale case study in China. *Resour. Conserv. Recycl.* https://doi.org/10.1016/j.resconrec.2018.01.033.

Havrysh, V., Kalinichenko, A., Brzozowska, A., and Stebila, J. (2021). Agricultural residue management for sustainable power generation: the Poland case study. *Appl. Sci. 11* (13): 5907. https://doi.org/10.3390/app11135907.

Hiloidhari, M. and Baruah, D.C. (2011a). Rice straw residue biomass potential for decentralized electricity generation: a GIS based study in Lakhimpur district of Assam. *India. Energy Sustain. Dev.* 15 (3): 214–222.

Hiloidhari, M. and Baruah, D.C. (2011b). Crop residue biomass for decentralized electrical power generation in rural areas (part 1): investigation of spatial availability. *Renew. Sust. Energ. Rev.* 15 (4): 1885–1892.

Hiloidhari, M., Das, D., and Baruah, D.C. (2014). Bioenergy potential from crop residue biomass in India. *Renew. Sust. Energ. Rev.* 32: 504–512.

IEA (2002). Sustainable Production of Woody Biomass for Energy A Position Paper Prepared by IEA Bioenergy. ExCo 2002:03, https://www.ieabioenergy.com/wp-content/uploads/2013/10/25_PositionPaper-SustainableProductionofWoodyBiomassforEnergy.pdf (accessed 13 April 2021).

IEA (2015). International Energy Agency, CCC World forest and agricultural crop residue resources for co-firing. ISBN 978–92–9029–571-6. https://www.usea.org/sites/default/files/042015_World%20Forest%20and%20agricultural%20crop%0residue%20resources%20for%20cofiring_ccc249.pdf (accessed 13 April 2021).

IEA CCC (2014). Author's analysis. IEA Clean Coal Centre, London, UK.

Jain, N., Pathak, H., and Bhatia, A. (2014). Sustainable management of crop residues in India. *Curr. Adv. Agric. Sci.* 6 (1): 1–9. ISSN: 0975-2315.

Jat, H.S., Hat, R.D., Nanwal, R.K. et al. (2020). Energy use efficiency of crop residue management for sustainable energy and agriculture conservation in NW India. *Renew. Energy 155*: 1372–1382. https://doi.org/10.1016/j.renene.2020.04.046.

Jha, P., Danewalia, S.S., Sharma, G., and Singh, K. (2018). Antimicrobial and bioactive phosphate-free glass–ceramics for bone tissue engineering applications. *Mater. Sci. Eng. C* 86: 9–17.

Jiang, D., Zhuang, D., Fu, J. et al. (2012). Bioenergy potential from crop residues in China: availability and distribution. *Renew. Sust. Energ. Rev.* 16: 1377–1382.

Joshi, P. and Visvanathan, C. (2019). Sustainable management practices of food waste in Asia: technological and policy drivers. *J. Environ. Manag. 247*: 538–550. https://doi.org/10.1016/j.jenvman.2019.06.079.

Jutidamrongphan, W. (2018). Sustainable waste management and waste to energy recovery in Thailand. In: *Advances in Biofuels and Bioenergy* (ed. M. Nageswara-Rao and J.R. Soneji). InTech https://doi.org/10.5772/intechopen.74988.

Kaithwas, A., Prasad, M., Kulshreshtha, A., and Verma, S. (2012). Industrial wastes derived solid adsorbents for CO_2 capture: a mini review. *Chem. Eng. Res. Des.* 90: 1632–1641.

Kan, A. (2009). General characteristics of waste management: a review. *Energy Educ Sci Technol Part A* 23: 55–69.

Karthikeyan, O.P., Trably, E., Mehariya, S. et al. (2018). Pretreatment of food waste for methane and hydrogen recovery: a review. *Bioresour. Technol.* 249: 1025–1039. https://doi.org/10.1016/j.biortech.2017.09.105.

Kaur, G., Pandey, O.P., Singh, K. et al. (2014). A review of bioactive glasses: their structure, properties, fabrication and apatite formation. *J. Biomed. Mater. Res. A* 102: 254–274.

Khan, M.A.I., Ueno, K., Horimoto, S. et al. (2009). CIELAB color variables as indicators of compost stability. *Waste Manag.* 29: 2969–2975.

Kibler, K.M., Reinhart, D., Hawkins, C. et al. (2018). Food waste and the food-energy-water nexus: a review of food waste management alternatives. *Waste Manag.* 74: 52–62. https://doi.org/10.1016/j.wasman.2018.01.014.

Kiran, E.U., Trzcinski, A.P., Ng, W.J., and Liu, Y. (2014). Bioconversion of food waste to energy: a review. *Fuel* 134: 389–399. https://doi.org/10.1016/j.fuel.2014.05.074.

Koch, W. (2010). Landfill Projects on the rise. https://usatoday30.usatoday.com/money/industries/energy/2010-02-24-landfill-energy_N.htm (accessed 17 March 2021.

Koerbitz, W. (1999). Biodiesel production in Europe and North America: an encouraging prospect. *Renew. Energy* 16: 1078–1083.

Kormi, T., Mhadhebi, S., Bel Hadj Ali, N. et al. (2018). Estimation of fugitive landfill methane emissions using surface emission monitoring and genetic algorithms optimization. *Waste Manag.* 72: 313–328. https://doi.org/10.1016/j.wasman.2016.11.024.

Kumar, S. (2000). Technology options for municipal solid waste-to-energy project. *TERI Inform. Monit. Environ. Sci.* 5: 1–11.

Kwapinski, W. (2010). Biochar from biomass and waste. *Waste Biomass Valoriz.* https://doi.org/10.1007/s12649-010-9024-8.

Lange, L.C., Hills, C.D., and Poole, A.B. (1996). Preliminary investigation into the effects of carbonation on cement-solidified hazardous wastes. *Environ. Sci. Technol.* 30: 25–30.

Lange, L.C., Hills, C.D., and Poole, A.B. (1997). The effect of carbonation on the properties of blended and non-blended cement solidified waste forms. *J. Hazard. Mater.* 52: 193–212.

Lee, K.C.L. (2018). Grocery shopping, food waste, and the retail landscape of cities: the case of Seoul. *J. Clean. Prod.* 172: 325–334. https://doi.org/10.1016/j.jclepro.2017.10.085.

Liu, C., Hotta, Y., Santo, A. et al. (2016). Food waste in Japan: trends, current practices and key challenges. *J. Clean. Prod.* 133: 557–564. https://doi.org/10.1016/j.jclepro.2016.06.026.

Maries, A. and Hills, C.D. (1986). Improvement in concrete articles. UK Patent 2,192,392.

Micales, J.A. and Skog, K.E. (1997). The decomposition of forest products in landfills. *Int. Biodeterior. Biodegrad.* 39: 145–158.

Miles, T.R. Sr. and Miles, T.R. Jr. (1992). Environmental implications of increased biomass energy use. National Renewable Energy Laboratory (NERL) Report. NREL{fP-230-4633 UC Category: 247, DE92001219.

Mortier, N., Velghe, F., and Verstichel, S. (2016). Organic recycling of agricultural waste today: composting and anaerobic digestion. In: *Biotransformation of Agricultural Waste and By-Products: The Food, Feed, Fibre, Fuel (4F) Economy* (ed. P. Poltronieri and O.F. D'Urso), 69–124. Elsevier https://doi.org/10.1016/B978-0-12-803622-8.00004-5.

Nakamura, T. (2009). Waste agriculture biomass convention, the 6th Biomass Asia Workshop in Hiroshima, 18–20 November 2009, IETC Osaka. http://www.biomassasiaworkshop.jp/biomassws/06workshop/presentation/25_Nakamura.pdf (accessed 13 April 2021).

National Environment Agency (2018). Waste-To-Energy Incineration Plants [WWW Document]. https://www.nea.gov.sg/our-services/waste-management/wastemanagement-infrastructure/waste-to-energy-and-incineration-plants (accessed 13 January 2022).

Ong, K.L., Kaur, G., Pensupa, N. et al. (2018). Trends in food waste valorization for the production of chemicals, materials and fuels: case study South and Southeast Asia. *Bioresour. Technol.* 248: 100–112. https://doi.org/10.1016/j.biortech.2017.06.076.

Otles, S. and Kartal, C. (2018). Food waste valorization. In: *Sustainable Food Systems from Agriculture to Industry* (ed. C.M. Galanakis), 371–399. Academic Press https://doi.org/10.11646/zootaxa.3726.1.1.

Paul, S., Dutta, A., Defersha, F., and Dubey, B. (2018). Municipal food waste to biomethane and biofertilizer: a circular economy concept. *Waste Biomass Valoriz.* 9: 601–611. https://doi.org/10.1007/s12649-017-0014-y.

Paula, L.E.d.R., Trugilho, P.F., Napoli, A., and Bianchi, M.L. (2011). Characterization of residues from plant biomass for use in energy generation. *Cerne* 17: 237–246.

Principato, L. (2018). *Food Policies to Tackle Food Waste: A Classification*, 35–42. Cham: Springer https://doi.org/10.1007/978-3-319-78887-6_3.

Raneses, A., Glaser, L., Price, J., and Duffield, J. (1999). Potential biodiesel markets and their economic effects on the agricultural sector of the United States. *Ind. Crop. Prod.* 9: 151–162.

Saratale, G.D., Saratale, R.G., Banu, J.R., and Chang, J.S. (2019). Biohydrogen production from renewable biomass resources. In: *Biohydrogen*, Biomass, Biofuels, Biochemicals, 2e (ed. A. Pandey, S.V. Mohan, J.-S. Chang, et al.), 247–277. Elsevier https://doi.org/10.1016/B978-0-444-64203-5.00010-1.

Schieber, A., Stintzing, F.C., and Carle, R. (2001). By-products of plant food processing as a source of functional compounds-recent developments. *Trends Food Sci. Technol.* 12: 401–413.

Sharma, K. and Garg, V.K. (2017). Vermicomposting: a green technology for organic waste management. In: *Waste to Wealth. Energy, Environment, and Sustainability* (ed. R. Singhania, R. Agarwal, R. Kumar and R. Sukumaran), 199–235. Singapore: Springer https://doi.org/10.1007/978-981-10-7431-8_10.

Sharma, R., Kamble, S.S., Gunasekaran, A. et al. (2020a). A systematic literature review on machine learning applications for sustainable agriculture supply chain performance. *Comput. Oper. Res. 119*: 104926. https://doi.org/10.1016/j.cor.2020.104926.

Sharma, G., Kaur, M., Punj, S., and Singh, K. (2020b). Biomass as a sustainable resource for value-added modern materials: a review. *Biofuels Bioprod. Biorefin. 14* (3): 673–695. https://doi.org/10.1002/bbb.2079.

Singh, K. and Bahadur, D. (1999). Characterization of SiO_2-Na_2O-Fe_2O_3-CaO-P_2O_5-B_2O_3 glass ceramics. *J. Mater. Sci. Mater. Med.* 10: 481–484.

Singh, J., Panesar, B.S., and Sharma, S.K. (2008). Energy potential through crop biomass using geographical information system – a case study of Punjab. *Biomass Bioenerg.* 32 (2008): 301–307.

Slade, R., Saunders, R., Gross, R., and Bauen, A. (2012). Energy from biomass: the size of the global resource. An assessment of the evidence that biomass can make a major contribution to future global energy supply. Imperial College Centre for Energ policy and Technology and UK Energy Research Centre, London. ISBN: 1 903144 108.

Smil, V. (1999). Nitrogen in crop production. *Glob. Biochem. Cycles* 13: 647–662.

Speratti, A.B., Johnson, M.S., Sousa, H.M. et al. (2018). Biochars from local agricultural waste residues contribute to soil quality and plant growth in a Cerrado region (Brazil). *Arenosol* 10: 272–286.

Statistical abstract of Punjab (2018). Economic Adviser to Government. Punjab.

Sumner, H.R., Hellwig, R.E., and Monroe, G.E. (1985). Harvesting cotton plant residue for fuel [1984]. *AGRIS* 27: 968–972. https://agris.fao.org/agris-search/search.do?recordID=US8534951.

The Economist (2018). https://www.economist.com/graphic-detail/2018/10/02/global-waste-generation-will-nearly-double-by-2050 (accessed 21 January 2022).

Tripathi, N., Hills, C.D., Singh, R.S., and Carey, P.J. (2017). Construction: use waste for building. *Nature* 550: 457.

Tripathi, N., Hills, C.D., Singh, R.S., and Atkinson, C.J. (2019). Biomass waste utilisation in low-carbon products: harnessing a major potential resource. *NPJ Clim. Atmos. Sci. 2* (1): https://doi.org/10.1038/s41612-019-0093-5.

Whitehead, K., Hills, C.D., MacLeod, C.L., and Carey P.J. (2003). A Field Application of ACT for the Remediation of an ex-Pyrotechnics Site: Monitoring Data and Microstructural Studies, WASCON 2003, Proc. 5th International Conference on the Environmental and Technical Implications of Construction with Alternative Materials, San Sebastian, Spain, 4–6th June 2003.

Yang, W., Nam, H., and Choi, S. (2007). Improvement of operating conditions in waste incinerators using engineering tools. *Waste Manag.* 27: 604–613.

Yevich, R. and Logan, J. (2003). An assessment of biofuel use and burning of agricultural waste in the developing world. *Glob. Biogeochem. Cycles* 17: 1095.

3

An Overview of Biomass Conversion from Agricultural Waste: Address on Environmental Sustainability

Palas Samanta[1], Tarakeshwar Senapati[2], Sukhendu Dey[3], and Apurba Ratan Ghosh[3]

[1] *Department of Environmental Science, Sukanta Mahavidyalaya, University of North Bengal, Dhupguri, West Bengal, India*
[2] *Department of Environmental Science, Directorate of Distance Education, Vidyasagar University, Midnapore, West Bengal, India*
[3] *Department of Environmental Science, The University of Burdwan, Burdwan, West Bengal, India*

CHAPTER OUTLINE

Agriculture Waste Management and Bioresource: The Circular Economy Perspective, First Edition.
Edited by Suruchi Singh, Pardeep Singh, Anu Sharma, and Moharana Choudhury.

3.1 Introduction

Energy crisis is now global constraint for future sustainability. Fossil fuels dependence, which accounts for 90% energy consumption, results in climate change *vis-a-vis* global warming and simultaneously exhausts natural sources (Guo et al. 2015). Accordingly, scientists are looking for alternative, clean, and renewable energy sources. Biofuels, in this regard, find themselves as an alternative energy source, reduce fossil energy dependency and greenhouse gas (GHG) emissions, and ensure energy security (Passoth and Sandgren 2019). Biofuels, in recent times, are generated using first-generation substrates like wheat grain, sugarcane, corn, vegetable oil, or palm oil. However, the excessive use of first-generation substrates indirectly enhanced food prices and led to global food crisis, since human consumed these substances as food stuff. Further, cultivation of these food crops leads to loss of natural ecosystems (Kluts et al. 2017). Hence, the attention has been shifted to second-generation substrates which have rich lignocellulose contents such as straw, wood residues, or agricultural residues (Guo et al. 2015). In addition to this, lignocellulosic crops solved the food versus fuel debate (Kluts et al. 2017). Currently, approximately 200 billion tonnes of lignocellulosic biomasses are generated annually by agricultural sector worldwide (Guo et al. 2010). Accordingly, lignocellulosic food crops serve as novel second-generation biofuels. In particular, agricultural residue biomass represents an ideal candidate for biofuel production as it is a co-product of food production. Nevertheless, higher food production did not impose any adverse effects on utilization of agricultural residue for biofuel production (Chandra et al. 2012). Further, agricultural residues are very cheap, readily available in vast amount, and easily biodegradable (Guo et al. 2015).

Agricultural residue includes both food-based part (contains simple carbohydrates) like sugarcane, corn, wheat, and beets and non-food-based parts (contains complex carbohydrates) like stalks, leaves, straw, corn stover's cobs, rice husk, and orchard trimmings (Chandra et al. 2012). Generally, agricultural residues are classified into three categories: direct agricultural waste, i.e. crop residues; livestock waste namely manure; and food waste (Guo et al. 2010). Crop residues include biomasses like stover, straw, peelings, stalks, cobs, bagasse, and other lignocellulosic residues (Lee et al. 2019). Maize, wheat, rice, and sugarcane, which are cultivated worldwide, are the prime sources of crop residues (Chandra et al. 2012). Livestock wastes, in contrast, are distinguished into three categories: liquid or slurry (urine) manure; solid manure (farm yard); and wastewater which contains processed water from farms, silage juices, feedlot run-off, disinfectants, bedding, and liquid manure (Guo et al. 2010). According to reports, in the United States, only 60 million tonnes manure can be utilized in 2030 for bioenergy production (Mohammed et al. 2018). Food waste also plays a crucial role in biofuel production as it has high energy potential. Further, food waste is highly biodegradable by microbial actions as it contains 75–85% moisture and 85–95% volatile solids, favouring microbial development (Guo et al. 2010).

Biofuel production from agriculture-based residues is growing rapidly worldwide considering economic and environmental concerns. The term biofuel includes biodiesel, biohydrogen, biogas, bioethanol, bioethers, biomethanol, synthetic biofuels, biohydrogen, and bio-oils. Figure 3.1 represents different second-generation biofuel generated from agricultural feedstocks. These biofuels are renewable and act as an alternative to supplement fossil fuels or become a new permanent energy source in near future. Further, biofuels have several advantages over conventional energy sources: (i) they are renewable and can be easily stored, (ii) they reduce GHGs emission, (iii) they reduce foreign fuel dependency, (iv) they reduce pollution level, (v) they solve waste management problems, (vi) they improve energy security and stabilize energy crisis, and (vii) they help to develop a new circular economy (Guo et al. 2010; Callegari et al. 2019). Considering the earlier-mentioned statements, this review paper is aimed to address the potential of agricultural residues for biofuel production. More specifically, the review chapter addressed comprehensive information about agricultural residue types, biofuel types, conversion technologies for biofuel production, factors controlling biofuel production, agriculture-based bioenergy potentials in India, and environmental and economic concerns of biofuel production.

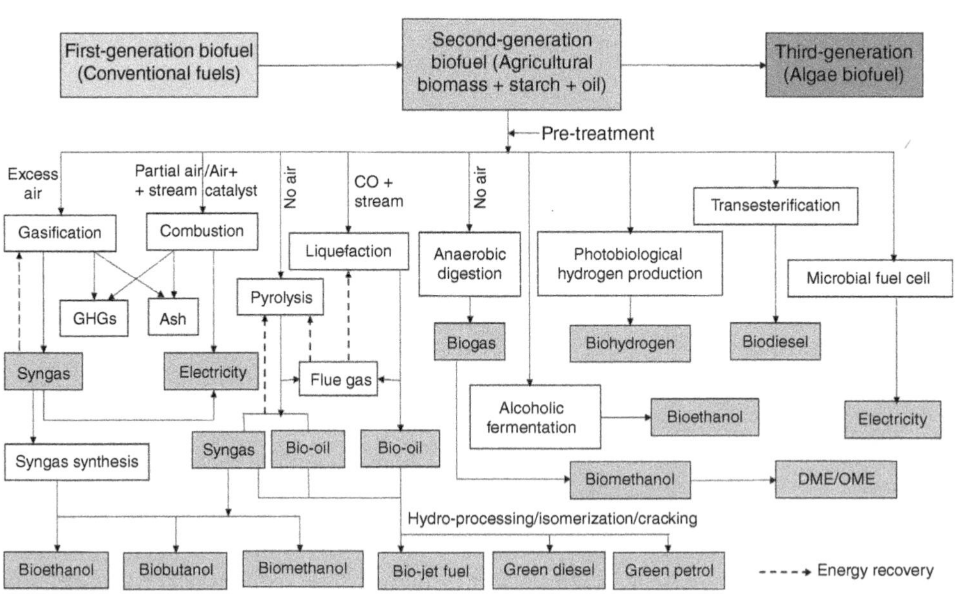

Figure 3.1 Different types of biofuel production with particular emphasis on second-generation biofuels from agricultural biomass and associated conversion technologies. *Source:* Modified after Verma et al. (2012), Fivga et al. (2019) and Lee et al. (2019).

3.2 Environmental Concerns of Renewable and Fossil Energy

Fossil fuel sources are mainly represented by coal, oil, and natural gas. They contribute approximately 90% of the world's primary energy demand today, because they are cheaply available and highly efficient (Chozhavendhan et al. 2020). Electricity generation from fossil fuels caused environmental consequences. Since 1900, the global CO_2 emissions from fossil energy have increased at a very alarming speed. China is in the driver seat to generate carbon, contributing up to 30% of global carbon emission; the United States accounts for 15% of the share; and India shares 8% of global CO_2 emissions (Boden et al. 2017). CO_2 emissions accelerated global phenomenon like global warming and climate change in long term. In addition to this, electricity generation from fossil fuels caused strain on water and land resources, while fossil fuel combustion emits pollutants like particulate matter (PM), nitrogen oxides (NOx), and sulphur dioxide (SO_2). Apart from this, annual capital expenditure on fossil fuels extraction, its transportation, and processing has increased several folds from 2000 to 2013 (World Energy Investment Outlook 2014). Additionally, ever-increasing population across the globe has aggravated much dependence on fossil fuel for energy generation (Awasthi et al. 2015).

This overdependence on fossil fuel for electricity generation might cause uncertainty on energy supply and politico-economic transitions, which in single word called 'energy trilemma'. Energy trilemma is defined by three terminologies, i.e. environmental sustainability, energy equity, and energy security. Energy security deals with continuous energy supply to masses, while energy equity denotes affordability of a particular energy. In contrast, concerns about environmental sustainability are to ensure low-carbon economy and reduction of GHGs emission to scale down global warming (Awasthi et al. 2015). Accordingly, to achieve environmental sustainability (i.e. environmental impacts reduction), shifting energy generation to renewable energy from fossil energy is major impetus.

Renewable energy is the energy that is naturally replenished with continuous supply. It contributes only 20% of global energy demand today, which is far below the expected global capacity (Musa et al. 2019). The major renewable energy sectors comprise solar, wind, hydropower, geothermal, tidal, and wave energy. Solar energy among them is most appealing alternate energy source. Renewable energy has a number of benefits over fossil fuel. Although renewable energy emits low GHGs amount and reduce conventional air pollution, they are unable to delineate all environmental issues. Renewable energy project installation contaminates water resources and disrupts land use, wildlife habitat, etc. (Musa et al. 2019). Accordingly, understanding the environmental impacts of renewable energy project is essential for utility operations through proper forecasting planning execution to mitigate or offset these problems.

Therefore, in quest for new renewable energy sources, policy-makers and governments moved to energy-efficient and cost-effective bio-based economy from fossil fuel-based economy. Biofuel industry, in this regard, could be a good solution to mitigate or offset these problems. Biofuels are generally developed from the wide variety of feedstock substances. Food crops are mainly used as feed material for first-generation fuels production (biodiesel and bioethanol). However, the major constrain of these first-generation biofuels is its fuel versus food debate. To address this concern, scientists developed biofuels from non-food crops and waste biomass. Recently, third-generation biofuels are developed by scientists to address the concern about feedstock's varied composition and associated conversion technologies of second-generation biofuels (Musa et al. 2019).

3.3 Agriculture and Biofuels

Agricultural residue biomasses are potential sources of renewable energy production. Green energy production from agricultural residues is considered as a principal contributor for country's economic development (Fivga et al. 2019). Apart from these, green energy, i.e. biofuel production, reduces environmental impacts such as carbon emission, global warming, and climate change phenomenon. Further, the produced biofuels are renewable in nature and production cost is very low as residues are cheaply available. A number of biofuels can be generated from agricultural residues based on physical state of biofuel: gaseous, liquid, solid, and other different chemical forms. Biogas, syngas, and biomethane are gaseous biofuels, while biodiesel, bioethanol, biobutanol, bio-jet fuel, bio-oil, dimethyl ether (DME)/poly (oxymethylene) dimethyl ethers (OME), green diesel, etc., are considered as liquid biofuels. Figure 3.2 represents replacement of traditional fuels by second-generation biofuel.

Biogas is a mixture of methane and carbon dioxide. It is an environment-friendly, very cheap, and clean versatile fuel. Synthesis gas, also called syngas, is a gasification or pyrolysis product containing H_2, CO_2, CO, and other ingredients. It is employed as an intermediate product during biomass conversion process (Schnürer 2016). Biomethanol, also known as wood alcohol, is generated from syngas or biogas and used in combustion engines. Biodiesel and bioethanol are widely used transport fuel worldwide. Biodiesel and fatty acid methyl esters (FAMEs) mixtures are developed from either non-edible or edible oils, while bioethanol is an alcoholic biofuel (CH_3CH_2OH) derived mainly from sugarcane. Both biodiesel and bioethanol are alternative of petro-diesel and dominating the market today. Bioethanol is developed from lignocellulosic materials using bioreactors as

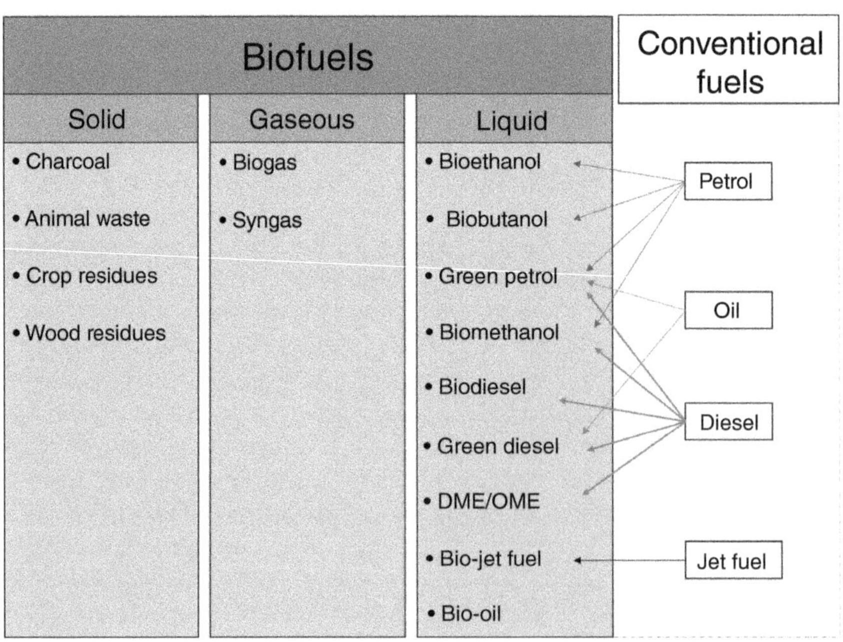

Figure 3.2 Biofuel types and replacement of traditional fuels by second-generation biofuels.

second-generation energy. Butanol is a renewable fuel and used as fuel for combustion engines but in less quantity than bioethanol and biodiesel (Callegari et al. 2019).

3.4 Agricultural Residue Composition and Properties

3.4.1 Composition of Agricultural Residues

The composition of agricultural residues varied in diverse quantity. For instance, plant-based residues mainly contain lignin, hemicellulose, cellulose with smaller quantity of pectin, protein, sugars, waxes, nitrogenous material, and inorganic minerals (Passoth and Sandgren 2019; Tursi 2019). In contrast, starch is the main ingredient for cereals, while protein is the prime ingredient of cattle manure. The major chemical constituents of agricultural residues are as follows:

3.4.1.1 Cellulose

Cellulose is a prime ingredient of plant cell wall containing linear polysaccharide polymer with maximum 10 000 D-glucose units, which are linked by β-1,4-glycosidic bonds (Tursi 2019). Cellobiose (4-o-β-D-glucopyranosyl-D-glucopyranose) constitutes the base of cellulose. It is an abundantly present organic molecule on earth either in crystalline or amorphous forms (Chandra et al. 2012). Cellulose is mainly consisted by crystalline cellulose with small percentage of amorphous cellulose. The cellulose chains are linked by van der Waals and hydrogen bonds, called 'elementary microfibrils'. These fibrils are associated as bundles and attached together by amorphous polymers (sugars), hemicelluloses, and pectin and covered by lignin. Although, cellulose structures are resistant to chemical as well as biological treatments, they are susceptible to enzymatic degradation in amorphous region for hydrolysis of molecules (Tursi 2019).

3.4.1.2 Hemicellulose

Hemicellulose is another major constituent of the plant cell wall (crosslink lignin and cellulose fibrils) and lignocellulosic biomass. Hemicelluloses are heterogeneous branched polysaccharides with xylan as a backbone and surrounded by glucuronic acid and arabinose. The hemicellulose backbone is linked either by β-1,3-glucan or by β-1,4-glucan bonds. The molecules, namely xylans, galactans, mannans, and arabinogalactans, predominantly played a crucial role in hemicellulose structure. Hemicellulose is less complex in structure, and its concentration in agricultural residues ranged between 25 and 35%. Generally, hemicelluloses are present as amorphous form and entirely composed of sugars with either five-carbon (arabinose and xylose) or six-carbon molecules (glucose, mannose, rhamnose, and galactose) with molecular weight <30 000 (Bonechi et al. 2017). Accordingly, they are easily decomposed (180–350 °C) compared with cellulose. Their structure and concentrations are dependent on plant types, tissue type, and even molecule structure (Tursi 2019).

3.4.1.3 Lignin

Lignin, a complex molecule, consisted of phenylpropanoid units (aromatic alcohols) in three-dimensional network. Generally, alkyl–alkyl, aryl–alkyl, and aryl–aryl ether bonds connected phenylpropanoid units. Coniferyl alcohol (guaiacyl propanol), sinapyl alcohol (syringyl alcohol), and coumaryl alcohol (p-hydroxyphenyl propanol) are the prime ingredients. Lignin contents are plant (from 25–30 to 50%) and age-dependent. Lignin provides structural strength, stiffness, and protection against pathogens (Passoth and Sandgren 2019). Further, lignin provides resistant to biological and chemical degradation (Passoth and Sandgren 2019). Lignin-derived value-added products are generally grouped as biofuels, aromatics, and macromolecules (Tursi 2019). Nevertheless, lignin is a recalcitrant constituent of lignocellulosic material to degrade (Bhatia et al. 2012).

3.4.1.4 Starch

Starch is the prime ingredient of carbohydrates. Generally, it is found as granules, and each granule contains millions of amylopectin and amylose molecules. Additionally, starch in nature is found in two forms: 25–27% soluble non-branched amylose (500–20 000 α-(1-4) D-glucose units) and 73–75% water-insoluble amylopectin. Amylopectin is consisted by linear α-D-glucopyranosyl units which are linked by α-(1 → 4) and α-(1 → 6) bonds (Tursi 2019). The empirical formula is $(C_6H_{10}O_5)_n$, where α-D-glucose units ($^4C_1$) are linked together (Edwards et al. 2003). Generally, amylose (without amylopectin) is produced by amylopectin hydrolysis with pullulanase, while amylopectin (without amylose) is produced from corn starch (Tursi 2019).

3.4.1.5 Minor Organic Substances and Fluid Matter

Agricultural residues sometimes contain organic compounds in trace amount. For example, 2–4% acetyls, 1–4% uronic acids, up to 10% proteins, and 2–40% lipids are detected in different agricultural biomass (Bhatia et al. 2012; Tursi 2019). Fluid material contains various anion and cation species. Residues contains 10–60% moisture normally, and they can reach as high as 80–90%.

3.4.2 Properties of Agricultural Residues

3.4.2.1 Proximate Properties

Generally, air-dried agricultural residue contains approx. 15–20% moisture and the remaining part is total solids. The solid mass is dominated by volatile dry solids and inorganic matter, i.e. ash in trace amount. In addition to this, this volatile dry solid undergoes energy conversion. Calcium, magnesium, sodium, potassium, magnesium, silicon, phosphorus, aluminium, and iron are common mineral elements found in ash (Alaswada et al. 2015). In addition to this, sulphates, chlorides, oxalates, carbonates, nitrates, and both inorganic and organic amorphous substances are found in ash (Tursi 2019). Sometimes, depending on conditions, the residues also contain carbon, oxygen, hydrogen, nitrogen, and sulphur as gaseous compounds (Tursi 2019).

3.4.2.2 Ultimate Properties

Carbon, hydrogen, and oxygen are the prime ingredients of agricultural residue. Generally, residue contains approx. 35–50% carbon and 40–45% oxygen on dry weight basis. The ultimate properties of residue governed the quality and quantity of product during energy production. Further, the elemental knowledge of residue helps in parameters adjustments during energy conversion process (Tursi 2019).

3.5 Agricultural Residue Pre-treatment Methods

Lignin, hemicellulose and cellulose provides protective barrier against biological and chemical degradation. Accordingly, hydrolysis of non-pre-treated residues is slow and produces lower yield (fermentable simple sugars). Therefore, pre-treatment is necessary for efficient biomass conversion to energy. Generally, pre-treatment enhances pore size and surface area, reduces degree of polymerization and crystallinity, increases reactivity to treatment methods, and produces higher yield (Bhatia et al. 2012; Chandra et al. 2012; Nwosu-Obieogu et al. 2016). Further, pre-treatment should fulfil the following criteria: (i) it should enhance simple sugar formation; (ii) it must destroy lignin shell; (iii) it must avoid loss of carbohydrate; (iv) it must minimize inhibitors production; and (v) its process must be cost-effective (Bhatia et al. 2012; Chandra et al. 2012). Pre-treatment methods are generally categorized into four types: (i) physical or mechanical method; (ii) chemical method; (iii) physicochemical method; and finally (iv) biological method. However, the pre-treatment method selection depends on biomass type.

3.5.1 Physical Pre-treatment Methods

Physical treatment, generally, is employed for size reduction through *milling/grinding* (hammer, two-roll, ball, colloid, and vibratory milling) or *chipping* to enhance the efficiency of downstream processing. Further, milling/grinding or chipping is used to reduce cellulose crystallinity. Although pre-treatment method is cost-effective and ecofriendly, a scanty of information in a collective way is still available. Chipping method yielded products in the 10–30 mm size range, while milling/grinding usually yielded 0.2–2 mm products. Among different milling techniques, vibratory milling is more effective to break down substrate's crystallinity and improve biomass digestibility (Bhatia et al. 2012). However, power requirement of *milling/grinding* or *chipping* methods depends on biomass type and particle size (Bhatia et al. 2012). Generally, high energy is required for *milling/grinding* or *chipping* methods.

Ultrasonic technique is another essential physical treatment method. *Ultrasonic* technique is generally employed to break down agricultural biomass into smaller molecules. In addition to this, *ultrasonic* technique shortened the extraction time, improved extractability of hemicelluloses, cellulose and lignin, and increased the yield (Bhatia et al. 2012). *Irradiation* through gamma rays or microwave or electron beam is proved to be another important technique to break down the lignin layer of agricultural residues. Generally, *irradiation* method breaks down β-1,4-glucan bonds of lignocellulosic materials to improve the surface area through size reduction and simultaneously lower lignocellulosic material's crystallinity (Tursi 2019).

Pyrolysis is generally employed to treat lignocellulosic materials. In *pyrolysis* technique, agricultural residues, mainly cellulose, are treated at higher temperatures (>300 °C) to produce gaseous by-products and residual char (Chandra et al. 2012). The decomposition of agricultural residues is much slower at lower temperatures and produced fewer volatile products. Other mechanical techniques such as *expansion* and *extrusion* are also used to decompose cellulose portion of lignocellulosic substances. The degradation of cellulose portion of lignocellulosic substances by these techniques produced fragile fibres and low-molecular-weight cellobiose and oligosaccharides. However, the major disadvantage of this method is higher cost (Lee et al. 2019; Tursi 2019).

3.5.2 Chemical Pre-treatment Methods

3.5.2.1 Acid Treatment

Acid pre-treatment generally uses diluted/concentrated acids to hydrolyze lignocellulosic biomass. Diluted sulphuric acid (H_2SO_4) is commonly used to treat wide variety of agricultural biomass (Lee et al. 2019). Generally, in diluted acid treatment, acid is sprayed over biomass residue for hydrolysis of cellulose, lignin, and hemicellulose under temperature range 160–220 °C. On contrarily, strong acids (H_2SO_4 and HCl) completely break down biomass components to sugars, but this process needs large volumes, and both methods produced furfural, an inhibitory by-product (Bhatia et al. 2012). The major advantage of this method is that the process does not need enzymatic hydrolysis, as acid treatment produced fermentable sugars (Nwosu-Obieogu et al. 2016). Sometimes, mixed acids such as H_2SO_4 and acetic acid are used for efficient treatment (Nwosu-Obieogu et al. 2016). However, the prime disadvantages are high acid cost and the need of neutralization.

3.5.2.2 Alkaline Treatment

Alkaline treatment generally uses bases namely potassium, sodium, calcium, magnesium or ammonium hydroxide, and ammonium sulphite for lignocellulosic biomass pre-treatment. Alkaline treatment aids to degrade ester and lignin's glycosidic chains, swelling and decrystallization of cellulose (Lee et al. 2019), and partial hemicellulosic solvation (Chandra et al. 2012). Additionally, alkaline treatment decreased cellulose's crystallinity and polymerization but enhanced accessible area (Chandra et al. 2012). For example, sodium hydroxide is extensively used to break down lignin structure and enhance enzyme's accessibility to hemicellulose and cellulose (Bhatia et al. 2012; Lee et al. 2019). Lime is also used for the pre-treatment of lignocellulosic feedstocks. Alkali efficacy followed the order as NaOH > KOH > $Mg(OH)_2$ and $Ca(OH)_2$. The technique is generally performed at ambient conditions, but it requires longer times and low temperatures. Among the different alkali treatments, lime treatment is very cost-effective. NaOH and $Ca(OH)_2$ are two commonly used alkali for biomass treatment. Particularly, $Ca(OH)_2$ treatment is more preferable due to cost-effectiveness, environment-friendly nature, and easy recovery than NaOH (Tursi 2019).

3.5.2.3 Oxidative Treatment

Oxidative treatment generally uses oxidizing agents, namely H_2O_2, oxygen, and ozone for cellulosic delignification. Oxidative treatment is distinguished as wet oxidation, oxidative agents, and ozonolysis. Wet oxidation uses oxygen as oxidizer. In wet oxidation, water is added during drying and milling of lignocellulosic biomass coated with dense wax (1 l water + 6 g biomass) to fractionate lignocellulosic material (particle size 2 mm) (Bhatia et al. 2012). Lignin is broken down to water, CO_2, and carboxylic acids during wet oxidation. H_2O_2 is the prime oxidizing agents and is used for delignification of cellulose. In ozonolysis technique, ozone is used for delignification of cellulose. However, the oxidative treatment in combination with other pre-treatment method showed higher efficacy for hydrolysis of lignocellulosic material. For example, wet oxidation in combination with alkaline pre-treatment (0.248 g $Ca(OH)_2$/g Alamo switch grass (*Panicum virgatum*) at 110 °C for 240 minutes and 6.89 bar O_2) reduced the formation of by-products (inhibitors) and enhanced the yield of sugar (26%) (Falls and Holtzapple 2011). Using oxidizing agents such as H_2O_2 or $C_2H_4O_3$ in alkaline pre-treatment (NaOH) under mild temperature condition enhanced the digestibility of crop residue than NaOH treatment alone (Nwosu-Obieogu et al. 2016).

3.5.2.4 Green and Organo-Solvents

Ionic liquids (ILs) and other solvents have gained considerable attention recently for treating lignocellulosic biomass due to tunable solvent chemistry. ILs are basically salts, contain both anionic (small) and cationic (large) part, and possess low vapour pressure. ILs have a great potential to dissolve cellulose-containing biomass (Bhatia et al. 2012). Cunha-Pereira et al. (2020) reported IL-treated cellulose significantly enhanced hydrolysis, and IL is recovered successfully. N-methylmorpholine N-oxide (NMMO), that has similar property with IL, is used to break down a wide range of lignocellulosic materials without chemical treatment, and simultaneously, greater than 99% of solvent recovery is possible because of reduced vapour pressure (Bhatia et al. 2012). Recently, acidic organic solvents (19.5% water, 80% ethylene glycol, and 0.5% HCl mixture at 178 °C for 90 minutes) are widely used for the pre-treatment of lignocellulosic materials (Yamashita et al. 2010). The major advantages of this method are higher potentiality of organic solvent recycling and recovery but require expensive high-pressure equipment.

3.5.3 Physicochemical Pre-treatment Methods

3.5.3.1 Liquid Hot Water (LHW) Treatment

Liquid hot water (LHW) (hydrothermal), which is heated at 150–240 °C temperature under high pressure, is used for hemicelluloses, cellulose, and lignin hydrolysis (Harmsen et al. 2010). It is considered as one of the prime treatment techniques. However, the reaction time (minutes to hour) depends on sugar formation types and amount (Yu et al. 2010; Nwosu-Obieogu et al. 2016). The technique is found to be advantageous from a number of points, such as no additives are required, required less expensive reactor, degradation products do not need neutralization, and no corrosion problems and no inhibitory products are generated.

3.5.3.2 Steam Explosion Treatment

Steam explosion is very common physicochemical method to break down lignocellulosic material (Chandra et al. 2012). Generally, steam explosion technique is based on hydrothermal method. Basically, lignocellulosic material is subject to highly pressurized saturated steam (0.69–4.83 MPa) at 160–260 °C under short duration (seconds to few minutes) after which system is depressurized rapidly. As a result, biomass undergoes rapid decompression to transform lignin and to degrade hemicellulose through disruption of cellulose microfibrils (Harmsen et al. 2010; Nwosu-Obieogu et al. 2016). Apart from these benefits, dilute acids are needed for the better pre-treatment of biomass. Further, high energy is needed for the efficient treatment. Additionally, factors namely chip size, residence time, temperature and moisture level adversely affect the process (Chandra et al. 2012; Nwosu-Obieogu et al. 2016).

3.5.3.3 Ammonia Fibre Explosion (AFEX) Treatment

Like steam explosion, the ammonia fibre explosion (AFEX) process is conducted under high pressures at 90–100 °C in the presence of liquid anhydrous ammonia and then rapidly depressurized. As a result, hemicellulose and lignin level reduced significantly but decrystallize cellulose (Harmsen et al. 2010). The major benefits of this technique are cost-effective, less energy required compared with stream explosion, low inhibitory products generation, and 100% material recovery. Environmental issue due to ammonia is a major disadvantage (Nwosu-Obieogu et al. 2016).

3.5.3.4 Ammonia Recycle Percolation (ARP) Treatment

Generally, ammonia recycle percolation (ARP) technique is used in association with AFEX process. In this process, aqueous ammonia (5–15%, wt%) is injected (5 ml/min) through packed bed reactor containing biomass to hydrolyze lignocellulosic substances. The process removes approx. 75–85% of lignin and solubilizes 50–60% of hemicellulose (Bhatia et al. 2012). The performance is comparatively better than AFEX method.

3.5.3.5 Supercritical Fluid (SCF) Treatment

Recently, supercritical fluid (SCF), a state where both gas and liquid coexist together, is used in treating lignocellulosic substances. It shows liquid like density and behaves like gas for viscosity and diffusivity (Chandra et al. 2012; Lee et al. 2019). As a result, they penetrate easily into crystalline lignocellulosic feedstock. Abrahamsson et al. (2015) reported that SC-CO_2-treated agricultural biomass (linseed) produced higher lipid yields. Alinia et al. (2010) observed that the combined supercritical CO_2 and steam method yielded more sugars than supercritical CO_2 alone. Sometimes, high-pressure CO_2 is used to degrade them. Actually, pressure reduction creates explosive decompression of biomass to improve the hydrolysis process (Tursi 2019).

3.5.4 Biological Pre-treatment Method

Biological method generally uses microbes (fungi, actinomycetes, and bacteria) and their enzymes for the selective degradation of substances (Nwosu-Obieogu et al. 2016). Particularly, brown-, white-, and soft-rot fungi are abundantly used to break down lignocellulosic substances (Bhatia et al. 2012). White-rot fungi proved to be more effective for biological decomposition. Among white-rot basidiomycetes *Penicillium chrysosporium*, *Phlebia radiata*, *P. floridensis*, and *Daedalea flavida* are widely used for lignin and cellulose treatment, while brown rots are mainly used to hydrolyze cellulose (Bhatia et al. 2012). Among bacteria, *Sphingomonas paucimobilis* and *Bacillus circulans* are very common to treat lignocellulosic substances (Bhatia et al. 2012). The major benefits of biological method are as follows: it needs very lower energy, it is environmentally mild in nature, it reduces degree of cellulose polymerization, and it produces minimal waste. However, the degradation rate of feedstock materials by biological method is very low.

Generally, the pre-treated hemicelluloses and celluloses are hydrolyzed to form soluble sugars using cellulases, ligninases, and hemicellulases. The enzymatic conversion eliminates corrosion problems, reduces maintenance costs, and gives high yields. Traditionally, three complementary enzymes are employed to break down cellulose: exoglucanases, β-glucosidases, and endoglucanases (Payne et al. 2015; Tursi 2019). A number of microorganisms (either anaerobic or aerobic and either thermophilic or mesophilic), namely fungi and bacteria, produce cellulases to break down lignocellulosic substances. Bacteria known to produce cellulases include genera of *Cellulomonas, Clostridium, Ruminococcus, Bacillus, Thermomonospora, Acetovibrio, Bacteroides, Erwinia, Streptomyces*, and *Microbispora* (Bhatia et al. 2012; Tursi 2019). Bacterial species namely *Clostridium thermocellum*, *Clostridium phytofermentans*, *Clostridium hungatei*, *Clostridium cellulovorans*, and *Clostridium papyrosolvens* are very common that produce cellulases. *Trichoderma reesei, Aspergillus niger, Phanerochaete chrysosporium, Sclerotium rolfsii,* and various *Schizophyllum* and *Penicillium* sp. are commonly known fungal species to produce glucanases (cellulases) (Passoth and Sandgren 2019).

Generally, hemicelluloses are hydrolyzed using thermal pre-treatment or acid treatments for the production of sufficient sugar monomers to aid enzymatic degradation. Due to their heterogeneous nature, hydrolysis of hemicelluloses needs different enzymes, including β-xylosidases, endo-xylanases, α-glucuronidases (GH family), and acetyl xylan esterases (CE family) (Passoth and Sandgren 2019). For example, in the first step, the xylan's β-1,4 linkages (major constituents of hemicelluloses) are hydrolyzed by xylanases for oligomers production, which are again hydrolyzed into xylose by β-xylosidase. In contrast, mannan-based or arabinofuranosyl-containing hemicelluloses need enzymes like α-L-arabinases, β-mannanases, or arabinofuranosidases for hydrolysis. Fungal species namely *Penicillium, Trichoderma, T. emersonii*, and *Aspergillus* have widely been known for the production of hemicellulases (Bhatia et al. 2012; Tursi 2019).

Fungi also play crucial role in lignin degradation by secreting enzymes known as ligninases. Generally, peroxidases (manganese peroxidase and lignin peroxidase) and phenol oxidase (laccase) are responsible for ligninases production (Passoth and Sandgren 2019). White-rot basidiomycetes such as *P. chrysosporium*, *Coriolus versicolor*, and *T. versicolor* are proved to be more efficient for the degradation of lignin (Bhatia et al. 2012).

3.6 Biomass Conversion Technologies

The production technologies of biofuel from agricultural residue biomass are generally classified into following three categories (Figure 3.3): thermochemical, biochemical, and physicochemical conversion technologies.

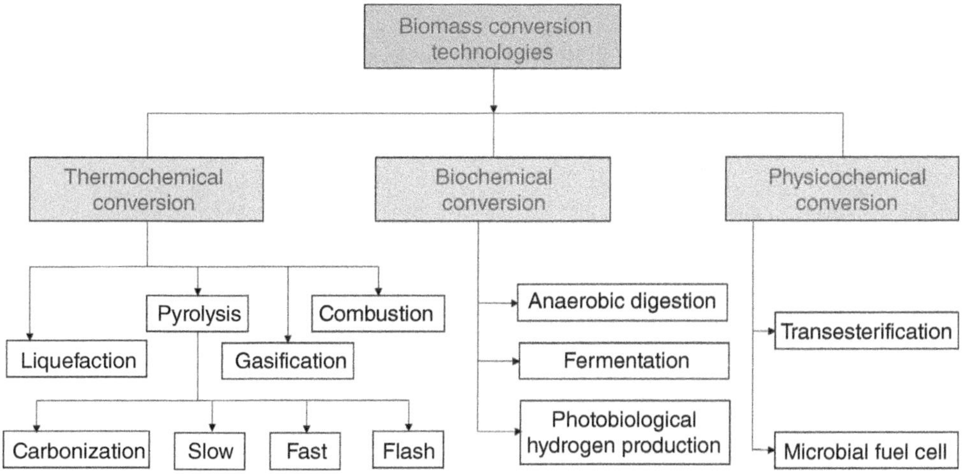

Figure 3.3 Different biomass conversion technologies for biofuel production.

3.6.1 Thermochemical Technological Conversion of Biomass

Generally, in thermochemical process, the electric or heat energy is produced by using chemical and heat process. The end products are mainly solid biochar, liquid bio-oil, and syngas. It is highly used conversion technologies due to industrial infrastructure availability as well as processes in very short time, reduces water usage, and has a capacity to acquire energy from plastics wastes (Lee et al. 2019). Generally, four thermochemical conservation techniques, namely pyrolysis, combustion, liquefaction, and gasification, are employed for biomass conversation.

3.6.1.1 Pyrolysis

Pyrolysis process has a great potential for all types of biomass conversions like into solid (charcoal), bio-oil, and combustible gas through combustion at higher temperatures (500 °C) under no oxygen level (Tursi 2019). Generally, pyrolysis process takes place in four stages namely biomass drying (100–120 °C) followed by distillation of outlet gases, exothermic reactions (280–500 °C), and removal of volatile compounds by evaporation. The process can be divided into four categories based on operating condition: carbonization, slow, and fast and flash pyrolysis (Table 3.1). It varies in heating rate, processing temperature, biomass particle size, and solid residence time, etc.

Carbonization, which is most ancient pyrolysis process, is used for complete conversion of biomass at temperature ranges from 300 to 500 °C under the absence of oxygen (Verma et al. 2012). ***Slow pyrolysis***, which is considered as a conventional method for biochar production, is generally employed at very low temperature (around 500 °C) and low heating rates. Additionally, slow pyrolysis process requires longer reaction time (5–30 minutes) and transformation time to convert biomass. Generally, slow pyrolysis is employed for bio-oil generation. However, the process requires higher energy input due to longer residence time (Tursi 2019). ***Fast pyrolysis***, which is generally employed for biofuel production, is conducted at medium–low temperature (500–650 °C), high heating rate (>200 °C/s), and short residence time (<2 seconds). Fast pyrolysis process yields approx. 60–75% oily products, 15–25% solids biochar, and 10–20% gaseous substance depending on raw materials. The prime advantages of this process are high heating rate, very short residence time, clear control over reaction temperature, and rapid cooling of yielded compounds (Tursi 2019). Fast pyrolysis method needs relatively low capital costs and has higher efficiency than other methods. In ***flash pyrolysis***, biomass conversion is conducted over 650 °C temperature under

Table 3.1 Characteristics of different pyrolysis techniques.

Production process	Temperature (°C)	Heating rate	Residence time	Biochar (%)	Bio-oil (%)	Syngas (%)
Carbonization	300–500	Slow to fast	Seconds to hours	50	50	–
Slow	~500	Slow (0.01–2 °C/s)	5–30 min	35	30	35
Fast	500–650	Fast (>200 °C/s)	Seconds (<2 s)	10	70	20
Flash	>650	Faster	Second (<1 s)	10–15	70–80	5–20

very short gas residence time (<1 seconds) and high reaction time (Tursi 2019). It is a very complex method for biofuel production (up to 75% bio-oil) from biomass (Tursi 2019). Moreover, this method has some technical fault, for example thermal instability and higher oil and solids corrosiveness in oil, enhanced viscosity, and so on (Kaltschmitt 2013; Tursi 2019).

3.6.1.2 Combustion

Combustion is another commonly accepted process for biomass transformation. It shares approx. 97% bioenergy production all around the world. Combustion is a traditional method for energy production from biomass for cooking and heating in some underdeveloped countries. Combustion process is highly acceptable technology due to cost-effectiveness and commercial availability (Kaushika et al. 2016). In combustion, generally three stages are involved to produce solid char and volatile gases: drying, pyrolysis, and reduction (Lebaka 2013). Volatile gas combustion contributes approx. 70% of total heat generation (Tursi 2019).

During combustion, oxygen is added to the biomass material to react with each other under high-temperature environment (800–1000 °C) to generate electricity. The process is conducted inside combustion chamber under <50% moisture level. Still, electricity generation depends on factors primarily on biomass type (Lee et al. 2019). The typical reactions of combustion reactions are described as follows:

$$\text{Biofuel} + \text{Oxygen} \rightarrow \text{Carbon Dioxide} + \text{Water} + \text{Heat}$$
$$CH_{1.44}O_{0.66} + 1.03\ O_2 \rightarrow CO_2 + 0.72\,H_2O + \text{Heat}$$

3.6.1.3 Liquefaction

Liquefaction is another important biomass transformation method carried out at 250–370 °C temperature under 10–25 MPa/700–3000 psi pressure. The process is totally different from pyrolysis with respect to operating condition and requires catalyst and final products. Liquefaction generates mainly liquid with small amounts of gaseous components in the presence of hydrogen (López Barreiro et al. 2013). Hydrothermal liquefaction (HTL) is widely recognized technique that utilizes subcritical water (SCW) under pressure range 40–220 bar at 250–374 °C to convert biomass into bio-oil. The end product has high heating value, low oxygen content, and chemically stable fuel. Apart from these, the prime aim of this technique is to produce high H/C ratio oil. The process is generally employed in two steps: in the first step catalysts hydrolyze biomass material into smaller molecules, and in the next step these smaller molecules are further degraded by dehydration followed by dehydrogenation, deoxygenation, and decarboxylation reactions (Verma et al. 2012; Lee et al. 2019). However, the process is commercially not successful, because process yield lower overall oil (between 20 and 55% w/w) compared with other commercial methods (Verma et al. 2012; Lee et al. 2019).

3.6.1.4 Gasification

The gasification technique converts biomass into syngas at temperature ≥800 °C under 33 bar atmospheric pressure with low/absent oxygen level. Syngas is used to generate electricity as well as ammonia, synthetic gasoline, methanol, etc. (Rahimpour et al. 2012; Lee et al. 2019). The technique is a very interesting method, which has an ability to evacuate all pollutants from syngas. However, the methodology is complex in nature compared with other thermochemical methods as it is conducted at air-tight closed chamber under low air pressure (Tursi 2019). Gasification of biomass is generally performed by following four stages (Portha et al. 2017; Lee et al. 2019): ***oxidation stage*** (exothermic in nature) is employed to generate electricity from biomass in the absence of oxygen. The typical reactions are as follows (Kaltschmitt 2013; Tursi 2019):

$$\text{Coal combustion} \quad C + O_2 \rightarrow CO_2 \quad \Delta H = -394\,\text{kJ/mol}$$

$$\text{Partial oxidation} \quad C + \tfrac{1}{2}\,O_2 \rightarrow CO \quad \Delta H = -111\,\text{kJ/mo}$$

$$\text{Hydrogen combustion} \quad H_2 + \tfrac{1}{2}\,O_2 \rightarrow H_2O \quad \Delta H = -242\,\text{kJ/mol}$$

Drying step (endothermic nature) removes moisture from biomass by converting moisture into steam at temperature 100–150 °C. Heat input is proportional to biomass moisture (Molino et al. 2016). In ***pyrolysis stage***, biomass is anaerobically decomposited to syngas, char, tars, and light hydrocarbons between 250 and 700 °C (biomass $\leftrightarrow H_2 + CO + CO_2 + CH_4 + H_2O$ (g) + Tar + Char). Pyrolysis stage is endothermic in nature (Roos 2010; Tursi 2019). ***Reduction phase***, also called gas phase reactions, is the last step of gasification process. In this step, end products of oxidation and pyrolysis are transformed to form final syngas. Reduction phase is endothermic in nature and favoured by temperature elevation. Temperature elevation is determinant of coal oxidation (high temperature indicates higher oxidation) and syngas composition and characteristics (Lee et al. 2019; Tursi 2019). The typical reactions of reduction phase are shown as follows (Zhang et al. 2013; Tursi 2019):

$$\text{Boudouard reaction}: C + CO_2 \leftrightarrow 2CO \quad \Delta H = 172\,\text{kJ/mol}$$

$$\text{Character reform}: C + H_2O \leftrightarrow CO + H_2 \quad \Delta H = 131\,\text{kJ/mol}$$

$$\text{Water gas displacement reaction}: CO + H_2O \leftrightarrow CO_2 + H_2 \quad \Delta H = -41\,\text{kJ/mol}$$

$$\text{Methanation}: C + 2H_2 \leftrightarrow CH_4 \quad \Delta H = -75\,\text{kJ/mol}$$

Recently, plasma gasification technique is used to convert biomass that is harmful. It is an allothermal process where generation of syngas, ash, and slug was conducted using exterior power. The technique is proved to be more efficient to convert all biomass types as process is conducted at higher temperature (Lee et al. 2019).

3.6.2 Biochemical Technological Conversion of Biomass

Biochemical methods utilized biological agents such as fungi, yeast, and/or specialized bacteria and their secreted enzymes to convert biomass into simpler forms. The widely used biochemical options are photobiological, fermentation, and anaerobic digestion (AD) techniques (Lee et al. 2019).

3.6.2.1 Anaerobic Digestion

AD is most widely accepted technique not only for renewable energy production but also for systematic waste management (Zamani 2015). AD takes place in four stages: biopolymer hydrolysis, acidogenesis, acetogenesis, and methanogenesis (Chandra et al. 2012; Passoth and Sandgren 2019). A series of chemical reactions happened in total oxygen-free environment through different metabolic pathways, governed by micro-organisms. During hydrolysis, agricultural biomass is converted into monomers, for example, amino acids, monosaccharides, and short-chain fatty acids by exoenzymes. In the next step, i.e. acidogenesis, monomers produced in hydrolytic phase are degraded mainly to volatile fatty acids (short-chain organic acids) and alcohols as well as C1–C5 molecules, ammonia, hydrogen, CO_2, and other compounds (Lee et al. 2019). In the acetogenesis step, the short-chain organic acids and alcohols are converted to CO_2, acetic acid, and hydrogen by homo-acetogenic micro-organisms. Finally, in methanogenesis, hydrogen and acetate are transformed into methane and CO_2 by methanogenic micro-organisms, called methanogenesis [mainly by acetoclastic methanogenesis ($C + 4H_2 \rightarrow CH_4 + 2H_2O$), hydrogenotrophic methanogenesis ($CH_3COOH \rightarrow CH_4 + CO_2$), and methyltrophic methanogenesis]. The first three steps are driven by bacterial consortia, but methanogenesis is performed by archaea (Schnürer 2016; Lee et al. 2019; Tursi 2019). Total chemical reactions break down biomass macromolecules to produce biogas and digestate.

3.6.2.2 Fermentation

Fermentation process is generally employed to ferment hydrolytic products (mannose, glucose, galactose, arabinose, and xylose) to form ethanol. The conversion of simple sugars to ethanol and CO_2 is generally governed by a number of biochemical reactions. The reaction is mainly conducted under anaerobic environment with the help of micro-organisms, mainly yeasts or bacteria. In particularly, *Saccharomyces cerevisiae* is important micro-organism during fermentation. In addition to this, different metabolic activities are generally responsible for this transformation into ethanol. For example, glycolysis or Embden–Meyerhof–Parnas (EMP) pathway is used for ethanol production from hexoses such as glucose, while pentose phosphate pathway is used for pentoses (xylose and arabinose) (Chandra et al. 2012; Callegari et al. 2019; Lee et al. 2019). Recently, genetically engineered bacteria such as *Rhodobacter* sp. is successfully used for the production of ethanol (Lee et al. 2019). The end product, that is ethanol, is finally distilled and dehydrated for obtaining chemical-grade alcohol (Tursi 2019). The typical reactions occurred during fermentation process are mentioned as follows (Tursi 2019):

$$\text{Sugars} \rightarrow \text{Ethanol} + CO_2 + \text{by} - \text{products}$$
$$C_6H_{12}O_6\left(\text{hexoses}\right) \rightarrow 2C_2H_5OH + 2CO_2$$
$$3C_5H_{10}O_5\left(\text{pentoses}\right) \rightarrow 5C_2H_5OH + 5CO_2$$

3.6.2.3 Photobiological Hydrogen Production

Photobiological hydrogen production method is generally employed for hydrogen production from biomass like algae. Generally, microalgal H_2 production is performed by two methods. During photosynthesis, microalgae produce H^+ and O_2 from water. In the next step, the produced H^+ is converted into H_2 gas by hydrogenase enzymes under anaerobic conditions. But the major problem is that the oxygen produced in this method inhibits hydrogenase enzymes. The second method utilizes a two-phase system for H_2 production. In the first step, algae are cultured, and in later phase continuously H_2 is generated under sulphur-deprived conditions (Lee et al. 2019). Actually, sulphur deprivation engages algae for their survival, and during algal survival, the generated H_2 provides energy to algal cells (Chandra et al. 2012; Lee et al. 2019).

3.6.3 Physicochemical Conversion Technique of Biomass

Physicochemical method generates high-density biofuels, mainly different types of vegetable oil and biodiesel by transesterification processes. Transesterification process converts oils and fats into glycerol and esters through the use of methyl or ethyl alcohol in the presence of catalysts such as NaOH or KOH (Tursi 2019). Generally, transesterification of triglycerides takes place at atmospheric pressure and at very low temperature (50–70 °C). The end product is neutralized and purified before its use (Leung et al. 2010):

$$\text{Oil} + \text{Short chain alcohol} \rightarrow \text{Biodiesel} + \text{glycerol}$$

3.7 Production of Biofuels from Agricultural Biomass

3.7.1 Bioethanol Production

Bioethanol production is most intensively investigated application of agricultural biomass. Generally, two processing routes are used to produce bioethanol from agricultural biomass: biochemical fermentation of sugar and syngas fermentation (Fivga et al. 2019; Passoth and Sandgren 2019). In biochemical fermentation, the pre-treated agricultural biomass (acid or enzymatic hydrolysis) produced fermentable sugar ($C_{12}H_{22}O_{11} \rightarrow C_6H_{12}O_6 + C_6H_{12}O_6$). In the next step, this fermentable sugar is converted into bioethanol ($C_6H_{12}O_6 \rightarrow 2C_2H_5OH + 2CO_2$) through the fermentation followed by product recovery (i.e. separation and conditioning). In the second pathway, the agricultural biomass is converted initially through gasification technique into syngas (CO_2, CO, H_2, CH_4, water vapour, and short-chain hydrocarbon gas mixture). In the next step, the syngas is fermented to produce bioethanol (Figure 3.4a). The bioethanol produced from fermentation of sugars or syngas is again used to produce green diesel and bio-jet fuel. The process generally involved in bio-jet fuel and green diesel production includes bioethanol dehydration followed by oligomerization, distillation, and hydrogenation. The typical reactions for bioethanol production from syngas fermentation are as follows:

$$4CO + 2H_2O \rightarrow CH_3COOH + 2CO_2$$

$$4CO + 3H_2O \rightarrow C_2H_5OH + 4CO_2$$

$$6H_2 + 2CO_2 \rightarrow C_2H_5OH + 3H_2O$$

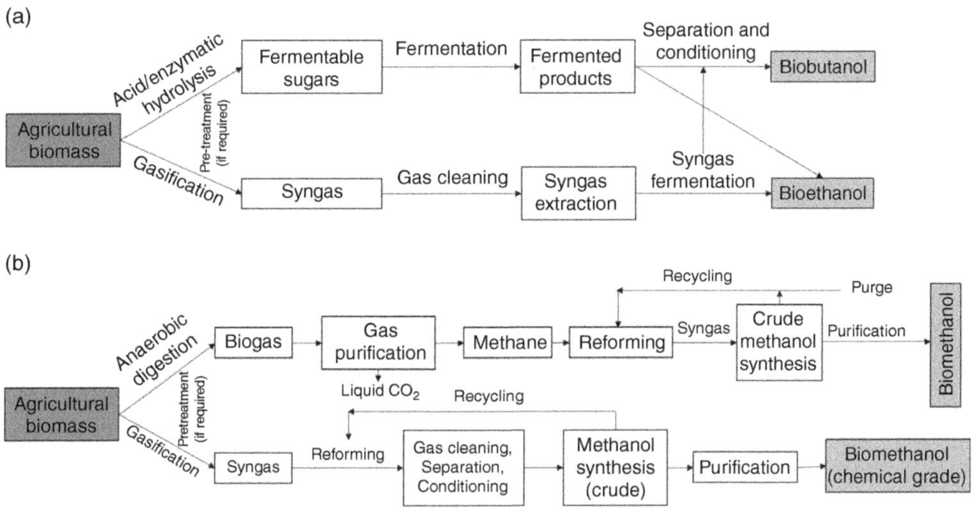

Figure 3.4 Different stages of biofuel production from agricultural biomass: (a) bioethanol and biobutanol, and (b) biomethanol. *Source:* Modified after Callegari et al. (2019) and Fivga et al. (2019).

3.7.2 Biobutanol Production

Biobutanol production from agricultural biomass can be produced by same process of bioethanol production, i.e. fermentation of sugars or syngas (Callegari et al. 2019; Fivga et al. 2019). Briefly, the biomass is first pre-treated and detoxified before the fermentable sugar production, and consequently, biobutanol production is carried out using acetone–butanol–ethanol (ABE, 3 : 6 : 1 ratio) process using the micro-organism *Clostridium acetobutylicum*. Lastly, the end product is purified and recovered by butanol, ethanol, and acetone distillation at their boiling points (118, 78, and 56 °C, respectively) (Figure 3.4a).

3.7.3 Biomethanol Production

Biomethanol production is another application of agricultural biomass (Callegari et al. 2019). Generally, biomethanol is produced predominantly by methane hydrogenation or syngas catalytic hydrogenation (usually zinc, copper, or chromium oxides) (Shamsul et al. 2014; Fivga et al. 2019). At first, methane is produced using AD from agricultural biomass, and then it is hydrogenated catalytically to produce biomethanol (Chandra et al. 2012). Figure 3.4b illustrates different stages of biomethanol production from agricultural biomass. AD takes place in four stages: biopolymer hydrolysis, acidogenesis, acetogenesis, and methanogenesis (Chandra et al. 2012; Passoth and Sandgren 2019). During hydrolysis, agricultural biomass is converted into monomers, for example amino acids, monosaccharides, and short-chain fatty acids by exoenzymes. In the next step, i.e. acidogenesis, monomers produced in hydrolytic phase are degraded mainly to volatile fatty acids (short-chain organic acids) and alcohols as well as C1–C5 molecules, ammonia, hydrogen, CO_2, and other compounds. In acetogenesis step, the short-chain organic acids and alcohols are converted to CO_2, acetic acid, and hydrogen by homo-acetogenic micro-organisms. Finally, in methanogenesis, hydrogen and acetate are transformed into methane and CO_2 by methanogenic micro-organisms, called methanogenesis [mainly by acetoclastic methanogenesis ($C + 4H_2 \rightarrow CH_4 + 2H_2O$), hydrogenotrophic methanogenesis ($CH_3COOH \rightarrow CH_4 + CO_2$), and methylotrophic methanogenesis]. The first three steps are driven by bacterial consortia, but methanogenesis is performed by archaea (Schnürer 2016). The produced methane is hydrogenated by catalytic conversion techniques to produce biomethanol. In the second pathway, the agricultural biomass is converted initially through gasification technique into syngas (CO_2, CO, H_2, CH_4, water vapour, and short-chain hydrocarbon gas mixture). In the next step, the syngas is hydrogenated by catalytic conversion technology (usually by using zinc, copper, or chromium oxides) to produce bioethanol. The typical reactions for biomethanol production from syngas or methane hydrogenation are as follows:

$$CO + H_2 \rightleftarrows CH_2O$$

$$CH_2O + H_2 \rightleftarrows CH_3OH$$

$$CO_2 + 3H_2 \rightarrow CH_3OH + H_2O$$

$$3H_2 + 2CO_2 \rightarrow CH_2OH + H_2O$$

3.7.4 Biodiesel Production

Biodiesel is generally derived from vegetative oils (second-generation biomass) through the transesterification/esterification reaction (Callegari et al. 2019; Fivga et al. 2019). Free fatty acids (FFAs) and triglycerides present in extracted vegetable oil react with methanol (sometimes, ethanol) in the presence of acidic/basic/enzymatic catalyst to form alkyl esters, FAME, and glycerol or water (Fivga et al. 2019). Upon completion of reaction, two layers namely glycerine layer or ethyl/methyl ester layer appeared, which are finally separated and purified (Figure 3.5). Before oil extraction, pre-treatment is mandatory. The techniques used for oil extraction include mechanical or chemical or enzymatic extraction, accelerated solvent extraction (ASE), supercritical fluid extraction (SFE), and microwave-assisted extraction (MAE) (Callegari et al. 2019). The simplified reactions are as follows:

$$\text{Triglycerides} + 3\,\text{Alcohol} \rightarrow 3\,\text{Ethyl esters} + \text{Glycerol}$$

$$\text{Fatty acids} + \text{Alcohol} \rightarrow \text{Ethyl ester} + \text{Water}$$

3.7.5 Syngas Production

Synthetic gas is produced by either pyrolysis or gasification technique. Generally, three transformations occur during syngas production: dehydration, pyrolysis, and partial oxidation (Guo et al. 2015). The dehydration phase is generally employed to heat up biomass rapidly at 700–1200 °C to reduce moisture level and then undergoes pyrolysis for char and vapour generation at elevated temperature >700 °C (Callegari et al. 2019). The partial oxidation step is employed to covert char into CO_2 and CO, and vapour to

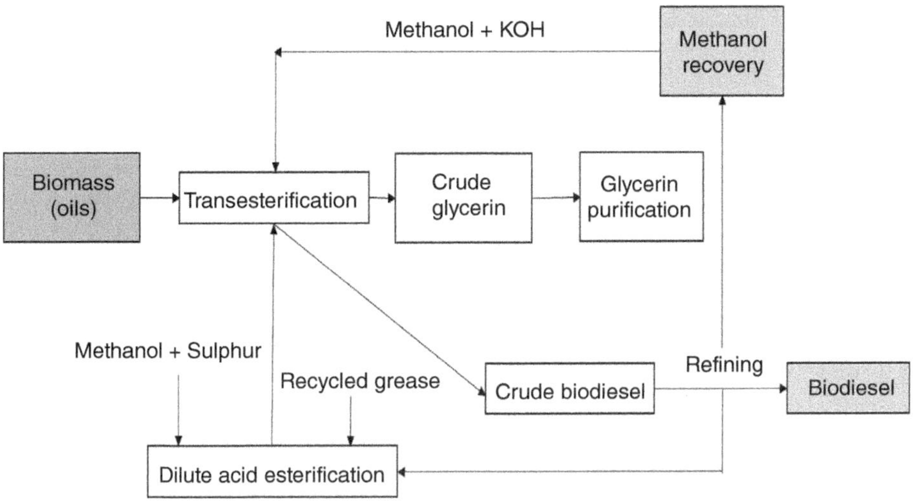

Figure 3.5 Different stages of biodiesel production. *Source:* Modified after Callegari et al. (2019).

CO_2 and H_2O. The char and CO_2 further react with each other to generate more H_2 and CO, while H_2O react with CO to generate CO_2 and H_2 (Guo et al. 2015). The whole process is conducted in the total absence of oxygen. The purified syngas can be further used for different applications such as synthesis of biomethanol, bioethanol, ammonia, Fischer–Tropsch (FT) processing to produce liquid hydrocarbons, and so on (Molino et al. 2018).

3.7.6 Bio-oil and Green Diesel Production

Bio-oil, a type of liquid biofuel, is developed from agricultural feedstock by thermochemical or biochemical process, particularly pyrolysis process (Fivga et al. 2019). Bio-oil is the prime product of pyrolysis process, which thermally decomposes lignocellulose biomass with rapid heating (250–850°C) in the complete absence of oxygen followed by rapid quenching of vapour products:

$$\text{Biomass pyrolysis} \rightarrow H_2 + CO_2 + CO + \text{hydrocarbons}(\text{liquid and gaseous})$$

$$\text{Biomass catalytic steam reforming} \rightarrow H_2 + CO_2 + CO$$

$$H_2 + CO(\text{Fisher} - \text{Tropsch synthesis}) \rightarrow \text{Hydrocarbons}(\text{liquid and gaseous})$$

Green diesel is synthesized from bio-oil or FT wax (Fivga et al. 2019). Generally, green diesel is derived by hydro-processing (HDO) technique followed by cracking or isomerization of bio-oil or FT wax instead of transesterification process. Its production cost is very high due to complex operating conditions but has higher cetane number (around 70–90) and calorific value, low level of oxygen, density, lubricity, and viscosity.

3.7.7 DME/OME Production

Dimethyl ether and OME are generally used as diesel substitutes and produced from biomethanol. DME production is conducted by biomethanol dehydration in catalytic (mainly acidic such as zeolites, alumina, $AlPO_4$, and CuO/ZrO_2) gas-phase reactor followed by purification and distillation (Bîldea et al. 2017). Recently, DME is directly produced from syngas in one-stage reactor under high temperature (250–400 °C) and pressure (>10 bar) using bifunctional catalyst to synthesize methanol followed by DME ($2CH_3OH \rightarrow CH_3OCH_3 + H_2O$) (Fivga et al. 2019).

OME production is done mainly in two stages. In the first stage, methanol is converted into formaldehyde by metal oxide catalytic process ($CH_3O + \frac{1}{2} O_2 \rightarrow CH_2O + H_2O$ or $CH_3O \rightarrow CH_2O + H_2$). In the next step, formaldehyde initially is converted into trioxane acid ($3CH_2O \rightarrow (CH_2O)_3$) in the presence of sulphuric acid, and finally methylal is produced by distillation in the presence of acidic heterogeneous catalyst ($2CH_3OH + CH_2O \rightleftarrows CH_3O\text{-}CH_2O\text{-}CH_3 + H_2O$) (Lautenschütz et al. 2016).

3.7.8 Bio-jet Fuel Production

Generally, different types of biofuels are used for aviation industry. The bio-jet fuels are produced in different pathways. First type of bio-jet fuel is hydro-processed renewable jet fuels (HRJs). HRJs are produced from triglyceride-based feedstock by hydrodeoxygenation (produced linear long-chain hydrocarbons) in the presence of hydrogen and catalyst at higher temperature and pressure, followed by isomerization and cracking (convert lineal long-chain hydrocarbons into bio-jet fuel) (Fivga et al. 2019). FT fuel is an another important bio-jet fuel and used extensively. FT fuel (C_8–C_{16}) is generated by catalytic conversion of syngas followed by cracking and isomerization. FT synthesis technology required high pressure and temperature (200–350 °C) and catalysts (iron- or cobalt-based catalysts for low-temperature FT fuel and iron-based catalysts for high-temperature FT fuel). Synthetic isoparaffin (SIP) fuel is produced by biochemical fermentation of carbohydrate sugars through purification and hydro-processing without producing ethanol. Alcohol to jet (ATJ) fuels are produced by two steps. In the first step, bioalcohol is generated by thermal and biochemical fermentation of hydrolyzed sugars. In the next step, the produced bioalcohol are dehydrated followed by oligomerized, hydrogenated, and distillated processes to form ATJ fuel (Fivga et al. 2019).

3.8 Factors Influencing Biofuel Production

The optimization of biofuel production depends on operating conditions (also called limiting factors) of different pre-treatment and conversion techniques. Accordingly, the factors that influence biofuel production are discussed as follows:

3.8.1 Feed Material

Feed material is the prime factor that affects economic feasibility of biofuel production. For example, approximately 80% of biodiesel production cost depends on feed material. Generally, substrates that contain fats, carbohydrates, and proteins can be used for biofuel production. In addition to this, feed material composition greatly influences biofuel production. For instance, biogas yield and its composition are greatly affected by feed material composition. Carbohydrates AD produced 886 L biogas (with 50% methane per kg volatile solids), while fat digestion produced 1535 L biogas (with 70% methane) and proteins digestion produced 587 L biogas (with 84% methane) (Chandra et al. 2012). In addition to this, water content is also critical factor for biofuel production. Generally, water content about 90% is considered as optimal, if feed material contains too much water, biofuel production decreases. On the other hand, if feed material contains too low water, acid accumulation occurs which inhibits biofuel production. For example, during biogas production through fermentation, too low water level leads to accumulation of acetic acid which inhibits biogas production (Chandra et al. 2012). In addition to this, water content significantly affects FFA level (Mathiyazhagan and Ganapathi 2011).

3.8.2 pH

The pH is the major factor regulating biofuel production. pH not only affects the biofuel production but also modifies by-product spectrum (Chandra et al. 2012). The pH significantly affects the growth of micro-organisms involved in conversion process (Guo et al. 2010). Generally, the pH range 5.0–8.5 is proved to be optimum for digester. The pH value outside this range hinders biofuel production; therefore, it should be maintained properly by feeding optimal organic load. For example, optimal biofuel production in case of food wastes takes place at pH 5.0–6.0, whereas for animal manure and crop residues a neutral pH is needed (Guo et al. 2010).

3.8.3 Reaction Temperature

Reaction temperature is very crucial factor that directly affects biofuel yield. For instance, higher reaction temperature enhanced reaction rate and reduced reaction time. However, temperature enhancement beyond optimum temperature yielded lower biofuel production (Mathiyazhagan and Ganapathi 2011). For example, optimal temperature for biodiesel production is 50–60 °C, but decreased biodiesel yield was noticed due to temperature elevation which accelerates triglycerides saponification (Mathiyazhagan and Ganapathi 2011). In addition to this, a small variation in operating temperature of digester significantly affects biological activity, especially anaerobic bacteria. This indicates that biological fermentation technique is directly dependent on reaction temperature. For example, during methane formation the optimal temperature for mesophilic bacteria is 30–35 °C, and it is 50–60 °C for thermophilic bacteria, while the activity of methane-forming bacteria is inhibited between 40 and 50 °C (Chandra et al. 2012).

3.8.4 Retention Time

Retention time is another most crucial parameter controlling biofuel production. Generally, higher retention time indicates higher biofuel yield. At the beginning, the biofuel production is very high, and after that it gradually declines over time (Chandra et al. 2012). High retention time enhances volume of digester, reduces amount of volatile substances, and provides protection against toxic compounds. In addition to this, high retention time improves acclimatization capacity of micro-organisms to toxic compounds (Mathiyazhagan and Ganapathi 2011; Chandra et al. 2012). Retention time greatly influences volatile solids conversion into gaseous products, but it should not exceed by 12 days (Mathiyazhagan and Ganapathi 2011). In contrast, retention time reduction decreases biofuel production cost due to enhanced digester volume.

3.8.5 Organic Loading Rate (OLR)

Organic loading rate (OLR) denotes amount of chemical oxygen demand (COD) or volatile solids fed per unit digester volume per day. Loading rate should be as per digester's size; otherwise, cost will be enhanced. For example, volatile solids should be in the range of 13–15% for dairy manure for maximum biogas yield (Chandra et al. 2012). However, organic load (25–30% total solids) indicates toxicity generation and limitations in trace nutrient. For example, during biogas production from sorghum and sorghum–cellulose mixture through thermophilic biomethanation (55 °C), organic loading yielded methane production about 5.7 (74.8%) and 7.5 l/kg/day (72.1%), respectively. Apart from this, they produce ammonia, which needs special attention. In addition to this, volatile acid formation enhances if ingestion of volatile solids to digester is higher, which ultimately shows alkalinity and pH effect. Accordingly, feed loading to digester should be at standard rate to reduce these problems (Chandra et al. 2012).

3.8.6 Free Fatty Acid (FFA) Content

FFA content is another critical factor for biofuel production. FFA generally contains 8–18 carbon chains, which include lauric, capric, myristic, caprylic, and oleic acids. Lauric acid among all is most toxic. Fatty acids content greater than 500 g/l caused toxicity during anaerobic treatments, while volatile fatty acid contents beyond standard value hinders methane yield (Chandra et al. 2012). On the contrary, low level of oleates and stearates inhibits thermophilic AD process during digestion of cattle manure. Further, oleate caused lower inhibitory effect when glycerol ester is added to it. Moreover, the FFA consumes catalysts of conversion technologies which lead to reduction of catalyst effect (Mathiyazhagan and Ganapathi 2011).

3.8.7 Catalyst Concentration

Biofuel production is greatly influenced by catalyst concentration. Generally, different catalysts are used for biofuel generation such as homogeneous/heterogeneous catalyst, enzymatic catalyst, or nanocatalyst. Alkalis and acids namely potassium hydroxide, sodium hydroxide, and sodium methoxide are commonly used homogeneous catalysts (Chandra et al. 2012). For example, during methane formation from Karanja oil, KOH excellently improves the yield of methyl esters above 90% (Boey et al. 2011). Although homogeneous catalyst has a number of benefits, the major disadvantages are separation problems, stable emulsion formation, catalysts regeneration not possible, and very sensitive to water and FFA. Accordingly, scientists are moved to alkali/acidic heterogeneous catalysts for biofuel production, because heterogeneous catalysts simplify the process, reduce the process cost, are recyclable and non-corrosive, have easy separation and purification process, have higher selectivity and long-life, and are acid-insensitive to FFA and water (Boey et al. 2011; Chozhavendhan et al. 2020). Recently, biocatalysts are gaining much attention for biofuel generation as biocatalysts perform equally with their chemical-based counterparts (Boey et al. 2011). Different biocatalysts are lipases, cellulases, and hemicellulases. Apart from this, biocatalysts are highly sensitive and can easily be immobilized into the feedstock biomass (Chozhavendhan et al. 2020). Recently, nanocatalysts such as nanosized alumina and nano-MgO are extensively used for biofuel production. For example, biodiesel is produced using nano-CaO, ferromagnetic zinc oxide nanocomposite, or nano-MgO with supercritical/subcritical methanol from various agricultural biomasses (Hsiao et al. 2011). However, the major constraint is too expensive.

3.8.8 C/N Ratio

The C/N ratio is very crucial parameter and regulates biofuel production directly. C/N ratio value varied according to the substrate. For example, cattle dung waste has average C/N ratio around 24.0 and human excreta waste has C/N ratio about 8.0, while plant materials have higher C/N ratio due to higher carbon content. Generally, the optimal C/N ratio for AD process is ranged from 20 to 30. Additionally, the C/N value 25 : 1 is considered as optimal for biogas production (Chandra et al. 2012). Biogas production hampered greatly if the ratio is too low or too high, leading to reduced gas formation by altering pH of biodigestate (Chandra et al. 2012). Accordingly, for biofuel production, a standard feedstock composition is very much necessary.

3.9 Biofuel Energy Potentials in India

India has to meet the target of blending 20% ethanol in gasoline and 5% biodiesel blending with diesel by 2030, as per requirement of National Policy on Biofuel. The fulfilment of blending target as per National Biofuel Policy (NBP) is very challenging till today as the biofuel industry in India is in developing stage. Presently, more than 300 oil seed bearing plants are documented in India. These species can be explored for the production of biofuel (Altenburg et al. 2009). Agriculture sector in India has an immense potential for biofuel production. Generally, bioethanol and biodiesel are considered as biofuels. Like biodiesel, bioethanol can also be produced from food (first generation) and non-food sources (second generation). In our country, sugarcane and wheat are the prime first-generation biofuel source. According to India Biofuels Annual Report (2020), India is currently producing 2.98 billion L ethanol yearly from sugarcane molasses against the installed capacity 5 billion L. It was estimated that 199.1 MMT of sugarcane is produced in India annually (Khalil et al. 2013). The country has produced 121 GJ ethanol in terms of energy used from sugarcane. But in reality, 70–80% produced sugarcane is used in sugar production (Timilsina and Shrestha 2011). As an alternative of sugarcane, sweet sorghum, sugar beet, and starch-containing materials, e.g. corn, rotten potatoes, damaged grains of rice, and wheat, are also taken into consideration for ethanol production (Singh 2018). In this regard, potato is the most important food crop around world, and India is second largest potato producer after

China. In our country, out of total potato production, 68% is consumed freshly as vegetable, 7.6% is used in other food processing, 8.5% is used as seed, and 16% of total production is wastage (Kumar et al. 2020). Generally, when potato production is high, market price is very low, so farmers dump large amounts of potato in the field. Due to low market price, farmers sometimes dump huge amount of crops in cold storages also. These waste/surplus potatoes can be utilized as feedstock materials for ethanol production through proper scientific procedure.

Agriculture residues, e.g. leaves and stalks of plant left over after harvesting, can be sustainable alternative to first-generation biofuel sources. India is the country with huge potential for biofuel production from crop residues (lignocellulosic materials). Many abundantly available cellulosic agricultural residues, such as rice straw, rice husk, sugarcane bagasse, wheat straw, sugarcane tops (SCTs), cotton stalks, leftover bamboo, and soft bamboo, are considered as potential biofuel sources in India as well as across the globe. Pulp of cashew apple, coffee pulp (Shenoy et al. 2011), and banana peels (Bhatia and Paliwal 2010) are also considered as potential biofuel sources in India. Purohit and Fischer (2014) have estimated that 150 Mt agricultural residues were available for energy production in 2010–2011 and 28 BL ethanol can be produced yearly. It was assumed that 20% collected residues may be lost during collection, storage, and transportation (Singh 2015). These residues amount can yield 214 lge/tonne dry matter (tDM) for cellulosic ethanol production as per estimation of IEA (2010). As per Purohit and Dhar (2015), it can be predicted that ethanol production per tDM will increase up to 250 lt/tDM in 2020–2021, 275 lt/tDM by 2025–2026, and 300 lt/tDM by 2030–2031 (Table 3.2). According to the estimation, net ethanol procurement is approximately 37 BL in 2020–2021, and it will be 50 BL by 2030–2031 (Table 3.2). Only one-fourth of the total residues availability was taken into consideration for this estimation.

Agricultural residues are sometimes used in some other purposes, e.g. paper making, production of power, and fodder, so there may be conflict of interest. In this situation, the production of biodiesel (another biofuel) from non-food oil seeds is very sustainable option. Non-edible oil seeds like *Jatropha curcas and Pongamia pinnata* have a great potential to fulfil the biofuel requirements in India. India has vast area of unutilized land, waste land across the country. *Jatropha* cultivation is an important option to utilize this land. This may bring boom in biofuel sector in India. According to Deshmukh and Marathe (2016), 3.75 Mt/Ha seeds can be produced from *Jatropa* plantation which can yield 1.2 Mt oil. In India, nearly 400 000 ha land has the potential to produce 0.48 MMt oil and 1.02 MMt oil cakes from which ethanol can be produced (Deshmukh and Marathe 2016). *Ricinus communis* (castor bean), which has a strong adaptability in different weather conditions and ability to grow in marginal soil, is to be considered as a potential feedstock for biodiesel production as they can supply up to 60% non-edible oil (McKeon 2016). *Pongamia pinnata* is a leguminous tree generally planted along roads, highways, and canals to control soil erosion. It can also be grown on marginal lands, thus making it a useful species for agroforestry. Uncountable numbers of these trees exist across India. This plant has immense potential to support large-scale biodiesel industry through its non-edible oil seeds (Kesari and Rangan 2010). It was estimated that annually 50 000 tonne oil could be produced from its seeds across the country. Commercial application of this oil is very limited. But continuous research and awareness will make this plant an alternative source of biodiesel in near future (Dwivedi et al. 2011).

Table 3.2 Biofuel potential from agricultural residues.

Crop residue	Uses of agricultural residue (%)			Net biofuel availability (Mt)			Net ethanol availability (BL)		
	Fodder	Fuel	other	2010/2011	2020/2021	2030/2031	2010/2011	2020/2021	2030/2031
Rice straw and husk	80.8	11.1	8.00	13.8	15.8	17.8	3.00	4.00	5.3
Wheat straw	86.4	0	13.6	0	0	0	0	0	0
Jowar stalk	100	0	0	0	0	0	0	0	0
Bajra straw	89.8	0	10.2	0	0	0	0	0	0
Maize stalk and cobs	81	19	0	7.4	8.5	9.7	1.6	2.1	2.9
Other cereals stalk	100	0	0	0	0	0	0	0	0
Gram waste	0	100	0	0	0	0	0	0	0
Tur shell and waste	3.5	48.5	48	2.9	3.1	3.3	0.6	0.8	1
Lentil shell and waste	3.5	48.5	48	1.0	1.3	1.4	0.2	0.3	0.4
Other pulses – shell and waste	3.5	48.5	48	6.3	6.4	6.9	1.3	1.6	2.1
Groundnut waste	0	13.2	86.8	1.8	1.9	2.1	0.4	0.5	0.6
Rape and mustard waste	0	100	0	11.8	13.9	15.9	2.5	3.5	4.8
Other oilseed waste	0	100	0	23.1	27.8	32.2	4.9	7	9.7
Cotton seeds and waste	0	100	0	14.1	15.3	16.2	3	3.8	4.9
Cotton gin and trash	0	100	0	0.3	0.3	0.4	0.1	0.1	0.1
Jute and mesta wastes	0	100	0	2.2	2.6	2.8	0.5	0.7	0.9
Sugarcane bagasse and leaves	11.8	41	47.2	35.9	42.6	48.2	7.7	10.7	4.5
Total				130.2	149.3	166.8	27.9	37.8	50.1

Source: Purohit and Dhar, 2015 / With Permission of Elsevier.

3.10 Environmental and Economic Concern of Agriculture-Based Biofuel Production

From environmental and economic perspective, biofuel production from cheap sources like agricultural wastes represents most promising alternatives. At economical level, biofuel industry provides employment and environmental protection simultaneously. There are still several drawbacks of using biomass residues and waste as second-generation biofuels, because biofuel production from agricultural residues is still not cost-effective, considering current technological advancement and policy considerations. This can be achieved through a variety of means as follows: (i) need R&D investment for research to find out optimal characteristics of feedstocks to maximize yield; (ii) need R&D investment to check economical potentiality of different cultivation areas; (iii) need R&D investment for improving feedstock pre-treatment technologies and process integration for conversion of agricultural residue into biofuel energy; and (iv) need R&D investment for understanding the conversion process to maximize co-product's value.

In addition to this, manufacturing cost and conversion efficiency of processes and capital investment should be taken into consideration importantly to assess the economic viability. Generally, the electricity generation cost is around 0.03–0.24 USD/kWh, while the cost is around 0.13–0.99 USD/L for biofuel production (World Energy Investment Outlook 2014; Deng et al. 2015). Nevertheless, the conventional fuel prices are still competitive compared with their bio-based counterparts. Despite this, biofuel production from agriculture-based wastes has several environmental advantages over conventional and alternative energy sources:

- avoid environmental damage stemming from stubble burning;
- reduce GHG emissions into atmosphere, which ultimately mitigates global environmental problems such as climate change and global;
- biomass is nature origin and renewable that means it will not exhaust;
- solve the waste management problem; and finally
- reduce land, water, and air contamination.

3.11 Future Perspectives

Replacement of fossil energy with bioenergy carriers namely biofuel as sustainable fuel cannot be underestimated because of their presumed future potential. In addition to this, they can revolutionize energy sector and can bring about positive impacts, in particularly, ensure economic sustainability, environmental security, and ecosystem health. However, there is still lack for efficient production of biofuel from agriculture biomass residue by using existing conversion technologies. This is mainly distinguished by two reasons: first one, diverse composition of agriculture biomass residue and second one, lack of agriculture residue-specific conversion technology. Accordingly, developing agriculture residue-specific technological practices and policies require R&D investment and smart public policy to increase productivity and reduce capital costs. Cost reduction can be achieved through multiple ways such as efficient biomass pre-treatment, specific activity improvement of recombinant micro-organisms, and development of co-generation system. In this regard, process engineering plays vital role for design, analysis, and technology improvement of employed bioprocesses for large-scale industrial production and provides significant outcomes during searching efficient biofuel production technologies. In addition to process improvement, the future of biofuel energy generation lies in integration of second-generation substrates with fossil fuel carriers such as cofiring coal with solid biomass, synthesis of gas mixing, and mixing natural gas with landfill gas or biogas. Recently, algae are used as feedstock for third-generation biofuel generation to address the concern of second-generation biofuel, i.e. diverse heterogenous composition, which reduced the biofuel generation efficiency.

3.12 Conclusions

The paper addressed the potential of agricultural residue biomass as second-generation biofuel production feedstock. The prime concern of biofuel production from agricultural biomass is acquiring environmental sustainability and ecological development. Additional research is needed to understand the impact of substrate composition on biofuel production performances. Nevertheless, biological pre-treatments method should be encouraged and employed to treat agricultural biomass because of their environment-friendly nature. Biomass conversion technologies for biofuel and electricity generation depend on feedstocks and desired end product. Particularly, thermochemical technique is restricted by biomass waste composition for biofuel production compared with biochemical strategies. However, the performance of conversion technologies and by-product redirection are directly dependent on operating conditions. Accordingly, operating conditions like medium pH, high temperature, medium reaction time, low FFA contents, high C/N ration of biomass, and acclimated microbial communities are recommended for optimum biofuel production. Apart from these, biofuel production from agricultural biomass is still considered as robust with regard to material handling and biomass conversion technology in comparison with traditional food crops-based biofuels. Additionally, agriculture-based biofuel production ensures economic sustainability and environmental security towards global climate change. Still, much more efforts at a large scale should be given for the development of cost-effective and eco-friendly technologies to fill up inadequacies of existing technologies to fulfil sustainable development goals (SDGs), i.e. Affordable and Clean Energy by 2030 set by US General Assembly.

Acknowledgements

Authors gratefully acknowledge Department of Environmental Science of Sukanta Mahavidyalaya, Vidyasagar University, and The University of Burdwan for allowing to work from home during the lockdown period.

References

Abrahamsson, V., Rodriguez-Meizoso, I., and Turner, C. (2015). Supercritical fluid extraction of lipids from linseed with on-line evaporative light scattering detection. *Analytica Chimica Acta* 853: 320–327.

Alaswada, A., Dassisti, M., Prescotta, T., and Olabia, A.G. (2015). Technologies and developments of third generation biofuel production. *Renewable and Sustainable Energy Reviews* 51: 1446–1460.

Alinia, R., Zabihi, S., Esmaeilzadeh, F., and Kalajahi, J.F. (2010). Pretreatment of wheat straw by supercritical CO_2 and its enzymatic hydrolysis for sugar production. *Biosystems and Engineering* 107 (1): 61–66.

Altenburg, T., Dietz, II., Hahl, M. et al. (2009). *Biodiesel in India – Value Chain Organisation and Policy Options for Rural Development.* Bonn, Germany: German Development Institute.

Awasthi, P., Shrivastava, S., Kharkwal, A.C., and Varma, A. (2015). Biofuel from agricultural waste: a review. *International Journal of Current Microbiology and Applied Sciences* 4 (1): 470–477.

Bhatia, L. and Paliwal, S. (2010). Banana peel waste as substrate for ethanol production. *International Journal of Biotechnology and Bioengineering Research* 1 (2): 213–218.

Bhatia, L., Johri, S., and Ahmad, R. (2012). An economic and ecological perspective of ethanol production from renewable agro waste: a review. *AMB Express* 2: 65. http://www.amb-express.com/content/2/1/65.

Bîldea, C.S., Győrgy, R., Brunchi, C.C. et al. (2017). Optimal design of intensified processes for DME synthesis. *Computers & Chemical Engineering* 105: 142–151.

Boden, T.A., Marland, G., and Andres, R.J. (2017). National CO_2 emissions from fossil-fuel burning, cement manufacture, and gas flaring: 1751-2014 (V. 2017). U.S. Department of Energy, 10.3334/CDIAC/00001_V2017.

Boey, P.L., Maniam, G.P., and Hamid, S.A. (2011). Performance of calcium oxide as a heterogeneous catalyst in biodiesel production: a review. *Chemical Engineering Journal* 168: 15–22.

Bonechi, C., Consumi, M., Donati, A. et al. (2017). Biomass: an overview. In: *Bioenergy Systems for the Future: Prospects for Biofuels and Biohydrogen* (ed. F. Dalena, A. Basile and C. Rossi), 3–42. London: Elsevier Publishing.

Callegari, A., Bolognesi, S., Cecconet, D., and Capodaglio, A.G. (2019). Production technologies, current role, and future prospects of biofuels feedstocks: a state-of-the-art review. *Critical Reviews in Environmental Science and Technology* 10.1080/10643389.2019.1629801.

Chandra, R., Takeuchi, H., and Hasegawa, T. (2012). Methane production from lignocellulosic agricultural crop wastes: a review in context to second generation of biofuel production. *Renewable and Sustainable Energy Reviews* 16: 1462–1476.

Chozhavendhan, S., Singh, M.V.P., Fransila, B. et al. (2020). A review on influencing parameters of biodiesel production and purification processes. *Current Research in Green and Sustainable Chemistry* 1-2: 1–6.

Cunha-Pereira, F.D., Matte, C.R., Costa, T.M. et al. (2020). Treatment and characterization of biomass of soybean and rice hulls using ionic liquids for the liberation of fermentable sugars. *Anais da Academia Brasileira de Ciências* 92 (3): e20191258. 10.1590/0001-3765202020191258.

Deng, Y.Y., Koper, M., Haigh, M., and Dornburg, V. (2015). Country-level assessment of long-term global bioenergy potential. *Biomass and Bioenergy* 74: 253–267.

Deshmukh, M.P. and Marathe, A.B. (2016). Production of bioethanol from deoiled *Jatropha* cake. *International Journal of Innovative Science, Engineering and Technology* 3 (6): 449–454.

Dwivedi, G., Jain, S., and Sharma, M.P. (2011). *Pongamia* as a source of biodiesel in India. *Smart Grid and Renewable Energy* 2: 184–189.

Edwards, S., Chaplin, M.F., Blackwood, A.D., and Dettmar, P.W. (2003). Primary structure of arabinoxylans of ispaghula husk and wheat bran. *Proceedings of the Nutrition Society* 62 (1): 217–222.

Falls, M. and Holtzapple, M.T. (2011). Oxidative lime pretreatment of alamo switchgrass. *Applied Biochemistry and Biotechnology* 165 (2): 506–522.

Fivga, A., Speranza, L.G., Branco, C.M. et al. (2019). A review on the current state of the art for the production of advanced liquid biofuels. *AIMS Energy* 7 (1): 46–76.

Guo, X.M., Trably, E., Latrille, E. et al. (2010). Hydrogen production from agricultural waste by dark fermentation: a review. *International Journal of Hydrogen Energy* 35: 10660–10673.

Guo, M., Song, W., and Buhain, J. (2015). Bioenergy and biofuels: history, status, and perspective. *Renewable and Sustainable Energy Reviews* 42: 712–725.

Harmsen, P.F.H., Hujigen, W., Bermudez, L., and Bakker, R. (2010). *Literature Review of Physical and Chemical Pretreatment Processes for Lignocellulosic Biomass*. Wageningen UR Food & Biobased Research.

Hsiao, M.C., Lin, C.C., and Hung Charg, Y. (2011). Microwave irradiation-assisted transesterification of soybean oil to biodiesel catalyzed by nano powder calcium oxide. *Fuel* 90: 1963–1967.

IEA (2010). *Sustainable Production of Second-Generation Biofuels: Potential and Perspectives in Major Economies and Developing Countries*. Paris: International Energy Agency (IEA).

India Biofuels Annual Report (2020). Global Agricultural Information Network; GAIN Report Number IN2021-0072. Foreign Agricultural Service, U.S. Department of Agriculture: Washington, DC, USA.

Kaltschmitt, M. (2013). Renewable energy from biomass. In: *Renewable Energy Systems* (ed. M. Kaltschmitt, N.J. Themelis, L.Y. Bronicki, et al.). New York: Springer.

Kaushika, N.D., Reddy, K.S., and Kaushik, K. (2016). Introduction to sustainable energy and environment. In: *Sustainable Energy and the Environment: A Clean Technology Approach*, 1–15. Springer.

Kesari, V. and Rangan, L. (2010). Development of *Pongamia pinnata* as an alternative biofuel crop – current status and scope of plantations in India. *Journal of Crop Science and Biotechnology* 13: 127–137.

Khalil, H.P.S.A., Aprilia, N.A.S., Bhat, A.H. et al. (2013). *Jatropha* biomass as renewable materials, for biocomposites and its application. *Renewable and Sustainable Energy Reviews* 22: 667–685.

Kluts, I., Wicke, B., Leemans, R., and Faaij, A. (2017). Sustainability constraints in determining European bioenergy potential: a review of existing studies and steps forward. *Renewable and Sustainable Energy Reviews* 69: 719–734.

Kumar, D., Dutt, S., Raigond, P. et al. (2020) Potato as a bioenergy crop for fuel ethanol production: perspectives and challenges. 10.13140/RG.2.2.25813.86247.

Lautenschütz, L., Oestreich, D., Seidenspinner, P. et al. (2016). Physico-chemical properties and fuel characteristics of oxymethylenedialkyl ethers. *Fuel* 173: 129–137.

Lebaka, V.R. (2013). Potential bioresources as future sources of biofuels production: an overview. In: *Biofuel Technologies* (ed. V.K. Gupta and M.G. Tuohy), 223–258. Springer.

Lee, S.Y., Sankaran, R., Chew, K.W. et al. (2019). Waste to bioenergy: a review on the recent conversion technologies. *BMC Energy* 1: 4. 10.1186/s42500-019-0004-7.

Leung, D.Y.C., Wu, X., and Leung, M.K.H. (2010). A review on biodiesel production using catalyzed transesterification. *Applied Energy* 87 (4): 1083–1095.

López Barreiro, D., Prins, W., Ronsse, F., and Brilman, W. (2013). Hydrothermal liquefaction (HTL) of microalgae for biofuel production: state of the art review and future prospects. *Biomass Bioenergy* 53: 113–127.

Mathiyazhagan, M. and Ganapathi, A. (2011). Factors affecting biodiesel production. *Research in Plant Biology* 1 (2): 1–5.

McKeon, T.A. (2016). Castor (*Ricinus communis* L.). In: *Industrial Oil Crops*, 75–112. Amsterdam: Elsevier.

Mohammed, N.I., Kabbashi, N., and Alade, A. (2018). Significance of agricultural residues in sustainable biofuel development. In: *Agricultural Waste and Residues* (ed. A. Aladjadjiyan), 71–88. IntechOpen. 10.5772/intechopen.78374.

Molino, A., Chianese, S., and Musmarra, D. (2016). Biomass gasification technology: the state of the art overview. *Journal of Energy Chemistry* 25 (1): 10–25.

Molino, A., Larocca, V., Chianese, S., and Musmarra, D. (2018). Biofuels production by biomass gasification: a review. *Energies* 11 (4): 1–31.

Musa, M., Ayoko, G.A., Ward, A. et al. (2019). Factors affecting microalgae production for biofuels and the potentials of chemometric methods in assessing and optimizing productivity. *Cells* 8: 851. 10.3390/cells8080851.

Nwosu-Obieogu, K., Chiemenem, L.I., and Adekunle, K.F. (2016). Utilization of agricultural waste for bioethanol production- a review. *International Journal of Current Research and Review* 8 (19): 1–5.

Passoth, V. and Sandgren, M. (2019). Biofuel production from straw hydrolysates: current achievements and perspectives. *Applied Microbiology and Biotechnology* 103: 5105–5116.

Payne, C.M., Knott, B.C., Mayes, H.B. et al. (2015). Fungal cellulases. *Chemical Reviews* 115: 1308–1448.

Portha, J.F., Parkhomenko, K., Kobl, K. et al. (2017). Kinetics of methanol synthesis from carbon dioxide hydrogenation over copper-zinc oxide catalysts. *Industrial & Engineering Chemistry Research* 56 (45): 13133–13145.

Purohit, P. and Dhar, S. (2015). Biofuel roadmap for India – (November 2015). UNEP (United Nations Environment Program) – DTU (Denmark Technical University) Partnership.

Purohit, P. and Fischer, G. (2014). Second generation biofuel potential in India: sustainability and cost considerations, UNEP Risø Centre on Energy, Climate and Sustainable Development, Technical University of Denmark, Copenhagen.

Rahimpour, M.R., Arab Aboosadi, Z., and Jahanmiri, A.H. (2012). Synthesis gas production in a novel hydrogen and oxygen perm-selective membranes tri-reformer for methanol production. *Journal of Natural Gas Science and Engineering* 9: 149–159.

Roos, C.J. (2010). Clean heat and power using biomass gasification for industrial and agricultural projects. U.S. Department of Energy.

Schnürer, A. (2016). Biogas production: microbiology and technology. *Advances in Biochemical Engineering/Biotechnology* 156: 195–234.

Shamsul, N.S., Kamarudin, S.K., Rahman, N.A. et al. (2014). An overview on the production of bio-methanol as potential renewable energy. *Renewable and Sustainable Energy Reviews* 33: 578–588.

Shenoy, D., Pai, A., Vikas, R.K. et al. (2011). A study on bioethanol production from cashew apple pulp and coffee pulp waste. *Biomass Bioenergy* 35: 4107–4111.

Singh, J. (2015). Overview of electric power potential of surplus agricultural biomass from economic, social, environmental and technical perspective: a case study of Punjab. *Renewable and Sustainable Energy Reviews* 42: 286–297.

Singh, G. (2018). *Rotten potatoes: that's the government's answer to your rising fuel bill.* https://economictimes.indiatimes.com/news/et-explains/rotten-potatoes-thats-the-governments-answer-to-your-rising-fuel-bill/articleshow/65656298.cms (accessed 3 September 2018).

Timilsina, G.R. and Shrestha, A. (2011). How much hope should we have for biofuels? *Energy* 36: 2055–2069.

Tursi, A. (2019). A review on biomass: importance, chemistry, classification, and conversion. *Biofuel Research Journal* 22: 962–979.

Verma, M., Godbout, S., Brar, S.K. et al. (2012). Biofuels production from biomass by thermochemical conversion technologies. *International Journal of Chemical Engineering* 2012: 542426. 10.1155/2012/542426.

World Energy Investment Outlook (2014). Factsheet fossil fuels. www.worldenergyoutlook.org.

Yamashita, Y., Sasaki, C., and Nakamura, Y. (2010). Effective enzyme saccharification and ethanol production from Japanese cedar using various pretreatment methods. *Journal of Bioscience and Bioengineering* 110: 79–86.

Yu, G., Yano, S., Inoue, H. et al. (2010). Pretreatment of rice straw by a hot-compressed water process for enzymatic hydrolysis. *Applied Biochemistry and Biotechnology* 160 (2): 539–551.

Zamani, A. (2015). Introduction to lignocellulose-based products. In: *Lignocellulose-Based Bioproducts* (ed. K. Karimi), 1–36. Cham: Springer.

Zhang, J., Weng, X., Han, Y. et al. (2013). Effect of supercritical water on the stability and activity of alkaline carbonate catalysts in coal gasification. *Journal of Energy Chemistry* 22 (3): 459–467.

4

Agriculture Wastes: Generation and Sustainable Management

Sumi Handique, Ankita Saha, Kundil Kumar Saikia, and Nirmali Gogoi

Department of Environmental Science, Tezpur University, Tezpur, Assam, India

CHAPTER OUTLINE

4.1 Introduction

Agriculture, or the cultivation of plants and livestock, has played an important role in the evolution of human civilization. Agriculture, like everything else, has advantages and disadvantages. With a burgeoning world population, the agricultural sector is under tremendous pressure to deliver food for all. Agriculture also requires vast land areas – globally, one-third of the land is under agriculture, which is approximately five billion hectares (Tubiello et al. 2021). Moreover, agricultural activities deplete the soil quality and consume about two-thirds of the freshwater resources. The increase in agricultural production is also associated with the generation of a considerable quantity of waste. Thus, agriculture has huge environmental and financial implications.

The United Nations has defined agricultural waste as 'waste produced as a result of various agricultural operations. It includes manure and other wastes from farms, poultry

Agriculture Waste Management and Bioresource: The Circular Economy Perspective, First Edition.
Edited by Suruchi Singh, Pardeep Singh, Anu Sharma, and Moharana Choudhury.

houses and slaughterhouses, harvest waste, fertilizer run-off from fields, pesticides that enter into water, air and soils, and salt and silt drained from fields' (Nagendran 2011). Agrowaste is the non-product output of agricultural production and processing (Obi et al. 2016; Ramírez-García et al. 2019). Among the numerous categories of agrowaste are plant waste, agro-industrial waste, waste from farming activities, aquaculture, livestock production, and horticultural farms (Prasad and Murugavelh 2019; Sharma and Garg 2019; Deka et al. 2021). According to Thangadurai et al. (2021), agrowaste is 'every type of waste material generated during agricultural practices that includes chemicals, pesticides, fertilizers, and unused crops along with non-edible plant parts'. Crop remains, plant stalks, leaves, bedding, animal manure, fruit and vegetable detritus, husk, and leaves are examples of agrowaste (Lim and Matu 2015; Kumar et al. 2022).

Agrowastes may be solid, liquid, or gaseous. They may potentially harm the environment as well as human health. Factors such as land use, cultural aspects, and a region's geography govern the quantity and composition of agricultural wastes (Tubiello et al. 2021). Contribution to greenhouse effect and climate change, eutrophication in water bodies, phosphorus and nitrogen pollution (i.e. nutrient pollution), air pollution due to stubble burning, and contribution to ozone layer depletion in the stratosphere are just a few of the major negative impacts of agrowastes on the environment (Adegbeye et al. 2020; Deka et al. 2021).

However, agricultural by-products are also rich in animal feed, antibiotics, antioxidants, biofertilizers, biofuels, enzymes, vitamins, and several other bioactive compounds. Therefore, they have the potential to be transformed into several valuable products. For example, antibiotics like oxytetracycline and enzymes like β-glucosidase and endoglucanase are obtained from different agrowastes (Kavitha and Nagarajan 2011; Mallek-Fakhfakh et al. 2016; Arevalos-Sánchez et al. 2021). Utilizing field residues properly with the application of appropriate technology can be environmentally beneficial as well as economically viable. In this chapter, we will discuss agrowaste with a special emphasis on its impacts on the environment, and its use as a resource towards achieving environmental sustainability.

4.2 Agrowastes and Its Composition

Agrowastes are classified as crop, animal, processing, and hazardous wastes (Agamuthu 2009). The composition of these wastes again varied based on the source (Table 4.1).

4.2.1 Crop Wastes

The crop wastes are produced after harvesting and processing. These wastes include leaves, stalks, seed pods, and stems. The food crop wastes of different types have varied utilization to human. It includes utilization of crop wastes as paper making pulp (from banana peel and sugarcane fibres), particleboards and softwood furniture (from oil palm stems and rubberwood), and mushroom cultivation (from husk and bagasse) (Agamuthu 2009), and also for production of biochar (Feng et al. 2020).

Table 4.1 Composition and sources of different agricultural wastes.

Waste type	Source	Composition	References
Crop wastes	Rice	Husk, bran, straw	(Goodman 2020)
	Wheat	Straw	(Goodman 2020)
	Sugarcane	Bagasse	(Goodman 2020)
	Maize	Bran	(Goodman 2020)
	Cotton	Stalks	(Goodman 2020)
Animal wastes	Cattle and piggery	Manure, urine	(Sorathiya et al. 2014)
	Meat	Suspended organic matter (fats and proteins), blood washings, urine	(Hansen and Cheong 2019)
Processing waste	Dairy	Milk and its various processing products such as cheese, butter, evaporated and powdered milk	(Hansen and Cheong 2019)
	Meat	Suspended organic matter (fats and proteins), blood washings, urine	(Hansen and Cheong 2019)
	Fruits and vegetables	Soil and organic matter, concentrated wastewater	(Hansen and Cheong 2019)
	Brewery and distillery waste	Dissolved solids, fermentation waste, and spent waste	(Hansen and Cheong 2019)
Hazardous waste	Agricultural solid wastes	Human and animal pathogens, medicines and feed additives, salts, meat processing waste, and leather tanning wastes	(Loehr 1978; Khan et al. 2020)
	Fertilizers and pesticides	Arsenic (As), cadmium (Cd), mercury (Hg), and lead (Pb)	(Loehr 1978)

4.2.2 Animal Wastes

Animal wastes primarily consist of solid and liquid manures. Other substances present are used bedding, spilled feed, animal carcass, etc. The composition and quantity of animal wastes depend on animal type, feed ration, health status, animal age, and climatic condition. They have potentiality of affecting the surface as well as groundwater (Gerba and Pepper 2009). Animal wastes are composed of nutrients, such as nitrogen, phosphorus, and potassium, which are essential elements for agricultural fields. Trace metals, organic solids, bacteria, viruses, and other micro-organisms can be present in animal wastes (Agamuthu 2009; Hansen and Cheong 2019).

4.2.3 Processing Waste

Food processing wastes result in the production of solid wastes and wastewater. Different type of food processing wastes includes the following:

i) Dairy processing waste: Dairy processing waste includes fluid milk in reception and bottling facilities. In addition, other processing facilities of production of cheese, butter, evaporated milk, condensed milk, powdered milk, etc., produce milk wastes with high biological oxygen demand (BOD) values (Hansen and Cheong 2019).
ii) Meat, poultry, and seafood processing waste: These categories of waste contain high dissolved and suspended organic matter, mostly proteins and lipids. They can carry a significant amount of pathogenic microbes. Slaughterhouses and meat packaging facilities produce strong and unpleasant wastes, faeces and blood washings from carcasses, urine, and undigested food from slaughtered animals (Hansen and Cheong 2019).
iii) Fruit and vegetable processing waste: These include polluted vegetable washing water containing organic matter and soil and concentrated wastewaters from food preparation and transformation facilities. Wastewater produced from various fruit and vegetable processing facilities contains highly suspended solids, colloidal, and dissolved organic matter (Hansen and Cheong 2019).
iv) Brewery and distillery waste: These wastes contain a high amount of dissolved solids. These may consist of high quantities of fermentation wastes and spent yeast, which are by-products of beer production, with extremely concentrated BOD levels (2000–15 000 mg/l), phosphate (20–140 mg/l), and total nitrogen (800–900 mg/l) (Hansen and Cheong 2019).

4.3 Agricultural Wastes as Source of Pollution

Various agricultural solid wastes that result in the formation of hazardous conditions to humans and animals include pathogens, medicines and feed additives, salts, meat processing wastes, and leather tanning wastes. These wastes contaminate the soil affecting its quality in and around the disposal area. Moreover, incorporating different transportation modes (e.g. run-off, percolation), these hazardous wastes affect the agricultural fields directly or indirectly (Loehr 1978). Hazardous wastes from applied pesticides and fertilizers in agricultural fields are the sources of toxic elements like DDT (dichlorodiphenyltrichloroethane), dieldrin, chlorinated hydrocarbons, organophosphates, and organochlorine pesticides (OCPs). Pesticides also release trace metals such as arsenic (As), cadmium (Cd), copper (Cu), lead (Pb), mercury (Hg), nickel (Ni), and zinc (Zn) in the environment that enters the human body via the food web. Some of these elements, like arsenic (As), lead (Pb), and cadmium (Cd), contribute to toxicity in the human body (Loehr 1978; Abbasi et al. 2014).

4.3.1 Air Pollution and Agriculture

Air pollution has an extremely negative impact on crop production both in its yield and the quality of the produce. Agricultural pollutants can include greenhouse gases, toxic chemicals, and other harmful airborne pollutants (Abbasi et al. 2014).

Different agricultural activities release pollutants into the atmosphere, which badly impacts the air quality. Figure 4.1 shows the effects of burning agricultural waste and

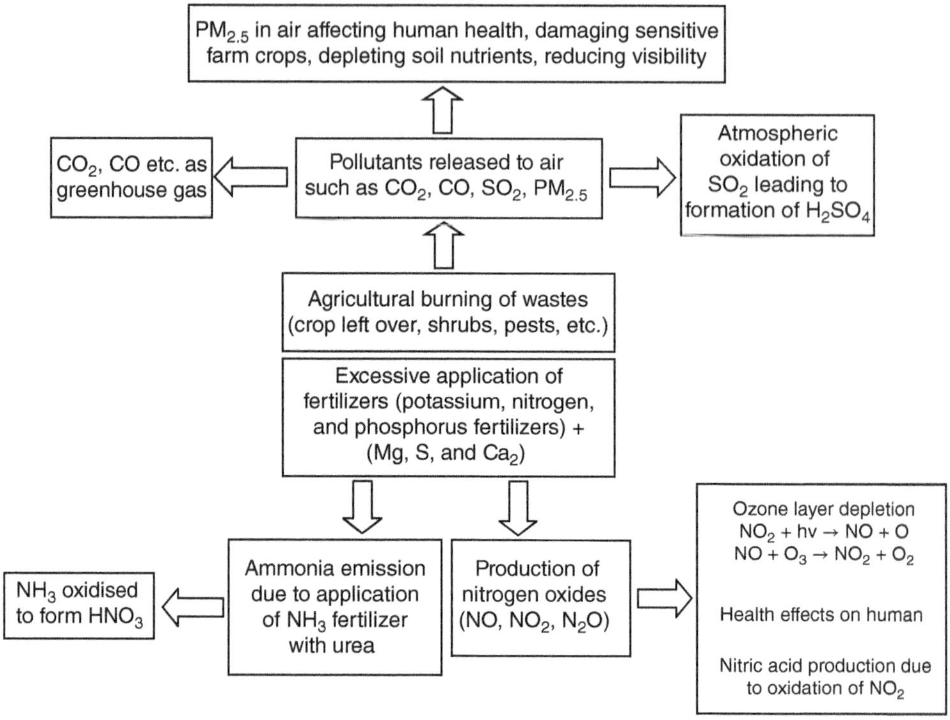

Figure 4.1 Sources of air pollutants from agriculture fields (Venkataraman et al. 2006; Abbasi et al. 2014).

fertilizers on the atmosphere. Some of the major farming activities that release pollutants to the atmosphere include (Amann et al. 2017):

a) Open burning of agricultural residues
The burning of crop residues involves the burning of various agricultural wastes that remain in the field after harvest. It is basically carried out for the clearing of land for the next crop. Some positive sides of burning include the quick release of nutrients and damage to soil-borne pathogens (Abbasi et al. 2014). However, burning processes release a massive amount of air pollutants such as particulate matter ($PM_{2.5}$), NO_x, ammonia (NH_3), volatile organic compounds (VOC), methane (CH_4), carbon dioxide (CO_2), carbon monoxide(CO), and sulphur dioxide (SO_2) (Abbasi et al. 2014) (Figure 4.1). Open biomass burning of residual wastes of rice and wheat results in the production of carbon monoxide, carbon dioxide, $PM_{2.5}$, and negligible amount of sulphur dioxide and a combination of black carbon and organic matter (Venkataraman et al. 2006).
b) Application of mineral fertilizers
Mineral fertilizers primarily consist of basic plant nutrients such as nitrogen (N), phosphorous (P), and potassium (K). Efficiently produced and transferrable nitrogen-rich fertilizers, such as ammonium or nitrate-based salts and urea, are used in masses around various agricultural fields across the world, thus contributing to the emission of N as NH_3 into the atmosphere. These nitrogen-rich fertilizers contain nitrogen

concentrations ranging between 33 and 46% (Amann et al. 2017). Application of these fertilizers in agricultural fields releases handful amount of NH_3 into the atmosphere, 5–40% of N lost as NH_3 compared to 0.5–5% for ammonium nitrate fertilizers and in case of urea about 5–11% of nitrogen loss as NH_3 (Schwenke et al. 2014).

Ammonia released into the atmosphere can be oxidized and form nitric acid, sulphuric acid, and form acid rain, resulting in damage to the vegetation and various organisms, leading to disturbance of the ecosystem (Shaviv 2001; Amann et al. 2017). Excessive application of fertilizers in agricultural fields also results in the production of nitrous oxide (N_2O) and other greenhouse gases such as water vapour, carbon dioxide, methane, and hydrogen sulphide with chlorofluorohydrocarbons (Figure 4.1) (Abbasi et al. 2014).

c) Application of manures from livestock
Livestock management is an essential part of the human development and growth; it contributes a large portion among the various interferences to the global nitrogen cycle. The requirement of protein to the growing human population results in a very higher demand of protein-rich foods. The N-rich foods are excreted as urea from the animal body. After exposing into the environment and in the presence of enzyme, urease would easily get transformed into NH_3 (Amann et al. 2017).

4.3.2 Water Pollution and Agriculture

Agricultural fields pollute the water bodies in varied ways. Water pollution caused by agricultural activities involves the application of chemical fertilizers, livestock and poultry breeding, aquaculture, and rural population (Figure 4.2). Pollutants released by agricultural activities into water body cause an increase in chemical oxygen demand (COD), along with high loads of nitrogen, ammonia, and phosphorus (Wu et al. 2013). The pressure led to agricultural fields by increasing food demand due to population growth and dietary

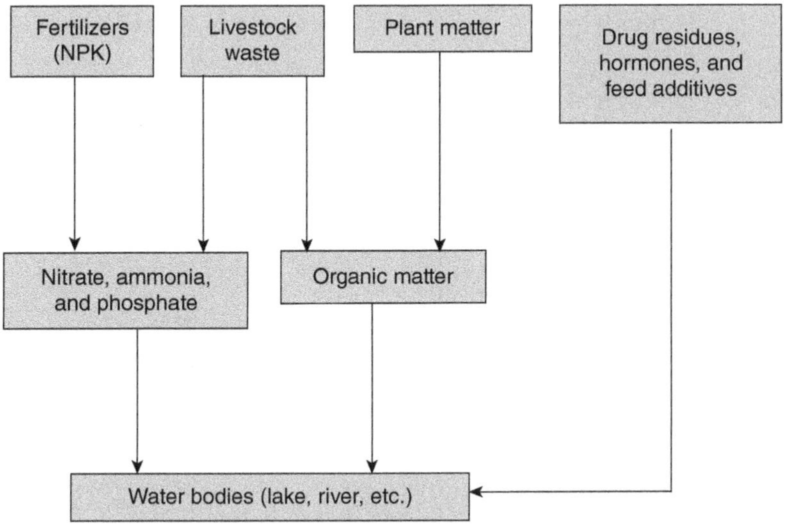

Figure 4.2 Flow of water pollutants from agricultural field (Wick et al. 2012).

patterns is indirectly responsible for the deterioration of water quality of both surface and groundwater. Agricultural pollution has manifold aspects of polluting the water bodies that resulted in waterlogging, erosion, salinization, and desertification (Savci 2012). Nitrate is the most common chemical contaminant released from agriculture and contaminate the groundwater aquifers (Wick et al. 2012). The flow of agricultural pollutants from fields to various water bodies such as rivers, wetlands, and ponds has been shown in Figure 4.2.

Agricultural pollutants that contribute to the water pollution are nutrients (Sharpley et al. 2015), pesticides, salts, sediments (Mateo-Sagasta et al. 2018), organic matter (Wen et al. 2017), pathogens (Zandaryaa and Mateo-Sagasta 2018), metals, and drug residues (Mateo-Sagasta et al. 2018).

a) Nutrients: Nutrients are largely used in agricultural fields in the form of fertilizers, and in livestock or aquaculture, they are used as feeds. In agricultural fields, the application of excess fertilizers results in the release of excess nitrogen and phosphates (Sharpley et al. 2015), which could be leached into the groundwater or transported to surface water via surface run-off (Mateo-Sagasta et al. 2018).
b) Pesticides: The prime use of pesticides in the agricultural field is to control weeds, pests, rodents, etc. Generally, the effects of pesticides are not species specific and raise residual issues even after a long duration of application. Various natural factors could lead to the flow of pesticides from agricultural fields into water bodies, including surface run-off. Harmful pesticides could cause injurious algal blooms (Kumar and Kumar 2019) and harm aquatic species (Helfrich et al. 2009).
c) Salts: Improper irrigation or poor drainage with other natural factors causes agricultural soils to accumulate leachable salts (Nabiollahi et al. 2017). These fractions of salts could get carried away by drainage water leading to salinization of water bodies (Mateo-Sagasta et al. 2018). The saline water aquifers that underlie the freshwater aquifers could affect the latter if the saline water is discharged onto the surface (Brindha and Schneider 2019). This affects the water quality and the freshwater biota. Changes within species and community composition cause habitat loss leading to a loss in biodiversity and migration (Lorenz 2013; Mateo-Sagasta et al. 2018).
d) Sediments: Sedimentation is a primary concern of water pollution. Excessive and improper land usage, tillage, and poor soil management in agriculture may increase soil erosion and sediment run-off into water bodies (Mateo-Sagasta et al. 2018). Soil erosion causes the degradation of waterways and releases suspended sediments into the water bodies (StenfertKroese et al. 2020). These suspended solids degrade the water quality of streams by deteriorating the physical, biological, and chemical properties of water (Owens et al. 2005; Horowitz 2008).
e) Organic matter: They consists of the components from animal excreta and feed and wastes from animal processing industries that contribute to water pollution (Mateo-Sagasta et al. 2018). BOD level is a common way to measure pollution caused by organic matter (Zandaryaa and Mateo-Sagasta 2018). Contamination with organic matter causes an increase of BOD levels in water bodies. Wastewater discharge from livestock farms is the leading organic pollutant load carriers into rivers (Meybeck 2003). Organic matter pollution from livestock farming has raised more concern than the organic matter pollution from urban discharge (Wen et al. 2017).

f) Pathogens: Livestock farming produces livestock excreta that could contain several types of multicellular parasites and zoonotic micro-organisms that could be harmful to human beings (Mateo-Sagasta et al. 2018). The transfer of these micro-organisms could be possible by using the contaminated water for irrigation purposes. Various zoonotic waterborne pathogens such as *Salmonella enteric, Escherichia coli O157:H7, Campylobacter* spp., *Cryptosporidium parvum*, and *Giardia uodenalis* are present in livestock and their excreta (Gannon et al. 2012).
g) Drug residues and other emerging pollutants: Chemicals such as vaccines, growth promoters, antibiotics, and hormones used in agriculture can harm the environment (Mateo-Sagasta et al. 2018). They could reach water bodies through leaching or surface run-off from livestock farming and aquaculture farms (OECD 2012). Other pollutants such as heavy metals in agricultural pesticides and animal feeds have raised severe issues to the biosphere's health.

Water pollution due to agricultural activities has adverse effects on human health, aquatic ecosystems, and terrestrial ecosystems. Negative impacts on human health include blue baby syndrome caused by high level of nitrates in water, various other illnesses, acute and chronic health effects are caused due to pesticide accumulation in water and in the food chain (Knobeloch et al. 2000). In aquatic ecosystems, eutrophication occurs due to excess nutrient accumulation affecting the aquatic organisms and their whole biodiversity (Mateo-Sagasta et al. 2018).

4.3.3 Soil Pollution Due to Agricultural Activities

Agriculture is responsible for soil pollution by acting as a source of solid wastes that are deposited on land and various chemicals applied on the agricultural fields. Agricultural solid wastes are mainly composed of biodegradable organics when disposed of or sometimes used in fields. This results in an inadequate supply of oxygen, high ammonia concentrations, etc., that leads to hazardous conditions to grow crops (Loehr 1978). Moreover, excessive use of fertilizers, pesticides, herbicides, insecticides etc., leads to soil pollution.

4.3.3.1 Effect of Fertilization

Chemical fertilizers added to the agricultural fields for maintaining good health of the crops indirectly affect the chemical, biological, and physical properties of soil. Application of fertilizers for a long duration causes reduction of soil pH, indirectly affecting the soil microbiological environment. Microbial activities and the nutrients solubility and availability depend on pH (Gentili 2018). Long-term fertilization leads to deterioration of soil microbial biomass by increasing the enzyme dehydrogenase activity (Upadhyay et al. 2020). Damage to soil structure, alteration in soil pH, and other physiochemical parameters can be increased when fertilizers containing a high amount of sodium and potassium are applied to the agricultural fields (Savci 2012). Cd accumulation is also observed due to the application of phosphate fertilizers. Application of N- based fertilizers accelerates soil organic matter (SOM) decomposition and adversely affects the soil health (Upadhyay et al. 2020). This type of fertilizer increases the activities of microbial decomposition. SOM is the composition of wide range of compounds organic materials in soil. These compounds include large amount of electron donating groups primarily comprising of oxygen (O) and nitrogen (N) (Matocha 2005).

4.3.3.2 Effect of Agrochemicals

Large-scale applications of selective agrochemicals caused the pests to adapt to these chemicals and insist on the development of new agrochemicals leading to more side effects on the environment (Carvalho 2006). Degradation of pesticides produces various by-products that enter the environment, soil, rivers, wetlands, or any type of water bodies, which leads to the accumulation of toxic substances. Most common pesticides include DDT (dichlorodiphenyltrichloroethane), dieldrin, chlorinated hydrocarbons, organophosphates, and OCPs, which is accumulated in soil particles. These chemicals contaminate the crops and enter the human body by ingestion of the contaminated plant parts, which in turn results in the accumulation of toxic substances in the human body, causing different diseases (Abbasi et al. 2014; Taufeeq et al. 2021). Long-term application of pesticides and chemical fertilizers could cause long-term changes in the soil properties such as bulk density (Yargholi and Azarneshan 2014).

Applied chemicals create a challenge against food safety and sustainability. For example, fungicides inhibit the activity of dehydrogenase, urease, catalase, acid phosphatase, and alkaline phosphate of microbes (Upadhyay et al. 2020). The application of pesticides also has some negative impacts on soil quality. Pesticides form some harmful products called transformation products (TPs) (Barceló and Hennion 1997). These pesticides and TPs are divided into two groups: (i) Hydrophobic, persistent, and bioaccumulable pesticides, which are firmly bound to soil particles and (ii) polar pesticides, which represents mainly herbicides and also includes carbamates, fungicides, and some organophosphorous insecticide TPs (Upadhyay et al. 2020). The formation of these TPs could be through photodegradation of pesticides (Agüera et al. 2014). Adsorption of TP particles affects the capability of soil to hold positively charged minerals. This results in decrease of the soil pH. With decreasing pH, adsorption increases for ionizable pesticides (e.g. 2,4,5-T, atrazine, etc.) thereby replacing the positively charged minerals with ionizable pesticides (Andreu and Pico 2004).

4.4 Realization of Sustainable Development from Agrowaste

In 1987, the World Commission on Environment and Development published the Brundtland report, popularly known as 'Our Common Future'. This report defined sustainable development as 'the idea that human societies must live and meet their needs without compromising the ability of future generations to meet their own needs. Sustainable development seeks to meet the needs and aspirations of the present without compromising the ability to meet those of the future' (World Commission on Environment and Development and Brundtland 1987). Thus, sustainable development is a strategy for a society's efficient organization to exist in the long term. It involves preservation of environmental elements, such as natural resources, along with social and economic fairness.

Issues related to environmental sustainability due to the development of human society, including industrial development, have been indicated by various economic theoreticians, including Adam Smith, during the eighteenth century. Philosopher Karl Marx and classical economists Malthus, Ricardo, and Mill debated several aspects of sustainable development in the nineteenth century (Willis 2005).

In 2015, at the UN Conference on Sustainable Development held in New York, the present challenges of sustainable development were reviewed, and a new resolution 'Transforming our world: the 2030 agenda for Sustainable Development' with seventeen new Sustainable Development Goals or SDGs was undertaken. Of these, SDG 2 for 'Zero Hunger' and SDG 12 for 'Responsible consumption and production' directly involve the agricultural sector. These goals are targeted to be achieved by the year 2030 (UN 2015; UNDP 2015). Hence, it is clear that reaching these SDG targets is simply not possible without a sustainable agriculture sector.

Agricultural wastes or agrowastes can be a source of sustainable development as they can be used to derive clean energy, used as resources for various industrial purposes, and can aid the improvement of soil health and boost food production. Crop residues can be used as animal fodder, can be composted, and used as resources for deriving bioenergy and biofertilizers. Animal dung based biogas plants may help to reduce fossil fuel dependency for cooking gas. Apart from waste reduction and management, the major advantage of utilizing agricultural residues is that many of these aforementioned products can also be extracted from non-food parts of the crop plant. This does not hamper food availability. Moreover, it aids economically as the generation of by-products, along with crops, will provide more revenue. It has been estimated that roughly 140 Gt of biomass waste is generated globally (Tripathi et al. 2019). Figure 4.3 illustrates the major agrowastes and technologies to obtain useful products from agrowastes.

The highest quantity of agricultural waste is generated from rice straws, wheat stalks, corn stalks, bagasse, etc. (Sadh et al. 2018). Although agrowaste seems to be a giant source for renewable energy, it is not always easy to collect the waste from their generation point and move to a power plant. Transportation of agrowaste from source to power plant is among the main challenges regarding utilization of agrowastes for sustainability. Additional energy extraction or bio-conversion processes may require specialized equipment or infrastructure, skilled human resources, specialized decontamination units, etc. These limitations may limit the process's net economic viability (Koul et al. 2021).

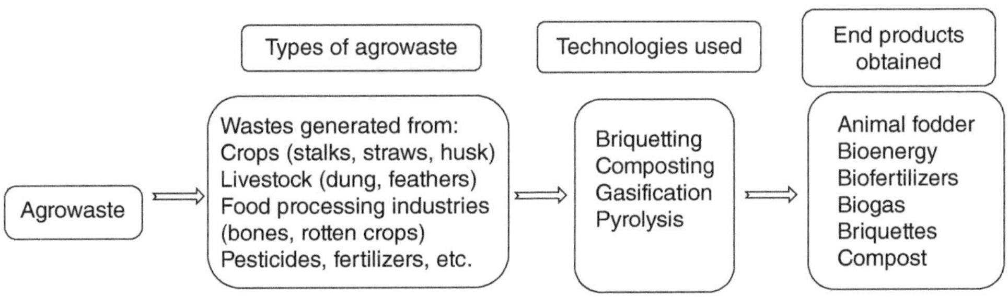

Figure 4.3 Agrowastes and technologies used to obtain valuable products from agrowastes (Ge et al. 2021; Gupta and Mondal 2020; El Haggar 2005).

4.4.1 Clean Energy

Clean energy refers to energy derived from renewable sources with zero carbon emission. Crop biofuel is considered to be carbon neutral as the quantity of carbon released upon combustion balances the amount of carbon consumed by the crop during its growth.

In recent years, the generation of vast quantities of agrowaste has raised the interest in using it as a fuel. The latent energy in waste is used for power generation. It serves various purposes apart from solving the issue of agricultural waste management, like reducing the dependence on fossil fuels for energy. The cost of transportation of agricultural biomass is greatly reduced, and thus helps in gaining overall environmental quality.

Energy extracted from agricultural waste can be termed as biofuel, which are further classified as bio-ethanol, biodiesel, green diesel, biogas, etc. Biofuels provide sustainability, reduce greenhouse gas emissions, and provide employment opportunities as renewable energy sources. The major biofuels obtained from agrowastes are given in Table 4.2.

Based on the type of feedstock used, biofuels are classified into different generations as demonstrated in Figure 4.4.

Thermochemical combustion and biochemical combustion are the two commonly used techniques to generate energy from agrowaste (Woods and Hall 1994). Thermochemical combustion refers to the thermal decomposition of agrowaste in order to produce fuel, whereas biochemical combustion refers to the enzymatic degradation of agrowaste through augmentation of microbial activity. There has been remarkable advancement in energy utilization technologies in recent years (Wei et al. 2020) some of which are mentioned in Figure 4.5.

The criteria for selecting the appropriate technology to convert a given agricultural waste material into energy have been presented in Table 4.3. The Central Public Health and Environmental Engineering Organisation (CPHEEO), which operates under the Ministry of Urban Development in India, established these standards.

Table 4.2 Biofuels obtained from agrowastes.

Biofuels	Agrowastes	References
Bio-ethanol	Corn, sugarcane, wheat	(Swain and Mohanty 2018)
Biodiesel	Jatropha, canola, soybean, palm oil, animal fats	(Zahan and Kano 2018)
Green diesel	Algae, plant sources	(Demirbas and Demirbas 2010)
Biogas	Digested organic matter	(Kolář et al. 2008)

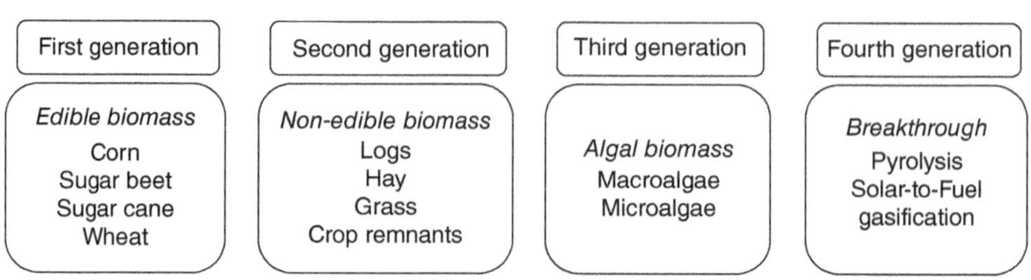

Figure 4.4 Classification of biofuel generations based on feedstock (Alalwan et al. 2019).

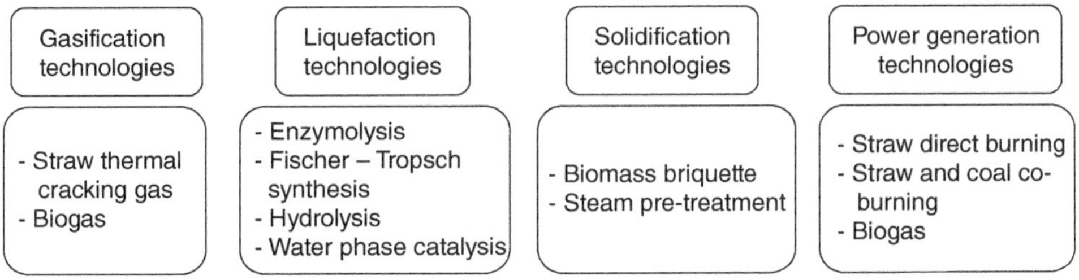

Figure 4.5 Technologies used to convert agrowaste biomass to energy.

Table 4.3 Criteria for selection of energy extraction technology (CPHEEO 2000).

Technology	Criteria	Component	Reference
Biochemical combustion	Moisture content	>50%	
	Volatile organic matter	>40%	
	C/N ratio	25–30 : 1	(CPHEEO 2000)
Thermochemical combustion	Moisture content	<45%	
	Volatile organic matter	>40%	
	Fixed carbon	<15%	
	Total carbon	<35%	
	Calorific value	1200 kCal/kg	

Screening agrowastes based on specific characteristics is essential before using these for energy extraction. Agrowastes with high density and a higher proportion of biodegradable contents are not considered suitable for thermal combustion.

4.4.1.1 Biochemical Combustion

Techniques associated with the biochemical combustion of agrowaste are biomethanation or anaerobic digestion. The process of converting agricultural biowaste into biogas with the help of biological agents such as micro-organisms is called biomethanation (Angelidaki et al. 2011). In this process, the production of biogas takes place in a closed container called a biogas digester, where bacteria decompose organic matter in anaerobic conditions.

Biomethanation, also known as anaerobic digestion, is a multi-stage biological and chemical process that includes (i) hydrolysis, (ii) acidogenesis, (iii) acetogenesis, and (iv) methanogenesis.

4.4.1.1.1 Stage – I: Hydrolysis

Complex organic matter in the biomass, such as organic polymers, proteins, lipids, and carbohydrates, is broken down into simpler molecules like amino acids, fatty acids, and simple sugars by anaerobic bacteria at this step (Angelidaki et al. 2011). Fermentative micro-organisms break down the long-chain fatty acids that are longer than acetate. The products at this stage are carbon, hydrogen, ammonia, carbon dioxide, simple sugars, fatty acids, etc. Methanogens may consume some hydrolysis products, such as hydrogen and acetate, further in the anaerobic digestion process. The majority of the remaining complex molecules must be further broken down in the acidogenesis process before being used to produce methane (Lin et al. 2021).

4.4.1.1.2 Stage II: Acidogenesis

In the second stage of fermentation, the agricultural biomass and organic products are further broken down by acidogenic micro-organisms such as *Bacteroides succinogens*, *Clostridium* sp., and *Eubacterium limosumz*, (Angelidaki et al. 2011; Azman et al. 2015; Tukanghan et al. 2021). While releasing ammonia, hydrogen, carbon dioxide, hydrogen sulphide, and other volatile fatty acids and organic acids, these fermentative bacteria generate an acidic environment in the digestive tank. Acetic acid, butyric acid, propionic acid, etc. are among the principal acids produced (Angelidaki et al. 2011; Azman et al. 2015; Lin et al. 2021).

4.4.1.1.3 Stage III: Acetogenesis

Acetogens like *Methanosarcina* spp. and *Methanosaeta* spp. dominate this stage by producing acetate, a derivative of acetic acid, from carbon and energy sources. Many of the chemicals produced during acidogenesis are metabolized into acetic acid, carbon dioxide, and hydrogen by these bacteria (Angelidaki et al. 2011). Acetogens break down biomass to the point that methanogens can use a large portion of the leftover material to produce methane (Lin et al. 2021).

4.4.1.1.4 Stage IV: Methanogenesis

In the final stage of anaerobic digestion, methanogens produce methane from the end products of acetogenesis, along with some of the intermediate products generated during hydrolysis and acidogenesis (Prabhudessai et al. 2013; Achinas and Euverink 2016; Lyu et al. 2018). There are two groups of micro-organisms following two pathways to produce methane:

i) *Methonabacterium* spp. and *Methanospirilium* spp. produce methane by either reducing carbon dioxide or oxidizing hydrogen (Wang et al. 2014), and
ii) *Aceticlastic* methanogens like *Methanosarcina* spp. break acetate to form methane (Angelidaki et al. 2011).

The process of biomethanation may be summarized as presented in the flowchart in Figure 4.6:

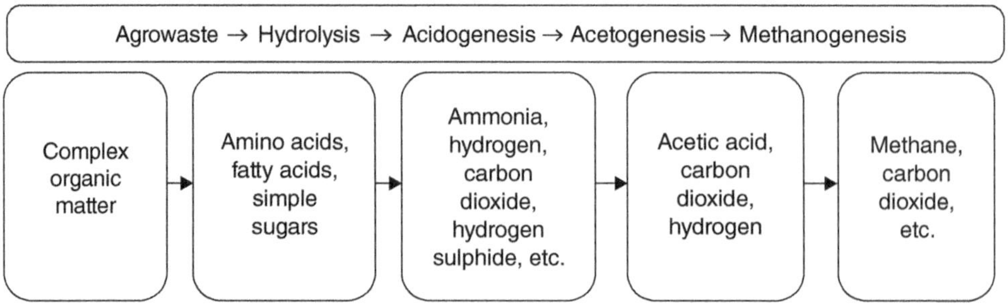

Figure 4.6 Flowchart showing various steps of biomethanation process (Beniche et al. 2021; Liu et al. 2021).

Different factors affect the production of biogas. Some of these factors are discussed as follows:

i) Temperature – One of the most crucial factors influencing biogas production is temperature. A temperature above 30 °C is best suited for fermentation. In the case of mesophilic flora, 30–40 °C is the optimal temperature range, whereas for thermophilic flora, 50–60 °C is the best temperature range for fermentation (Sambo et al. 1995).
ii) pH – Low pH hinders methanogenic bacteria growth and gas production and is frequently caused by overloading. A neutral pH, i.e. between 6.0 and 8.0, is the most efficient for digestion to take place (Sambo et al. 1995).
iii) Nutrients – Carbon (C) and nitrogen (N) are two of the most important nutrients, and the net C/N ratio is a crucial criterion for selecting raw materials. Excess nitrogen causes the creation of ammonia, which can prevent microbes from growing further. Dilution or lowering the load may help reduce ammonia toxicity (McCarty and McKinney 1965). Maintaining a C/N ratio near to 30 : 1 is critical for achieving an optimal rate of digestion. Combining materials with low carbon content with those with high nitrogen content, or vice versa, can help precisely manage the C/N ratio (McCarty and Mckinney 1965).
iv) Toxic Materials – Anaerobic digestion can be hampered by a range of contaminants found in wastes and biodegradable wastes. Salts of chromium, copper, mercury, nickel, and zinc are all common toxic substances (Raja and Wazir 2017). Salts of sodium, potassium, calcium, and magnesium, on the other hand, may have stimulatory or toxic effects, depending on whether they are connected with the cation or the anionic component of the salt. In the process, pesticides and synthetic detergents may operate as deterrents. Processes such as dilution or the inclusion of shredded bagasse or straw can help to minimize ammonia toxicity and hence improve the C/N ratio (Raja and Wazir 2017).

4.4.1.2 Biogas Reactors

Biogas is a potential source of renewable energy. Its efficient utilization will help us with the various emission reduction goals. It is a wonderful waste management tool. It reduces the need for landfills and minimizes pollution due to disposal of agricultural waste (Thrän and Fritsche 2016). Two major types of digester are floating drum and fixed drum biogas plants. An airtight digesting chamber with inlet and outlet and biogas collection chamber are among the most critical features of biogas reactors. Different types of biogas reactors such as fixed dome reactor and floating drum reactors are used (Nakamya et al. 2020).

The installation of new biogas plants also helps generate employment. Although biofuels are carbon neutral, they do pollute the environment to a small extent, but they are still the better alternatives to using fossil fuels for energy extraction (Srivastava et al. 2021). With improved scientific technologies, the drawbacks can be overcome, while better storage facilities and purification techniques may be developed to enhance this attractive fuel further, so that our goals for sustainable development may be achieved within a shorter span.

4.4.1.3 Thermochemical Combustion

The major techniques associated with thermochemical combustion include incineration, pyrolysis, and gasification (Saidi et al. 2020). In the incineration method, combustion is controlled to reduce solid and liquid biomass primarily into carbon dioxide, water vapour, and other gases and some non-combustible residues that ultimately remain as flying ash. This process, however, also releases polychlorinated biphenyls and sulphur-containing gases that are highly corrosive in nature along with partially combusted char (Khan et al. 2009). Organic volatiles burn at temperatures of 750–1200 °C, and the remaining inert products and char burn at temperatures exceeding 1200 °C (Samolada and Zabaniotou 2014). The incinerator undergoes cooling and condensation after all the agricultural waste has been burnt. Fly ash and other residues are collected and disposed of or reused further. A constant air supply is extremely crucial for incineration processes as it will control the amount of toxic gases produced. The final waste biomass is reduced by 85–90% (Sun et al. 2016).

Agricultural biomass, such as stalks and husks, consists of approximately 43% cellulose, 37% lignin, and 10–15% hemicellulose (Ruan et al. 2019). In pyrolysis, thermal decomposition of agrowaste is done at high temperatures of up to 1000 °C in the absence of oxygen (Prasad and Murugavelh 2019; Zhang et al. 2019). In this process, the cellulose chains undergo depolymerization and break down into sugars, methane, carbon dioxide, carbon monoxide, and hydrogen. Pyrolysis also generates commercially useful by-products such as acetic acid and acetone (Yanik et al. 2007), whereas gasification is the process of controlled partial oxidation of carbonaceous agrowaste. Controlled oxidation refers to the reduced supply of oxygen than what is stoichiometrically required for the complete combustion (Tripathi and Shukla 2015). It converts the biomass in the agricultural waste into combustible gas that mainly consists of ash, carbon dioxide, carbon monoxide, hydrogen, and water vapour. This combustible gas has a high calorific value, equivalent to 25% of natural gas, thereby rendering it an attractive candidate for production of clean energy (Zainal et al. 2001). The process of gasification involves drying, pyrolysis (devolatilization), gasification, and secondary reactions like cracking of tar.

Theoretically, most agricultural plant residues, such as wheat stalks and straws, have moisture content less than 30% and are ideal for gasification. But in practice, gasification experiments have been conducted on only a handful of agrowastes (Chang et al. 2000). Apart from moisture content, other important parameters such as calorific value, ash content, volatile matter content, bulk density, charring properties, help determine whether a feedstock is suitable for gasification. Most experiments conducted involve coconut shells and maize cobs and have been proven to be efficient fuels for gasification as feedstock (Arun et al. 2020). Cereal straws and rice husks constitute a significant portion of agrowaste, but have ash contents above 10 and 20%, respectively (Mansaray and Ghaly 1999; Vargas et al. 2012; Arun et al. 2020). Such high ash content causes problems of slagging, which can be corrosive in nature (Wang et al. 2017).

Gasification is a promising process for the extraction of clean fuel and power generation because this process produces the economically important SynGas. Synthesis gas or SynGas consists of carbon monoxide (35–45%), carbon dioxide (10–15%), hydrogen (27–30%), water vapour (15–25%), and other gases (0.2–1%) (Wang et al. 2009; Tinoco-Caicedo et al. 2021). Fischer–Tropsch synthesis is a gasification technique named after the chemist duo Franz Fischer and Hans Tropsch, who devised the reaction. SynGas can be used to make liquid transportation hydrocarbon fuels and a variety of other chemical products (van de Loosdrecht et al. 2013).

$$CO_2 + 2H_2 \xrightarrow{\text{Catalyst (Fe or Co)}} \text{-}(CH_2)\text{-} + H_2O$$

Based on this reaction, several types of feedstock are being experimented upon for the extraction of biofuels. Based on the different properties of the feedstock, there is a range of gasifier designs such as updraught, downdraught, crossdraught and fluidized bed. Major disadvantages of gasification are its high cost and the high technical expertise required to perform the process impeccably for each type of agrowaste feed (Pinto et al. 2016; Arun et al. 2020). Another disadvantage is the presence of an undesirable product – tar, which is produced during the production of syngas (Arun et al. 2020; Ratchahat et al. 2021).

4.4.2 Resources for Industries

Apart from biofuel, agricultural wastes serve as source of different resources for the production of industrially useful products, such as biofertilizers, bioethanol, biodiesel, bioenzymes, briquettes, bioplastic, biochar, bio-compost, and animal feed. A brief overview of some economically important resources has been discussed later.

4.4.2.1 Biochar

Biochar is a carbon-rich form of charcoal that is produced by the thermochemical degradation of plant biomass in anaerobic conditions. Biochar serves to improve soil quality by improving the water retention capacity of soil, improving soil structure and air circulation, increasing retention of nutrients, and preventing leaching. It also helps mitigate global climate change by adding carbon back to the soil by enhancing soil carbon sequestration (Giri et al. 2017; Kumar et al. 2021).

4.4.2.2 Biocompost

Biocomposting denotes the degradation of plant organic matter by the action of micro-organisms under controlled aerobic conditions. They consist of bacteria that can solubilize nitrogen, phosphate, and other fungi such as *Trichoderma viridea*. Micro-organisms present in biocompost also provide protection against various soil-borne diseases and help to improve soil fertility, thus boosting crop growth (Rai et al. 2021).

4.4.2.3 Bioenzymes

Enzymes are biological catalysts that are used in a variety of industries, from cleaning to cooking to the creation of pulp for the paper industry (Singhania et al. 2015). Enzymes such as α-amylase, cellulase, pectinase, and protease have been extracted from agricultural residues such as soybean meal, sunflower meal, maize bran, maize pericarp, olive oil cake, and wheat bran using *Bacillus* as the biological degradation agent (Chen et al. 2020; Kukreti et al. 2022). Bioenzymes are preferred over chemical enzymes due to their high efficiency of substrate-specific binding efficiency. Technologies such as recombinant DNA techniques and genetic engineering have led to manifold improvement in bioenzyme production. Solid-state fermentation is one of the best methods to convert agrowaste into valuable bioproducts (Sivakumar et al. 2022).

Xylanase is an economically important bioenzyme for industries like food, biomedical, animal feed, and bioethanol. Xylanase is extracted from lignocellulosic food waste such as wheat straws and cereal stalks (Kukreti et al. 2022; Sharmila et al. 2022). It employs fungi *Aspergillus niger* and *Aspergillus terreus* strains to extract the enzymes (Rehman et al. 2014). Inulinase is another commercially important bioenzyme used in the food and pharmaceutical industry and in the production of bioethanol, citric acid, lactic acid, etc. and hydrolysis of inulin to fructose (Silva et al. 2013; Singh 2014). It is synthesized by the action of micro-organisms like *Actinomyces viscosus, Chrysosporium pannorum, Kluyveromyces fragilis, Penicillium* sp., and *Streptococcus salivarius* on lignocellulosic substrates (Ravindran et al. 2018). Other important bioenzymes are amyloglucosidase from *Aspergillus niger*, amylase from *Aspergillus oryzae*, dextranase from *Chaetomium erraticum*, glucose isomerase from *Streptomyces murinus*, etc. (Abdel-Azeem et al. 2021; Kukreti et al. 2022; Sivakumar et al. 2022).

4.4.2.4 Biofertilizers

Biofertilizers are defined as 'biological products containing living micro-organisms that, when applied to seed, plant surfaces, or soil, promote growth by several mechanisms such as increasing the supply of nutrients, increasing root biomass or root area, and increasing nutrient uptake capacity of the plant' (Vessey 2003). Agricultural wastes, especially fruits and vegetable wastes, are subjected to solid-state fermentation to cultivate micro-organisms that can serve as efficient biofertilizers (Vessey 2003). The role of microbial interactions on sustainable agriculture has been demonstrated in Figure 4.7.

4.4.2.5 Bioplastic

Biodegradable or bio-based plastics are also called as bioplastics. These can be made from plant starch and have a variety of applications. The bio-based monomers are usually made by chemical conversion of agricultural biomass. Polylactic acid is the most common biopolymer derived by the bioactivity of bacteria on agrowaste such as sugarcane, sugar beet, and cane (Chan et al. 2021). Rawoof et al. (2021) have compared several substrates and found that food waste, broken rice, sugarcane molasses, etc. have a high yield of lactic acid with good purity. The micro-organisms incorporated for the extraction of lactic acid are *Lactobacilli amylophilus, Lactobacilli salivarius, Lactobacilli acidophilus,* etc. The lactic acid thus extracted undergoes oligomerization to form lactide, where a catalyst, such as stannous alkoxide, poly L-lactide, or poly D-lactide, initiates the polymerization process (Chan et al. 2021). While technologies are rapidly evolving for the extraction of better quality biopolymers, the major hindrance in their commercialization is the cost-effective production processes and their relatively inferior quality compared to plastics derived from fossil fuels (Chan et al. 2021; Riera and Maldonado 2021).

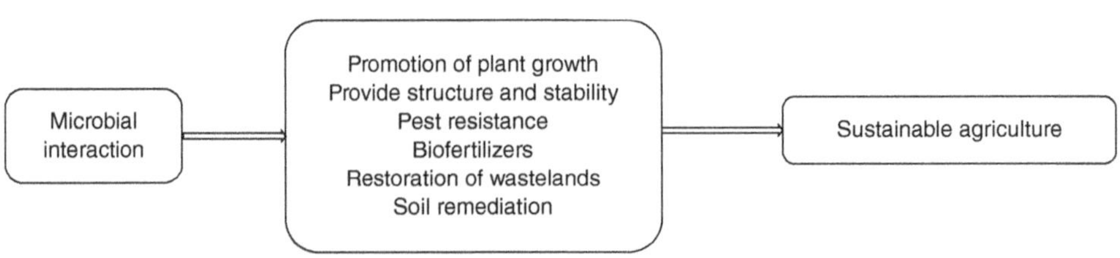

Figure 4.7 Various ways microbial interactions support sustainable agriculture (Gupta et al. 2021; Ke et al. 2021)

4.4.2.6 Briquettes

Using simple compression techniques, briquettes are 'bricks' made from agricultural waste materials. Briquettes can be used as fossil fuel or firewood substitute in small-scale operations. Cow dung cakes or straw briquettes are good examples of this technique (Stewart et al. 2021). The bulk of the agrowaste is compressed, and the high calorific value of the briquette ensures high energy value. Agricultural wastes such as hay, sugarcane bagasse, corn stalk, coconut shells and leaves, groundnut shells, rice husk, sawdust, and other agricultural wastes are carbonized, crushed, dried, compacted into bricks, and further compressed to ensure biomass briquettes have a long life span (Jagtap and Kalbande 2021; Thekedar et al. 2021).

4.4.2.7 Animal Feed

Agricultural wastes such as fruits and vegetable wastes, sugarcane molasses and the peel, husk or straws of cereals such as corn, rice and, wheat are used as feed for cattle and other livestock (Koul et al. 2021). This helps to reduce the cost of animal fodder. Table 4.4 shows some agricultural wastes and their end products along with the conversion processes:

Table 4.4 Methods used to convert agricultural wastes to economically useful products. (Yalinkilic et al. 1998; Khedari et al. 2003; Sud et al. 2008; UNEP 2009; Karade 2010; Rodríguez et al. 2010; Raut et al. 2011; Mohammad et al. 2013; Asim et al. 2015).

Agricultural wastes	Conversion process	Product
Pineapple leaves, sugarcane residues	Bio-reduction	Animal feed, industrial absorbents, additives for beverages, etc.
Wheat straw, cotton fibres		Wheat straw polypropylene, pelletized feedstock, fertilizers
Rice husk	Bio-refinery system	Water-soluble oil, synthetic lubricants, etc.
Sugarcane bagasse Wheat straw		Lumber materials, panel boards
Basil leaves		Fabric, ropes, fibre craft, cellulose, and speciality papers
Coconut husk	Decortication	Coconut fibre, coir pads, rope and twine, brooms and brushes, doormats, rugs, mattresses, upholstery, etc.
Bamboo fibres, sugarcane bagasse	Hot melt process	Paper and packaging materials
Sugar mill boiler ash from bagasse	Hydro-separation	Filtration materials and absorbent products
Oil palm fruit residues, sugarcane bagasse	Moulding	Biodegradable packaging materials, paper and paper wares
Jute and hemp fibres	Pulping	Soundproofing systems, thermal insulators
Coconut coir	Twining	Coconut twines

4.5 Conclusion and Future Prospects

Transformation of agrowastes to valuable products and their proper handling can safeguard the environment as well as can be a source of clean energy. This will reduce the dependence on fossil fuels and help in generating employment and thereby improve the economy of a country. However, technological interventions are needed to receive maximum benefits from agrowaste. Awareness about handling of agrowaste to the general public is also essential to reduce its loss at the sources. Thus, with technological advancement and knowledge to common masses, agrowaste can not only reduce environmental pollution but also help in attaining the sustainable development goals.

References

Abbasi, A., Sajid, A., Haq, N. et al. (2014). Agricultural pollution: an emerging issue. In: *Improvement of Crops in the Era of Climatic Changes* (ed. P. Ahmad, M.R. Wani, M.M. Azooz and L.-S.P. Tran), 347–388. Springer. 10.1007/978-1-4614-8830-9_13.

Abdel-Azeem, A.M., Abu-Elsaoud, A.M., Nahas, H.H.A. et al. (2021). Biodiversity and industrial applications of Genus Chaetomium. In: *Industrially Important Fungi for Sustainable Development* (ed. A.M. Abdel-Azeem, A.N. Yadav, N. Yadav and Z. Usmani), 147–206. Cham: Springer.

Achinas, S. and Euverink, G.J.W. (2016). Theoretical analysis of biogas potential prediction from agricultural waste. *Resource-Efficient Technologies 2* (3): 143–147.

Adegbeye, M.J., Reddy, P.R.K., Obaisi, A.I. et al. (2020). Sustainable agriculture options for production, greenhouse gasses and pollution alleviation, and nutrient recycling in emerging and transitional nations – an overview. *Journal of Cleaner Production* 118319. 10.1016/j.jclepro.2019.118319.

Agamuthu, P. (2009). Challenges and opportunities in agro-waste management: an Asian perspective. In *Inaugural Meeting of First Regional 3R Forum in Asia*. Tokyo Japan, pp. 11–12. https://www.uncrd.or.jp/content/documents/Session2_Agamuthu.pdf.

Agüera, A., Ramos, M.d.M., and Fernández-Alba, A. (2014). Transformation products of pesticides in the environment: analysis and occurrence. In: *Transformation Products of Emerging Contaminants in the Environment: Analysis, Processes, Occurrence, Effects and Risks* (ed. D.A. Lambropoulou and L.M.L. Nollet), 385–412. Wiley. 10.1002/9781118339558.ch13.

Alalwan, H.A., Alminshid, A.H., and Aljaafari, H.A. (2019). Promising evolution of biofuel generations. Subject review. *Renewable Energy Focus 28*: 127–139.

Amann, M., Holland, M., Maas, R. et al. (2017). Costs, benefits and economic impacts of the EU Clean Air Strategy and their implications on innovation and competitiveness. *IIASA report: International Institute for Applied Systems Analysis (IIASA): Laxenburg, Austria*, pp. 1–59. http://gains.iiasa.ac.at.

Andreu, V. and Pico, Y. (2004). Determination of pesticides and their degradation products in soil: critical review and comparison of methods. *Trends in Analytical Chemistry 23*: 772–789. https://doi.org/10.1016/j.trac.2004.07.008.

Angelidaki, I., Karakashev, D., Batstone, D.J. et al. (2011). Biomethanation and its potential. *Methods in Enzymology 494*: 327–351.

Arevalos-Sánchez, M.M., Maynez-Perez, A.O., Rodríguez-Almeida, F.A. et al. (2021). in vitro assessment of two novel Cellulases from *Trabulsiella odontotermitis* for agricultural waste utilization. *BMC Biotechnology* 21 (1): 1–10. https://doi.org/10.1186/s12896-021-00687-6.

Arun, K., Ramanan, M.V., and Mohanasutan, S. (2020). Comparative studies and analysis on gasification of coconut shells and corn cobs in a perforated fixed bed downdraft reactor by admitting air through equally spaced conduits. *Biomass Conversion and Biorefinery* 12 (4): 1–13.

Asim, N., Emdadi, Z., Mohammad, M. et al. (2015). Agricultural solid wastes for green desiccant applications: an overview of research achievements, opportunities and perspectives. *Journal of Cleaner Production 91*: 26–35.

Azman, S., Khadem, A.F., Van Lier, J.B. et al. (2015). Presence and role of anaerobic hydrolytic microbes in conversion of lignocellulosic biomass for biogas production. *Critical Reviews in Environmental Science and Technology 45* (23): 2523–2564.

Barceló, D. and Hennion, M. (1997). *Trace Determination of Pesticides and their Degradation Products in Water*, 3. Amsterdam: Elsevier.

Beniche, I., El Bari, H., Siles, J.A. et al. (2021). Methane production by anaerobic co-digestion of mixed agricultural waste: cabbage and cauliflower. *Environmental Technology 42* (28): 4550–4558.

Brindha, K. and Schneider, M. (2019). Impact of urbanization on groundwater quality. In: *GIS and Geostatistical Techniques for Groundwater Science* (ed. S. Venkatramanan, P.M. Viswanathan and S.Y. Chung), 179–196. Elsevier.

Carvalho, F.P. (2006). Agriculture, pesticides, food security and food safety. *Environmental Science & Policy* 9 (7): 685–692. 10.1016/j.envsci.2006.08.002.

Central Public Health & Environmental Engineering Organisation (CPHEEO) (2000). Chapter 15, Energy recovery from municipal solid waste, manual on solid waste management. http://cpheeo.gov.in/upload/uploadfiles/files/chap15(1).pdf (accessed 11 August 2021).

Chan, J.X., Wong, J.F., Hassan, A., and Zakaria, Z. (2021). Bioplastics from agricultural waste. In: *Biopolymers and Biocomposites from Agro-Waste for Packaging Applications* (ed. N. Saba, M. Jawaid and M. Thariq), 141–169. Elsevier.

Chang, C.F., Chang, C.Y., and Tsai, W.T. (2000). Effects of burn-off and activation temperature on preparation of activated carbon from corn cob agrowaste by CO_2 and steam. *Journal of Colloid and Interface Science* 232 (1): 45–49. 10.1006/jcis.2000.7171.

Chen, Y., Wang, W., Zhou, D. et al. (2020). Biodegradation of lignocellulosic agricultural residues by a newly isolated Fictibacillus sp. YS-26 improving carbon metabolic properties and functional diversity of the rhizosphere microbial community. *Bioresource Technology* 310: 123381.

Deka, P., Handique, S., Kalita, S., and Gogoi, N. (2021). Recycling of agro-wastes for environmental and nutritional security. In: *Input Use Efficiency for Food and Environmental Security* (ed. A. Hossain, R. Bhatt and R.S. Meena), 605–626. Singapore: Springer.

Demirbas, A. and Demirbas, M.F. (2010). *Algae Energy: Algae as a New Source of Biodiesel, Green Energy Technology*. Springer.

El Haggar, S.M. (2005). Rural and Developing Country Solutions. In: *Environmental Solutions* (ed. F.J. Agardy and N.L. Nemerow), 313–400. Academic Press. 10.1016/B978-012088441-4/50015-0.

Feng, Z., Yuan, R., Wang, F. et al. (2020). Preparation of magnetic biochar and its application in catalytic degradation of organic pollutants: a review. *Science of The Total Environment* 765: 142673. https://doi.org/10.1016/j.scitotenv.2020.142673.

Gannon, V., Grace, D., and Atwill, E.R. (2012). Zoonotic waterborne pathogens in livestock and their excreta: interventions. In: *Animal Waste, Water Quality and Human Health* (ed. A. Dufour, J. Bartram, R. Bos and V. Gannon), 115–156. IWA Publishing.

Ge, S., Yek, P.N.Y., and Cheng, Y.W. (2021). Progress in microwave pyrolysis conversion of agricultural waste to value-added biofuels: a batch to continuous approach. *Renewable and Sustainable Energy Reviews* 135: 110148.

Gentili, R. (2018). Effect of soil pH on the growth, reproductive investment and pollen allergenicity of *Ambrosia artemisiifolia* L. *Frontiers in Plant Science* 1–12. https://doi.org/10.3389/fpls.2018.01335.

Gerba, C.P. and Pepper, I.L. (2009). Wastewater treatment and biosolids reuse. In: *Environmental Microbiology*, 2 (ed. R.M. Maier, I.L. Pepper and C.P. Gerba), 503–530. Elsevier. 10.1016/B978-0-12-370519-8.00024-9.

Giri, B.S., Goswami, M., and Singh, R.S. (2017). Review on application of agro-waste biomass biochar for adsorption and bioremediation dye. *Chemosphere* 99: 19–33.

Goodman, A.B. (2020). Utilization of waste straw and husks from rice production: a review. *Journal of Bioresources and Bioproducts* 5 (3): 143–162. 10.1016/j.jobab.2020.07.001.

Gupta, G.K. and Mondal, M.K. (2020). Bioenergy generation from agricultural wastes and enrichment of end products. In: *Refining Biomass Residues for Sustainable Energy and Bioproducts* (ed. R.P. Kumar, E. Gnansounou, J.K. Raman and G. Basker), 337–356. Academic Press.

Gupta, R., Anand, G., Gaur, R., and Yadav, D. (2021). Plant–microbiome interactions for sustainable agriculture: a review. *Physiology and Molecular Biology of Plants* 27: 1–15.

Hansen, C.L. and Cheong, D.Y. (2019). Agricultural waste management in food processing. In: *Handbook of Farm, Dairy and Food Machinery Engineering* (ed. M. Kutz), 673–716. Academic Press. 10.1016/b978-0-12-814803-7.00026-9.

Helfrich, L.A., Weigmann, D.L., Hipkins, P.A., and Stinson, E.R. (2009). *Pesticides and Aquatic Animals: A Guide to Reducing Impacts on Aquatic Systems.* Virginia Polytechnic Institute and State University. https://pubs.ext.vt.edu/420/420-013/420-013.html.

Horowitz, A.J. (2008). Determining annual suspended sediment and sediment-associated trace element and nutrient fluxes. *Science of the Total Environment 400* (1): 315–343. https://doi.org/10.1016/j.scitotenv.2008.04.022.

Jagtap, A. and Kalbande, S.R. (2021). Densification technologies for agro waste management. In: *Bioenergy Engineering* (ed. M.S. Seveda, P.D. Narale and S.N. Kharpude), 231–251. CRC Press.

Karade, S.R. (2010). Cement-bonded composites from lignocellulosic wastes. *Construction and Building Materials 24* (8): 1323–1330.

Kavitha, S. and Nagarajan, P. (2011). Production of endoglucanase by *Aspergillus niger* using agro residue. *Journal of Engineering Resource 3*: 89–96.

Ke, J., Wang, B., and Yoshikuni, Y. (2021). Microbiome engineering: synthetic biology of plant-associated microbiomes in sustainable agriculture. *Trends in Biotechnology 39* (3): 244–261.

Khan, A.A., de Jong, W., Jansens, P.J. et al. (2009). Biomass combustion in fluidized bed boilers: potential problems and remedies. *Fuel Processing Technology 90* (1): 21–50.

Khan, F.A., Tomar, A., Agarwal, Y.K., and Shukla, H.O. (2020). Agricultural solid waste management: an approach to protect the environment and increase agricultural productivity. In: *Handbook of Solid Waste Management: Sustainability through Circular Economy* (ed. C. Baskar, S. Ramakrishna, S. Baskar, et al.), 1–27. Springer.

Khedari, J., Charoenvai, S., and Hirunlabh, J. (2003). New insulating particleboards from durian peel and coconut coir. *Building and Environment 38* (3): 435–441.

Knobeloch, L., Salna, B., Hogan, A. et al. (2000). Blue babies and nitrate-contaminated well water. *Environmental Health Perspectives* 108 (7): 675–678.

Kolář, L., Kužel, S., Peterka, J. et al. (2008). Agrochemical value of organic matter of fermenter wastes in biogas production. *Plant, Soil and Environment* 54 (8): 321–328. https://doi.org/10.17221/412-PSE.

Koul, B., Yakoob, M., and Shah, M.P. (2021). Agricultural waste management strategies for environmental sustainability. *Environmental Research* 206: 112285.

Kukreti, N., Kag, S., Ruhal, R. et al. (2022). A sustainable biorefinery approach to valorize corn waste to valuable chemicals. In: *Zero Waste Biorefinery* (ed. Y.K. Nandabalan, V.K. Garg, N.K. Labhsetwar and A. Singh), 269–292. Springer.

Kumar, V. and Kumar, P. (2019). Pesticides in agriculture and environment: Impacts on human health. *Contaminants in Agriculture and Environment: Health Risks and Remediation* 1: 76, https://doi.org/10.26832/AESA-2019-CAE-0160-07.

Kumar, P.S., Gayathri, R., and Rathi, B.S. (2021). A review on adsorptive separation of toxic metals from aquatic system using biochar produced from agro-waste. *Chemosphere* 285: 131438.

Kumar, V., Mitra, D., Rani, A. et al. (2022). Bio-inoculants for biodegradation and bioconversion of agrowaste: status and prospects. In: *Bioremediation of Environmental Pollutants*, 351–367.

Lim, S. and Matu, S.U. (2015). Utilization of agro-wastes to produce biofertilizer. *International Journal of Energy and Environmental Engineering* 6: 31–35. https://doi.org/10.1007/s40095-014-0147-8.

Lin, C.Y., Chai, W.S., Lay, C.H. et al. (2021). Optimization of hydrolysis-acidogenesis phase of swine manure for biogas production using two-stage anaerobic fermentation. *Processes 9* (8): 1324.

Liu, S., Cao, L., Xu, F. et al. (2021). Integration of algae cultivation to anaerobic digestion for biofuel and bioenergy production. *Advances in Bioenergy* 6 (1): 199–300.

Loehr, R.C. (1978). Hazardous solid waste from agriculture. *Environmental Health Perspectives* 27: 261–273. 10.2307/3428887.

van de Loosdrecht, J., Botes, F.G., Ciobîcă, I. et al. (2013). Fischer–Tropsch Synthesis: Catalysts and Chemistry. In: *Comprehensive Inorganic Chemistry II*, 2 (ed. J. Reedijk and K. Poeppelmeier), 526. Elsevier https://doi.org/10.1016/B978-0-08-097774-4.00729-4.

Lorenz, J.J. (2013). A review of the effects of altered hydrology and salinity on vertebrate Fauna and their habitats in Northeastern Florida Bay. *Wetlands* 34 (1): 189–200. 10.1007/s13157-013-0377-1.

Lyu, Z., Shao, N., Akinyemi, T., and Whitman, W.B. (2018). Methanogenesis. *Current Biology* 28 (13): R727–R732.

Mallek-Fakhfakh, H., Fakhfakh, J., Masmoudi, N. et al. (2016). Agricultural wastes as substrates for β-glucosidases production by *Talaromyces thermophilus*: role of these enzymes in enhancing waste paper saccharification. *Preparative Biochemistry & Biotechnology* 47. 10.1080/10826068.2016.1252928.

Mansaray, K.G. and Ghaly, A.E. (1999). Determination of kinetic parameters of rice husks in oxygen using thermogravimetric analysis. *Biomass and Bioenergy 17* (1): 19–31.

Mateo-Sagasta, J., Zadeh, S.M., and Turral, H. eds. (2018). More people, more food, worse water?: a global review of water pollution from agriculture.

Matocha, C.J. (2005). Oxidation–reduction of contaminants. *Encyclopaedia of Soils in the Environment* 133–140. 10.1016/B0-12-348530-4/00185-5.

McCarty, P.L. and McKinney, R.E. (1965). Salt toxicity in anaerobic digestion. *Water Pollution Control Federation* 33 (4): 399–415.

Meybeck, M. (2003). Global analysis of river systems: from Earth system controls to Anthropocene syndromes. *Philosophical Transactions of the Royal Society of London. Series B, Biological Sciences* 358 (1440): 1935–1955. https://doi.org/10.1098/rstb.2003.1379.

Mohammad, M., Yaakob, Z., and Abdullah, S.R.S. (2013). Carbon derived from Jatropha seed hull as a potential green adsorbent for cadmium (II) removal from wastewater. *Materials 6* (10): 4462–4478.

Nabiollahi, K., Taghizadeh-Mehrjardi, R., Kerry, R., and Moradian, S. (2017). Assessment of soil quality indices for salt affected agricultural land in Kurdistan Province, Iron. *Ecological Indicators 83* (September 2016): 482–494. https://doi.org/10.1016/j.ecolind.2017.08.001.

Nagendran, R. (2011). Agricultural waste and pollution. In: *Waste* (ed. T.M. Letcher and D.A. Vallero), 341–355. Academic Press.

Nakamya, J., Tumuhairwe, J.B., Sabiiti, E.N. et al. (2020). Influence of biogas digesters on faecal indicator organisms in digestate and around homesteads in Ethiopia. *Biomass and Bioenergy 142*: 105746.

Obi, F., Ugwuishiwu, B., and Nwakaire, J. (2016). Agricultural waste concept, generation, utilization and management. *Nigerian Journal of Technology* 35 (4): 957. https://doi.org/10.4314/njt.v35i4.34.

Organisation for Economic Co-operation and Development (2012). *New and Emerging Water Pollutants Arising from Agriculture*. OECD Publishing.

Owens, P., Batalla, R.J., Collins, A.J. et al. (2005). Fine-grained sediment in river systems: environmental significance and management issues. *River Research and Applications* 21 (7): 693–717. https://doi.org/10.1002/rra.878.

Pinto, F., André, R., Miranda, M. et al. (2016). Effect of gasification agent on co-gasification of rice production wastes mixtures. *Fuel 180*: 407–416.

Prabhudessai, V., Ganguly, A., and Srikanth, M. (2013). Biochemical methane potential of agro wastes. *Journal of Energy* 2013: 350731. 7 pages. https://doi.org/10.1155/2013/350731.

Prasad, K.M. and Murugavelh, S. (2019). Bio oil production from Agro waste residue: thermochemical conversion to improve oil quality. In: *IOP Conference Series: Earth and Environmental Science*, vol. 312(1) (ed. S.K. Natarajan and P. Karthikeyan), 012031. IOP Publishing.

Rai, S., Solanki, M.K., Anal, A.K.D. et al. (2021). Emerging frontiers of microbes as agro-waste recycler. In: *Waste to Energy: Prospects and Applications* (ed. B.K. Kashyap, M.K. Solanki, D.V. Kamboj, et al.), 1. Springer.

Raja, I.A. and Wazir, S. (2017). Biogas production: the fundamental processes. *Universal Journal of Engineering Science* 5 (2): 29–37.

Ramírez-García, R., Gohil, N., and Singh, V. (2019). Recent advances, challenges, and opportunities in bioremediation of hazardous materials. In: *Phytomanagement of Polluted Sites* (ed. V.C. Pandey and K. Bauddh), 517–568. Elsevier.

Ratchahat, S., Srifa, A., and Koo-amornpattana, W. (2021). Syngas production with low tar content from cellulose pyrolysis in molten salt combined with Ni/Al_2O_3 catalyst. *Journal of Analytical and Applied Pyrolysis 158*: 105243.

Raut, S.P., Ralegaonkar, R.V., and Mandavgane, S.A. (2011). Development of sustainable construction material using industrial and agricultural solid waste: a review of waste-create bricks. *Construction and Building Materials 25* (10): 4037–4042.

Ravindran, R., Hassan, S.S., Williams, G.A., and Jaiswal, A.K. (2018). A review on bioconversion of agro-industrial wastes to industrially important enzymes. *Bioengineering 5* (4): 93.

Rawoof, S.A.A., Kumar, P.S., and Vo, D.V.N. (2021). Production of optically pure lactic acid by microbial fermentation: a review. *Environmental Chemistry Letters 19* (1): 539–556.

Rehman, S., Aslam, H., Ahmad, A. et al. (2014). Production of plant cell wall degrading enzymes by monoculture and co-culture of *Aspergillus niger* and *Aspergillus terreus* under SSF of banana peels. *Brazilian Journal of Microbiology* 45 (4): 1485–1492.

Riera, M.A. and Maldonado, S. (2021). Agro-industrial waste as an option for the sustainable development of bioplastic. In: *Innovations in Biotechnology for a Sustainable Future* (ed. N.R. Maddela and L.C. García), 117–132. Cham: Springer.

Rodríguez, A., Sánchez, R., Requejo, A., and Ferrer, A. (2010). Feasibility of rice straw as a raw material for the production of soda cellulose pulp. *Journal of Cleaner Production 18* (10–11): 1084–1091.

Ruan, Z., Wang, X., Liu, Y., and Liao, W. (2019). Corn. In: *Integrated Processing Technologies for Food and Agricultural By-Products* (ed. Z. Pan, R. Zhang and S. Zicari), 59–72. Academic Press.

Sadh, P.K., Duhan, S., and Duhan, J.S. (2018). Agro-industrial wastes and their utilization using solid state fermentation: a review. *Bioresourses and Bioprocessing* 5: 1. https://doi.org/10.1186/s40643-017-0187-z.

Saidi, M., Gohari, M.H., and Ramezani, A.T. (2020). Hydrogen production from waste gasification followed by membrane filtration: a review. *Environmental Chemistry Letters* 18: 1529–1556.

Sambo, A.S., Garba, B., and Danshehu, B.G. (1995). Effect of some operating parameters on biogas production rate. *Renewable Energy 6* (3): 343–344.

Samolada, M.C. and Zabaniotou, A.A. (2014). Comparative assessment of municipal sewage sludge incineration, gasification and pyrolysis for a sustainable sludge-to-energy management in Greece. *Waste Management 34* (2): 411–420.

Savci, S. (2012). An agricultural pollutant: chemical fertilizer. *International Journal of Environmental Science and Development* 3: 77–80. 10.7763/IJESD.2012.V3.191.

Schwenke, G.D., Manning, W., and Haigh, B.M. (2014). Ammonia volatilisation from nitrogen fertilisers surface-applied to bare fallows, wheat crops and perennial-grass-based pastures on Vertosols. *Soil Research* 52 (8): 805–821.

Sharma, K. and Garg, V.K. (2019). Vermicomposting of waste. In: *Sustainable Resource Recovery and Zero Waste Approaches* (ed. M.J. Taherzadeh, K. Bolton, J. Wong, et al.), 133–164. Elsevier.

Sharmila, V.G., Tamilarasan, K., Kumar, M.D. et al. (2022). Trends in dark biohydrogen production strategy and linkages with transition towards low carbon economy: an outlook, cost-effectiveness, bottlenecks and future scope. *International Journal of Hydrogen Energy* 47 (34): 15309–15332.

Sharpley, A.N., Bergstrom, L., Aronsson, H. et al. (2015). Future agriculture with minimized phosphorus losses to waters: research needs and direction. *Ambio 44*: S163–S179. https://doi.org/10.1007/s13280-014-0612-x.

Shaviv, A. (2001). Advances in controlled-release fertilizers. *Advances in Agronomy* 71: 1–49. 10.1016/S0065-2113(01)71011-5.

Silva, M.F., Golunski, S.M., Rigo, D. et al. (2013). Fructooligosacharides production using Inulinase of *Aspergillus niger* after treatment with fluid pressurized in two different means. In *III Iberoamerican Conference on Supercritical Fluids Cartagena de Indias (Colombia)* Cartagena de Indias, Colombia (1–5 April 2013), (pp. 1–7).

Singh, R.S. (2014). Industrial biotechnology: an overview. In: *Advances in Industrial Biotechnology*, 1–35. India: IK International Publishing House Pvt. Ltd.

Singhania, R.R., Patel, A.K., Thomas, L. et al. (2015). Industrial enzymes. In: *Industrial Biorefineries & White Biotechnology* (ed. A. Pandey, R. Höfer, M. Taherzadeh, et al.), 473–497. Elsevier.

Sivakumar, D., Srikanth, P., Ramteke, P.W., and Nouri, J. (2022). Agricultural waste management generated by agro-based industries using biotechnology tools. *Global Journal of Environmental Science and Management* 8 (2): 281–296. 10.22034/gjesm.2022.02.10.

Sorathiya, L.M., Fulsoundar, A.B., Tyagi, K.K. et al. (2014). Eco-friendly and modern methods of livestock waste recycling for enhancing farm profitability. *International Journal of Recycling Organic Waste in Agriculture* 3: 50. https://doi.org/10.1007/s40093-014-0050-6.

Srivastava, R.K., Shetti, N.P., Reddy, K.R. et al. (2021). Biomass utilization and production of biofuels from carbon neutral materials. *Environmental Pollution 276*: 116731.

StenfertKroese, J., Batista, P.V.G., Jacobs, S.R. et al. (2020). Agricultural land is the main source of stream sediments after conversion of an African montane forest. *Scientific Reports-UK 10* (1): 1–15. https://doi.org/10.1038/s41598-020-71924-9.

Stewart, G.J., Nelson, B.S., Acton, W.J.F. et al. (2021). Emissions of intermediate-volatility and semi-volatile organic compounds from domestic fuels used in Delhi, India. *Atmospheric Chemistry and Physics* 21 (4): 2407–2426.

Sud, D., Mahajan, G., and Kaur, M.P. (2008). Agricultural waste material as potential adsorbent for sequestering heavy metal ions from aqueous solutions – a review. *Bioresource Technology 99* (14): 6017–6027.

Sun, X., Li, J., Zhao, X. et al. (2016). A review on the management of municipal solid waste fly ash in American. *Procedia Environmental Sciences 31*: 535–540.

Swain, M. and Mohanty, S. (2018). Bioethanol production from corn and wheat: food, fuel, and future. In: *Bioethanol Production from Food Crops* (ed. R.C. Ray and S. Ramachandran), 45–59. Academic Press https://doi.org/10.1016/C2017-0-00234-3.

Taufeeq, A., Baqar, M., Sharif, F. et al. (2021). Assessment of organochlorine pesticides and health risk in tobacco farming associated with River Barandu of Pakistan. *Environemental Science and Pollution Research* 28: 1–18.

Thangadurai, D., Naik, J., Sangeetha, J. et al. (2021). Nanomaterials from agrowastes: past, present, and the future. In: *Handbook of Nanomaterials and Nanocomposites for Energy and*

Environmental Applications (ed. O.V. Kharissova, L.M.T. Martínez and B.I. Kharisov), 471–487. Springer https://doi.org/10.1007/978-3-030-36268-3_43.

Thekedar, K., Karale, S., and Awari, G. (2021). A systematic review on the future potential of tea waste and other admixtures in bio briquetting from rice husk. *AIP Conference Proceedings,* Chennai, India (24–26 April 2021), (Vol. 2417, No. 1, p. 020008). AIP Publishing LLC.

Thrän, D. and Fritsche, U. (2016). Standards for biobased fuels and resources–status and need. In *IEA Bioenergy Conference 2015 Proceedings,* Berlin, Germany (27–29 October 2015) (pp. 148–158).

Tinoco-Caicedo, D.L., Mero-Benavides, M., Santos-Torres, M. et al. (2021). Simulation and exergoeconomic analysis of the syngas and biodiesel production process from spent coffee grounds. *Case Studies in Thermal Engineering* 28: 101556.

Tripathi, A. and Shukla, S.K. (2015). Comparative studies of gasification potential of agro-waste with wood and their characterization. *Advances in Energy Research 3* (3): 181.

Tripathi, N., Hills, C.D., Singh, R.S., and Atkinson, C.J. (2019). Biomass waste utilisation in low-carbon products: harnessing a major potential resource. *Npj Climate and Atmospheric Science* 2 (1): 1–10. https://doi.org/10.1038/s41612-019-0093-5.

Tubiello, F., Conchedda, G., de Santis, G. et al. (2021). Land use statistics and indicators statistics. Global, regional and country trends 1990–2019. *FAOSTAT Analytical Brief Series No. 28.*

Tukanghan, W., Hupfauf, S., Gómez-Brandón, M. et al. (2021). Symbiotic bacteroides and clostridium-rich methanogenic consortium enhanced biogas production of high- solid anaerobic digestion systems. *Bioresource Technology Reports* 14: 100685.

United Nations (2015). Resolution “Transforming our world: the 2030 Agenda for Sustainable Development”. https://sustainabledevelopment.un.org/content/documents/21252030%20Agenda%20for%20Sustainable%20Development%20web.pdf (accessed 19 August 2021).

United Nations Development Programme (2015). Sustainable Development Goals (SDGs). https://www.undp.org/sustainable-development-goals (accessed 18 August 2021).

United Nations Environmental Programme (2009). Converting Waste Agricultural Biomass into a Resource: Compendium of Technologies.

Upadhyay, P., Vaishampayan, A., and Jaiswal, S.K. (2020). Soil pollution caused by agricultural practices and strategies to manage it. In: *Plant Responses to Soil Pollution* (ed. P. Singh, S.K. Singh and S.M. Prasad), 119–132. Springer.

Vargas, F., González, Z., Sánchez, R. et al. (2012). Cellulosic pulps of cereal straws as raw material for the manufacture of ecological packaging’. *BioResources 7* (3): 4161–4170.

Venkataraman, C., Gazala, H., Kadamba, D. et al. (2006). Emissions from open biomass burning in India: Integrating the inventory approach with high-resolution Moderate Resolution Imaging Spectroradiometer (MODIS) active-fire and land cover data. *Global Biogeochemical Cycles* 20. 10.1029/2005GB002547.

Vessey, J.K. (2003). Plant growth promoting Rhizobacteria as Biofertilizers. *Plant and Soil* 255: 571–586. 10.1023/A:1026037216893.

Wang, Z., Yang, J., Li, Z., and Xiang, Y. (2009). Syngas composition study. *Frontiers of Energy and Power Engineering in China* 3 (3): 369–372.

Wang, C., Zuo, J., Chen, X. et al. (2014). Microbial community structures in an integrated two-phase anaerobic bioreactor fed by fruit vegetable wastes and wheat straw. *Journal of Environmental Sciences 26* (12): 2484–2492.

Wang, L., Skjevrak, G., Skreiberg, Ø. et al. (2017). Investigation on ash slagging characteristics during combustion of biomass pellets and effect of additives. *Energy & Fuels 32* (4): 4442–4452.

Wei, J., Liang, G., Alex, J. et al. (2020). Research progress of energy utilization of agricultural waste in China: bibliometric analysis by citespace. *Sustainability* 12 (3): 812. https://doi.org/10.3390/su12030812.

Wen, Y., Schoups, G., and Van de Giesen, N. (2017). Organic pollution of rivers: combined threats of urbanization, livestock farming and global climate change. *Scientific Reports-UK 7* (1): 1–9. https://doi.org/10.1038/srep43289.

Wick, K., Heumesser, C., and Schmid, E. (2012). Groundwater nitrate contamination: factors and indicators. *Journal of Environmental Management* 111: 178–186. https://doi.org/10.1016/j.jenvman.2012.06.030.

Willis, K. (2005). *Theories and Practices of Development.* London, New York: Routledge p. 37, 164.

Woods, J. and Hall, D.O. (1994). Bioenergy for development – Technical and environmental dimensions - FAO Environment and energy paper 13. FAO - Food and Agriculture Organization of the United Nations; http://www.fao.org/3/t1804e/t1804e00.htm#Contents.

World Commission on Environment and Development and Brundtland, G.H. (1987). *Presentation of the Report of World Commission on Environment and Development to African and International and Non-Governmental Organizations. . . June 7, 1987, Nairobi, Kenya.* World Commission on Environment and Development.

Wu, M., Tang, X., Li, Q. et al. (2013). Review of ecological engineering solutions for rural non-point source water pollution control in Hubei Province, China. *Water, Air, and Soil Pollution* 224 (5): 1–18.

Yalinkilic, M.K., Imamura, Y., Takahashi, M. et al. (1998). Biological, physical and mechanical properties of particleboard manufactured from waste tea leaves. *International Biodeterioration & Biodegradation 41* (1): 75–84.

Yanik, J., Kornmayer, C., Saglam, M., and Yüksel, M. (2007). Fast pyrolysis of agricultural wastes: characterization of pyrolysis products. *Fuel Processing Technology 88* (10): 942–947.

Yargholi, B. and Azarneshan, S. (2014). Long-term effects of pesticides and chemical fertilizers usage on some soil properties and accumulation of heavy metals in the soil (case study of Moghan plain's (Iran) irrigation and drainage network). *International Journal of Agriculture and Crop Sciences* 7: 518–523.

Zahan, K. and Kano, M. (2018). Biodiesel production from palm oil, its by-products, and mill effluent: a review. *Energies* 11 (8): 2132. https://doi.org/10.3390/en11082132.

Zainal, Z.A., Ali, R., Lean, C.H., and Seetharamu, K.N. (2001). Prediction of performance of a downdraft gasifier using equilibrium modeling for different biomass materials. *Energy Conversion and Management 42* (12): 1499–1515.

Zandaryaa, S. and Mateo-Sagasta, J. (2018). 'Organic matter, pathogens and emerging pollutants' *More people, more food, worse water?: a global review of water pollution from agriculture.* Rome, Italy: FAO; Colombo, Sri Lanka: International Water Management Institute (IWMI). CGIAR Research Program on Water, Land and Ecosystems (WLE), pp. 125–138.

Zhang, Y., Cui, Y., Chen, P. et al. (2019). Gasification technologies and their energy potentials. In: *Sustainable Resource Recovery and Zero Waste Approaches* (ed. M.J. Taherzadeh, K. Bolton, J. Wong and A. Pandey), 193–206. Elsevier.

5

Microbiological Digestion of Agricultural Biomass: Prospects and Challenges in Generating Clean and Green Energy

Sangita Agarwal[1], *Soumendra Darbar*[2,3], *Srimoyee Saha*[4], *and Ravindra Pratap Singh*[5]

[1] *Department of Applied Science, RCC Institute of Information Technology, Kolkata, West Bengal, India*
[2] *Faculty Council of Science, Jadavpur University, Kolkata, West Bengal, India*
[3] *Research and Development Division, Dey's Medical Stores (Mfg.) Ltd., Kolkata, West Bengal, India*
[4] *Department of Life Science and Biotechnology, Jadavpur University, Kolkata, West Bengal, India*
[5] *Central Public Works Department, New Delhi, India*

CHAPTER OUTLINE

Agriculture Waste Management and Bioresource: The Circular Economy Perspective, First Edition.
Edited by Suruchi Singh, Pardeep Singh, Anu Sharma, and Moharana Choudhury.

5.1 Introduction

Before the practice of agriculture started, people were gatherers and nomads and were primarily engaged in the quest for food-collecting plants, which grew in forests and hunted wildlife. Approximately thousands of years ago, human beings started learning the art and science of growing grains and root crops and thus finally settling into a farming-based lifestyle. Over the centuries, growth in agriculture has contributed to the development of communities and nations (Flannery 1973).

As revealed from the report prepared by the Food and Agriculture Organization (FAO) in 2011, 30% of global energy is consumed by food production and supply chains. While growing bio-energy-supplied plants are criticized for competing with food plants, it threatens food security and biodiversity. Feeding an ever-growing population, an estimated 10.5 billion in the year 2050, would also require increasing food production, which inevitably would create more agricultural remnants.

Agricultural practices generate huge quantities of wastes that are derived from harvestable yield. Globally, the biomass waste generated annually is in the order of 140 Gt (UNEP 2009), which can have adverse impact on the health of environment. The biomass or the agricultural wastes consists of leaves, roots, fruit peels, seeds, and crop stalks, which is generally discarded or burnt.

Some crops like sugarcane produce secondary and tertiary waste along with primary biomass on harvesting (Schieber et al. 2001). It is assumed that a given crop would consistently generate same amount of by-product from a geographical region. In developing countries, the biomass is not treated or utilized but burnt openly or left to decompose. The agricultural land is projected to increase by 13% in developing countries from 1999 to 2030 by FAO (FAO 2012). Hence, agricultural residues are bound to increase that needs to be utilized to create wealth from waste.

The management of food and garbage from agriculture would be crucial in conserving the environmental resources in many countries, especially in developing countries.

5.1.1 Need for Renewable Energy

Energy is the driving force for fulfilling many basic human needs and services; it also encompasses social and economic development. **The progress of humankind is closely related to energy consumption, and the quantity of energy used is a determinant of the size of a nation in terms of economy.** Table 5.1 shows the primary energy consumption in various regions across the globe.

Humankind has been nonjudiciously using mineral-based resources like coal and petroleum for their growth and development. They are finite reserves in nature and hence would extinct in the future. The heavy usage of fossil fuels for power generation and the exponential rise in population have also resulted in significant global climate change (Asumadu-Sarkodie and Owusu 2016). The interest in alternate energy sources started in the latter part of 1990 when oil prices were hiked shockingly (Abbasi et al. 2011) and the need to achieve a sustainable future. As an alternative option, people explored renewable energy sources. **There are many sources of renewable energy, including wind, biomass, solar, water, and geothermal heat, which could be utilized for industrial energy**

Table 5.1 Primary energy consumption in various regions across the globe.

Sl. No.	Region	Primary energy consumption (TWh) in year		Increase in energy consumption over a period of 36 year
		1980	2016	
1.	Afghanistan	7.773	35.737	27.96
2.	Africa	1698.122	5105.052	3406.93
3.	Algeria	177.097	616.819	439.72
4.	Angola	16.953	104.001	87.05
5.	Antigua and Barbuda	1.76	3.067	1.31
6.	Argentina	460.339	993.388	533.05
7.	Asia Pacific	13683.237	64758.213	51074.98
8.	Australia	840.092	1634.105	794.01
9.	Bahrain	40.392	202.719	162.33
10.	Bangladesh	36.474	373.061	336.59
11.	Belgium	565.183	729.817	164.63
12.	Bhutan	0.033	24.753	24.72
13.	Brazil	1091.087	3310.703	2219.62
14.	Burma	21.195	150.811	129.62
15.	Canada	2643.899	3871.303	1227.40
16.	China	4856.292	35264.122	30407.83
17.	Egypt	210.263	1038.808	828.55
18.	Europe	20021.95	23305.822	3283.87
19.	India	1208.518	8352.07	7143.55
20.	Iraq	95.036	539.031	444.00
21.	Japan	4244.663	5179.953	935.29
22.	Malaysia	127.406	1168.961	1041.56
23.	Maldives	0.062	6.714	6.65
24.	Mauritius	4.048	25.14	21.09
25.	Nigeria	123.501	506.867	383.37
26.	Thailand	146.451	1488.844	1342.39
27.	Mexico	903.239	2162.618	1259.38
28.	Switzerland	300.304	307.959	7.66
29.	Zimbabwe	38.612	47.5	8.89
30.	World	77628.384	152944.938	75316.55

Source: https://ourworldindata.org/energy-production-consumption.

requirements, and biomass could be the alternative source for fuel production (Abbasi and Abbasi 2011; **Kaygusuz** 2012)**, which is in tune with the Sustainable Development Goals (SDGs) (Edenhofer et al.** 2014)**.**

The transformation of the current energy system holds the key to alleviating global climate changes (Mentis et al. 2017). The concept of sustainable energy is based on three key targets: affordability, reliability, and universal accessibility. Thus, searching and developing nonconventional energy sources is the need of this *era* and essential to sustain the life cycle of humans. The energy or fuel generated by agricultural waste can be the key to counter the shortage of fossil fuels worldwide. Several studies have shown that unutilized or residues from the agriculture sector, including livestock manure and crop straw, would significantly contribute to alternative energy in the future. There are abundant bioenergy resources in Asian countries, which would be of strategic importance in the global biomass energy atlas. Asian countries have started rapidly trading and using solid biomass for bioenergy, much after than in Europe and North America (Goh et al. 2013). Countries like Japan and South Korea use domestic biomass and import wood pellets and palm kernel shells from nations a part of the Pacific Rim (Aikawa 2016). The usage of residues of agriculture for combustion and anaerobic digestion (AD) of effluents of agriwaste has been rising.

Many countries have effectively utilized the waste generated from the agricultural sector and contributed to relieving the issues of global resource waste. Thus, the comprehensive utilization ratios of wastes from agricultural industries have shown a remarkable increase. The level of energy utilization of these wastes is not equitable and varies from country to country. In 2015, the biogas production of Europe was almost half of the total biogas production globally (Scarlat et al. 2018).

5.1.2 Agricultural Waste: Indian Perspective

The agricultural waste generated by India is quite significant as agriculture is the primary source of livelihood. With population rising, the quantity of waste generated would increase, and so is the case with agricultural waste. The agricultural waste generated by India every year is about 350 million tonnes, but this figure would be increasing with the rise in population. According to the Ministry of New and Renewable Energy report, **agricultural waste can be effectively used** to produce greater than 18 000 MW of power every year and environmentally friendly green fertilizer usage in agriculture (Reardon et al. 2019).

The biomass abandoned in the field after harvesting the agricultural produce is known as rural residues. There are vast amounts of these residues in various forms like woody stalks, sugarcane leaves, and cereal straws during harvesting generated every year. The milling of these farm products also generates a large number of residues. Although some of these residues are utilized as cooking fuel in rural homes and industries, as animal feed and thatching homes in rural settings, many of these residues remain unutilized and left in fields. Removing such significant residues is a big challenge, and the farmers burn them *in situ* to prepare their fields for the successive cropping. The burning of these residues is one of the primary causes of air pollution besides fossil fuels. As reported by a group of researchers, the burning of residues from the rice–wheat system by the farmers, especially in the Punjab and Haryana belt and western part of Uttar Pradesh, is adding to the greenhouse emission and loss of nutrients and other health hazards (Jain et al. 2014).

The Ministry of New and Renewable Energy (MNRE), Government of India, with the help from state governments, has been promoting biogas projects for energy production. **Several methods can be used to** convert biomass into electric power. The direct combustion of biomass material is a commonly used method wherein the starting materials are agricultural waste or woody materials. Proper and effective utilization can help minimize air pollution and generate wealth out of waste in **the form of bio-energy and green manure**.

5.1.3 Types of Agricultural Wastes

Agricultural operations produce various types of waste during their daily process at their premises (Weiland 2004). The garbage generated due to such farming practices can be solid or liquid, namely cattle manure, crop residue (for example, corn stalks), pesticides, and fertilizers (Weiland and Rieger 2005; Weiland 2006). Table 5.2 shows the **residues** generated by the processing of major crops.

5.1.3.1 Classification of Agricultural Wastes

5.1.3.1.1 Solid Agricultural Wastes

The farming activities generate various agricultural solid wastes that include production and other associated activities related to farming. Each phase of these associated farming activities produces solid waste, broadly classified under the headings mentioned in Table 5.3.

Table 5.2 Major agricultural crop wastes and processing residues.

Sl. No.	Crop	Waste products and processing residues
1.	Rice	Straw, hulls
2.	Vegetables	Rejects
3.	Sugarcane	Sugarcane tops, bagasse, mudpress
4.	Corn	Cobs, husks, leaves, stalks
5.	Coconut	Coconut milk, coconut husk (as coir dust)
6.	Banana	Trunk, leaves, reject fruits, peels
7.	Pineapple	Plant, crown, peel, core, pulp
8.	Abaca	Leaves
9.	Coffee	Pulp
10.	Mango	Rejects, peel, pulp
11.	Tobacco	Midribs, reject leaves
12.	Peanut	Plant, pods, reject beans
13.	Mung bean	Plant, pods, reject beans
14.	Cacao	Pods, core
15.	Cassava	Trunk, leaves, peel, washings

Table 5.3 Generation of solid wastes in various areas.

Sl. No.	Generation of solid wastes in various areas
1.	Solid wastes generated during crop production
2.	Solid wastes generated during the use of chemicals in farming
3.	Solid wastes generated during horticultural production
4.	Solid wastes generated during the production of livestock
5.	Solid wastes generated during industrial agricultural production
6.	Solid wastes generated during the processing of food and meat
7.	Solid wastes generated during the use of medicines on-farm

Solid wastes generated during crop production During agricultural activities, much solid waste is generated, which are associated with crop production. For example, crop residues, husks, and others fall under agricultural solid wastes (Karak et al. 2015).

Solid wastes generated during the use of chemicals in farming **The containers or bottles after usage, which are disposed of, of the chemicals that are used in farming come under the category of solid waste generated while using chemical wastes**. These are wastes generated when farmers use chemicals like herbicides, pesticides, and insecticides during farming activities. Often, ignorant farmers in developing countries mishandle the containers. After usage, the remnant chemicals in the containers are thrown into the environment, polluting the water body or the field, causing health risks (Wang et al. 2016).

Solid wastes generated during horticultural production **The solid wastes are left over after farming, upkeeping of horticultural plants, ornamentation, and beautification of the environment in homes, institutes, and other landscapes.** Grass cuttings, dead leaves, debris, and prunings are examples of this type of solid wastes (Adejumo et al. 2015). The entire procedure involves various operations such as pruning, earthing, training, and thinning of different plants that generate wastes.

Solid wastes generated during the production of livestock During the production of livestock, some solid wastes are generated, which constitute these types of wastes. Animal carcasses, water trough, and litter/bedding are some examples of solid waste generated during the production of animals (Oluseun et al. 2014).

Solid wastes generated during industrial agricultural production When solid wastes are generated during agricultural and livestock activities, the products are not used for dietary consumption. These products are used for large-scale industrial productions, such as paper production, and solid waste left over after the industrial production of finished goods would come under this category (Dien and Vong 2006).

Solid wastes generated during the processing of food and meat These are the categories of solid waste produced when animal products or crops are processed to consume. In meat processing, wastes like bones, hoofs, and feathers are the solid waste generated from slaughterhouses or abattoirs. In the case of food processing of bananas, agricultural wastes like banana peels are generated. Similarly, wastes from food and meat processing produce many of such solid wastes (Obi et al. 2016).

Solid wastes generated during the use of medicines on-farm The medical wastes generated when drugs or vaccines are used on animals for treating them are categorized as solid wastes produced during the use of medicines on-farm. Wrappers, containers, and disposable syringes are examples that are a part of this category (Klich 2007).

5.1.3.1.2 Liquid Agricultural Wastes

The different types of liquid agricultural waste are listed in Table 5.4 and discussed later.

Vegetable washing water The water obtained after washing vegetables is contaminated with soil and pieces of vegetables. This wastewater may also contain the insecticides, pesticides, chlorine, and disinfectants used to treat vegetables (Borgens 2019).

Silage liquor Silage is a fermented product made from fresh or dried grass or other green plants stored and used as a winter feed for domestic animals. In the production of silage, acid conditioners like sulphuric acid or formic acid are used.

The silage production process yields a brownish-yellow liquid rich in monosaccharides, alcohol, short length carbon chains, and silage conditioner. This silage liquor has a detrimental effect on the environment as it is one of the most polluting organic substances known (Agricultural Waste usatanksales.com 2013).

Dairy farming wastes Milk is a vital dairy product, but if present in wastewater can make the water very polluted because of its high organic content. Such pollutants are high oxygen demanding, rendering water to deoxygenate. The dissolved oxygen in water is reduced, making it difficult for the survival of aquatic life. Dairy waste also contains a lot of washdown water, waste of other animals, and chemical cleaning and disinfectant agents used for washing (Anukam et al. 2019).

Slaughtering waste Slaughtering wastes are obtained when the wastes from slaughter activities are mixed with wastewater. In such wastewater, the dissolved oxygen content is

Table 5.4 Generation of liquid wastes in various sectors.

Sl. No.	Generation of liquid wastes in various sectors
1.	Liquid waste obtained from washing vegetables – Vegetable washing water
2.	Liquid waste obtained from fermentation of dried grass – Silage liquor
3.	Liquid waste obtained after waste water gets contaminated by – Dairy farming wastes
4.	Liquid waste obtained from slaughter houses – Slaughtering waste
5.	Sol Liquid waste obtained from firefighting at farms – Firewater

further reduced because the organic content of these substances is relatively high. Thus, it can pollute the water body, more threatening the existence of aquatic species (Stagich 2017).

Firewater Firewater is very toxic to the environment and is produced from firefighting at farms. It is composed of herbicides, pesticides, fuel oil, and fertilizers. Fires occur more frequently on farms than at industrial premises (Spillages 2011).

5.1.3.1.3 Nature of Agricultural Waste

The agricultural waste can be divided into biodegradable and nonbiodegradable, depending on the ability of micro-organisms to degrade or break the farm waste.

Biodegradable wastes are agricultural wastes that micro-organisms can degrade because of their high organic content within a reasonable length of time. The waste from poultry farms, waste from dairy industries, sugar mills, food and starch processing, and many other wastes used as food by micro-organisms are classified as biodegradable waste.

Nonbiodegradable wastes are resistant to degradation by micro-organisms, which implies that these wastes either have meagre rates of degradation or do not decay by micro-organisms' actions. Some of the examples of garbage that are nonbiodegradable are woody plants, pouches, containers, plastic wrappings, cardboard, and others.

The circular economy concept like effective and efficient utilization of by-products can positively correlate between economy and problems associated with growing environmental issues of any country. Severe environmental problems can result from the waste generated from agricultural sectors; for instance, burning the stubble would cause air pollution. Feedstocks can be used for generation of energy (Jayathilakan et al. 2012). The world's growing economies like India, where the livestock capacity is increasing exponentially at 6.7% per annum can reap the benefit.

Underutilization or nonutilization of these by-products can be an added menace, increasing disposal costs and loss of prospective revenues.

5.1.4 Treatment Process

The by-products and waste generated from the food industry and agriculture sectors have been degraded by anaerobic micro-organism since a long time and this help in proper utilization of these wastes. Microbial digestion facilitates the stabilization of waste, controlling pollution, improving manure quality, and producing biogas.

There are many advantages of using AD, and it can also solve the waste disposal problem; consequently, utilizing this technique, one can convert waste to wealth. **The technique of AD is also useful** in generating precious fuel and manure rich in nutrients, substituting mineral fertilizers for organic farming. Thus, this process can play a pivotal role in waste treatment and be a good source of fuel production, and the biogas generated should contain at least 50% methane. Biomass from diverse sources, including wastes mainly organic in nature, manure, and slurry, can be converted to a valuable energy source by valorization (Hossain and Badr 2007; Holm-Nielsen et al. 2009; Ravindran et al. 2018).

There are practically two different treatments processes used for biological waste, namely AD and composting. High-tech plants that produce biogas in industrialized nations process thousands of tons of waste per year (Anon 2000). In the world's less industrialized and developing economies, biogas generation is done in small-scale low-tech plants. **AD**, followed by composting has been organized in industrialized countries at different technological levels, even in the backyard, from a simple heap or bin to technologically advanced vessels systems to expand the capacity for commercial production (Ayalon et al. 2000; Yadav et al. 2009).

With this background, various biological waste treatment strategies are evaluated and compared in terms of efficiency of fuel generation, the economy of the process, and environmental impacts.

5.2 Agricultural Waste: Generation to Proper Utilization of Agricultural Waste

5.2.1 Components of Agricultural Waste

Increased urbanization, industrialization, intense agricultural activities, and animal husbandry have resulted in the greater manufacturing and augmentating the stocks of waste materials.

The prolific usage of pesticides and various fertilizers in fodder production has impaired the integral and interconnected association between agriculture produce and livestock. The environmental imbalance is caused by soil enrichment due to the addition of waste containing organic materials. The components of agroindustrial wastes are carbohydrates, complex polysaccharides, proteins, polyphenolic compounds, and others (Marinari et al. 2000; Solomon et al. 2007).

The technique of integrated waste utilization can help in overpowering the environmental crisis. The efficient and effective utilization of waste from both agricultural and industrial sectors through recycling and clean technology can be helpful in these circumstances.

5.2.2 Waste Materials

Mostly, the generation of wastes from the various agricultural industries is a significant cause of pollution. These wastes can be subdivided based on the sources they emanate from (i) agricultural, (ii) forestry, and (iii) industrial activities (Figure 5.1).

The agricultural waste includes slurry of livestock, remains of the crop, manure, and waste from pruning.

The by-products of the food and agricultural industry consist of bagasse, coffee residues, degummed fruits, legumes, sludge from wool, milk serum, and cellulose fall under organic industrial waste. The problems associated with agricultural waste can be resolved if this waste could be further recycled and used in the production of citric acid, bioactive compounds, enzymes, pigment food, flavoring products, and fuel generation, to name a few (Figure 5.2).

The quantity of these agricultural and industrial wastes is increasing manifold, thus adding more to the existing environmental burden; many countries have passed laws and legislation to save the environment from the detrimental effects of these wastes, which

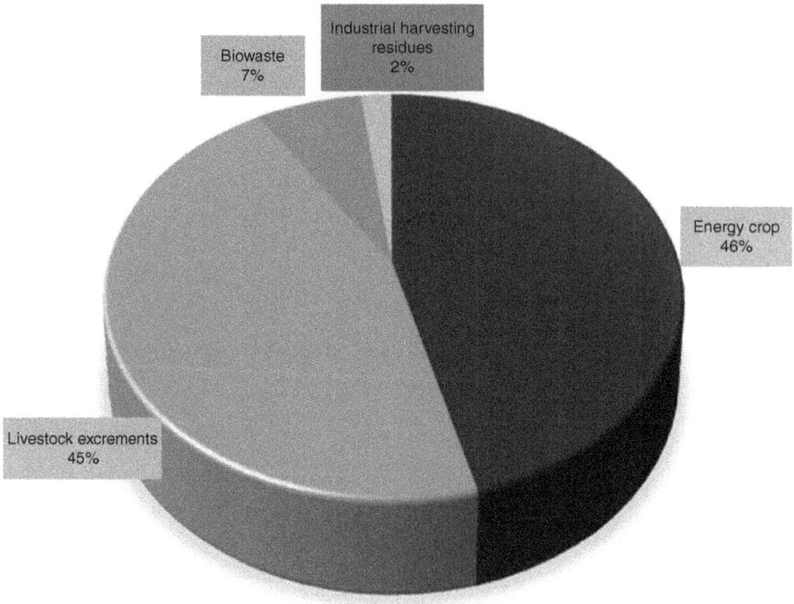

Figure 5.1 Agroindustrial waste materials.

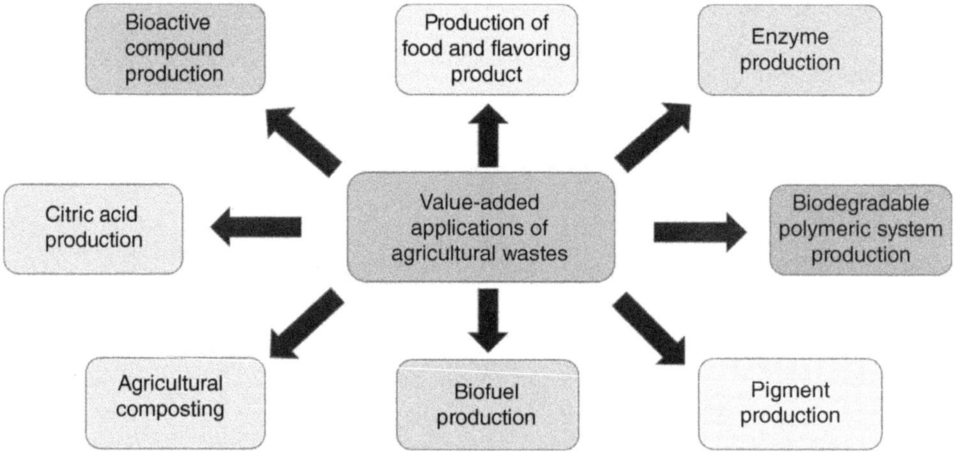

Figure 5.2 Value-added applications of agro-industrial wastes.

clearly defines **the waste and principles governing its management emphasizing on waste assessment and elimination in such a way so that minimum harm is caused to all the natural resources.** The principal concept is to safeguard water, air, soil, and our biotic community from any danger and regulate the usage of organic fertilizers in farming (Venglovsky et al. 2006; Bharathiraja et al. 2017). Implementing clean and green technology through proper organization and planning, assessment, and a feasibility study would help minimize industrial pollution. Figure 5.3 represents the process of manufacturing biofuel from organic material-based wastes.

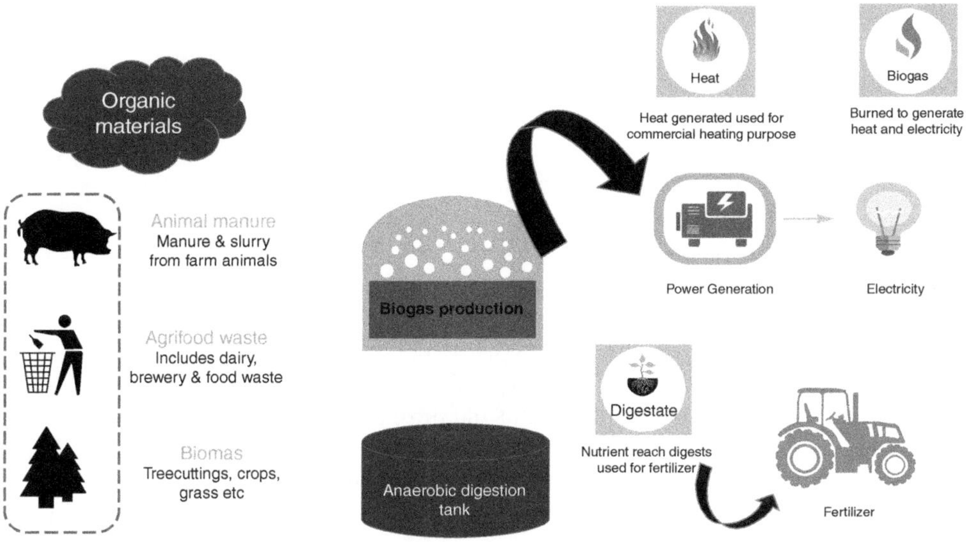

Figure 5.3 Biofuel generation from organic wastes.

5.2.3 Biofuel Production

The municipal solid waste and nonedible plant materials can serve as potential sources for biofuels production (Ramos et al. 2016). **Bioethanol can be produced** by utilizing agricultural wastes, which may be an adequate replacement for fossil fuels.

Kim et al. estimated the quantity of bioethanol **that could be generated** from the lignocellulosic biomass. They wasted total crop residues that could produce more than 490 billion L of bioethanol in a year (Kim and Dale 2004). Several hurdles that include challenges in transporting and handling biomass and its total **delignification** need to be solved. Bioethanol production can be carried out by several processes like enzymatic degradation, pyrolysis, fermentation, physical–chemical treatments, and others. An efficient pretreatment method can also increase the concentration of fermentable sugar to augment the efficiency of the process. Some new, economical, complementary, and unconventional technologies in fermentation need to be applied for converting glucose and xylose to ethanol (Jecu 2000; Schievano et al. 2011).

Not only ethanol but the production of other chemicals like carboxylic acids, butanol and farnesene from nonfood feedstocks can be used to add value and utility to biofuels (Eggleston and Lima 2015; Ramos et al. 2016; Wang et al. 2022).

5.3 Anaerobic Digestion

Different anaerobic processes and various types of digesters are utilized in AD. A particular digester would be most suitable only for a specific type of organic waste, and it is difficult to predict that digester would be most pertinent for the treatment of selected organic waste material. For example, decentralized plants should serve each farm separately to digest farm waste, thus making the entire process viable, feasible, and economical. Similarly, waste from a city should be considered a component in treating municipal wastes based on organic materials because such organic waste varies even from the exact geographical locations. The maximal conditions for substrate digestion can be determined if and **only** if the waste from each unit is studied carefully. It is highly advisable to run the process in a laboratory setting and scale up from pilot to actual output of full-scale digester due to the considerably high cost of the digester and when past data are not there to fall back on. One of the most significant challenges is that after construction, the digester does not perform as expected and thus its needs to be reconstructed (Logan and Visvanathan 2019; Kunatsa and Xia 2022).

The process of AD is segregated into two parts, namely solid digestion and wet digestion. The former acts as anaerobic composters as in this procedure, both the biomass and substrate, are present in presoaked solid form, which contains 22% dry matter and rests water. This process has many merits and challenges. Since the amount of water used is less, the reactor volume is reduced, which is one of the most significant advantages. A five times concentrated substrate means that the importance of the reactor has decreased by five times.

Transporting the substrate can be one of the most significant challenges of the reliable digestion process. In the solid digestion process, probably some inhibitors like ammonium can show less inhibitory effects. The energy requirement for transporting the substrate in solid form in and out of the digesters is more. The probability and possibility of air intrusion into the digesters cannot be ruled out and risky to the stability and safety of the process. Recently, the technology-based procedure has gained attention in a broader range, such as the Kompogas® process (Kompogas 2011).

The wet digestion process shows variability and could be operational in conventional areas. Various reactor technologies are utilized to carry out the digestion process anaerobically. The process of digestion can be continuous or can be split into batches. Further, the operation in constant processes may be carried out in a single-step or two-step process. **Mostly in all wet digestion, the micro-organism is mixed and suspended along with the substrate in the digester, making the separation of the substrate from the micro-organism impossible after the process. The process can be selected** based on the state of the substrate; for example, a flow-through method is used when the substrate is mainly dissolved organic liquid containing very little solid. **The micro-organisms in the granular form are suspended in liquid in this process; the micro-organism granules can be separated from the exhausted substrate.**

5.3.1 Dry and Wet Anaerobic Digestion

The dry AD process uses less fresh water and consumes less energy, with advantages over the wet process (Angelonidi and Smith 2015). The biomass containing lignocellulose has high TS content, while biomasses of the straw, corn silage, and grasses range between 25 and 89% (Chiumenti et al. 2018). In manure from livestock, the total solid content /challenging phase is in the range of 18–30%, depending on the pretreatment done (Møller et al. 2002; Nkoa 2014; Riya et al. 2018). Thus, the dry anaerobic process is more suitable for the digestion of agricultural waste than the wet process because of the high total solid content. In wet digestion, the digestate has a TS content of 23–30% after solid–liquid separation (Zhou et al. 2016), which is at par or marginally higher than the total solid content of the dry anaerobic digestate (complete substantial range in the reactor) (Han et al. 1997).

Accordingly, the cost of plant operation, maintenance, and energy consumption in dry AD are less because of reduced postdigestate treatment such as solid–liquid separation and treatment of liquid fraction compared to the wet process.

5.3.2 Essentials of Anaerobic Digestion

The anaerobic decomposing digestion method is defined as a controlled process wherein biodegradable materials are decomposed. The parameters like oxygen and temperature are controlled in digesters since it is an anaerobic process. **No oxygen is supplied, and temperature is selected depending on the anaerobic species (thermophilic or mesophilic anaerobic, facultative bacteria, and archaea) and their optimal temperature. This process is mainly used for the sludge obtained by the wastewater treatment plants and organic biodegradable wastes. It decreases the mass and volume of the starting material and produces biogas (Han et al.** 1997; **Zhou et al.** 2016**). Biogas is a mixture of CO_2 and methane with trace amounts of H_2S, NH_3, and H_2.** These micro-organisms are responsible for the conversion to biogas and whole digestate (0-62).

5.3.3 Biochemical Reactions in Anaerobic Digestion

The following are four main biochemical reactions of AD are represented in Figure 5.4 namely:

a) Hydrolysis,
b) Acidogenesis,
c) Acetogenesis, and
d) Methanogenesis.

Mainly the biomass consists of large organic compounds. Thus, macromolecular chains first need to break down into smaller molecules so that the micro-organism can access the potential chemical energy of the organic material. The organism can easily further process the monomers such as sugars.

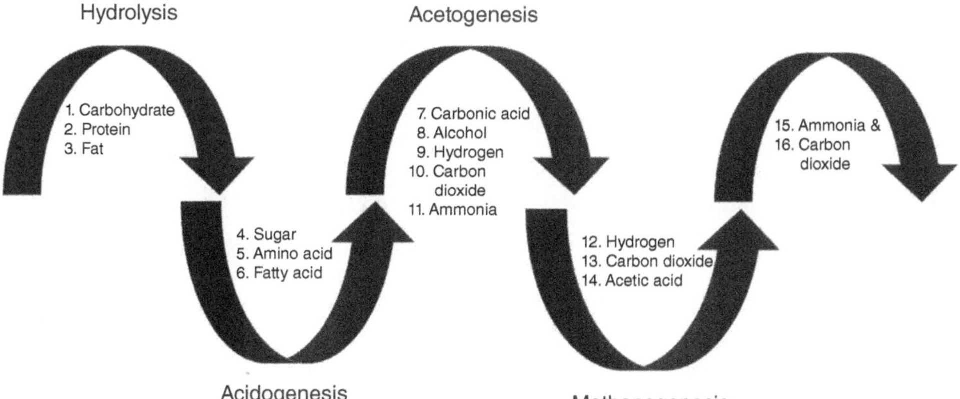

Figure 5.4 Biochemical reactions in anaerobic digestion.

5.3.3.1 Hydrolysis

The first and essential step in AD is hydrolysis. In this procedure, the chains are broken, and the tiny molecules are dissolved into solution. **Various pretreatment processes like mechanical, thermal, or chemical hydrolysis can be enhanced** to improve digestion. For breaking down and dissolving the smaller molecules, **biological or a combination of the biochemical, chemical, and physical processes can be used**. The natural process makes use of hydrolytic micro-organisms. In the biochemical process, extracellular enzymes are utilized, while catalytical reactions, thermal energy, and pressure **acidogenesis** are employed in chemical and physical processes, respectively.

5.3.3.2 Acidogenesis and Acetogenesis

Methanogens can directly use hydrogen and acetates, which are produced in the first stage. Molecules like VFAs (volatile fatty acids) having a series of chain lengths more than acetate are needed to break into the compounds that the methanogens can utilize directly, **which is followed by degradation of the remaining constituents by acidogenic bacteria (fermentative) through the biological process of acidogenesis**. This process generates VFAs along with gases like CO_2, NH_3, H_2S, and other by-products. This process is followed by acetogenesis, which is the next stage of digestion by anaerobic decomposing. The smaller molecules generated by the process of acidogenesis are finally acted upon by acetogens through acetogenesis to generate acetates or acetic acid as well as CO_2 and H_2 gas.

5.3.3.3 Methanogenesis

The main constituent of biogas is methane. Methanogenesis is the last phase of anaerobic decomposing digestion. In this process, the methanogenic archaea utilize the intermediate products generated in the earlier stages and transform them into carbon dioxide, methane, and water. The method of methanogenesis is sensitive to both low and high pH, and these bacteria function well in different pH concentrations between the pH ranges 6.6 and 8.1. The dead bacterial residues and the nondigestible mineral material and organic matter remain because microbes cannot feed upon comprises the solid digestate.

5.3.4 Factors Affecting Anaerobic Digestion

Table 5.5 summarizes the optimum conditions of the environment required for the effective operation of anaerobic decomposing digestion. There are several parameters on which microbial metabolism depends (Figure 5.5); thus, these need to be carefully controlled in a practical situation.

In waste degradation, the two-stage process has been evolved so that each group of micro-organisms (Figure 5.6) is provided with optimum conditions in an automated reactor for each stage. The two stages are as follows:

Stage 1: hydrolysis / acidification and
Stage 2: acetogenesis / methanogenesis

Table 5.5 Environmental requirements: optimal conditions for biogas production.

Parameter	pH	Temperature (°C)	C/N ratio	Redox potential (mV)	C:N:P:S ratio
Hydrolysis/ Acidogenesis phase	5.2–6.3	25–35	10 : 1–45 : 1	+400 to −300	500 : 15 : 5 : 3
Methanogenesis phase	6.7–7.5	32–42	20/1–30/1	< −250	600 : 15 : 5 : 3

Source: Adapted from Chandra et al. (2012).

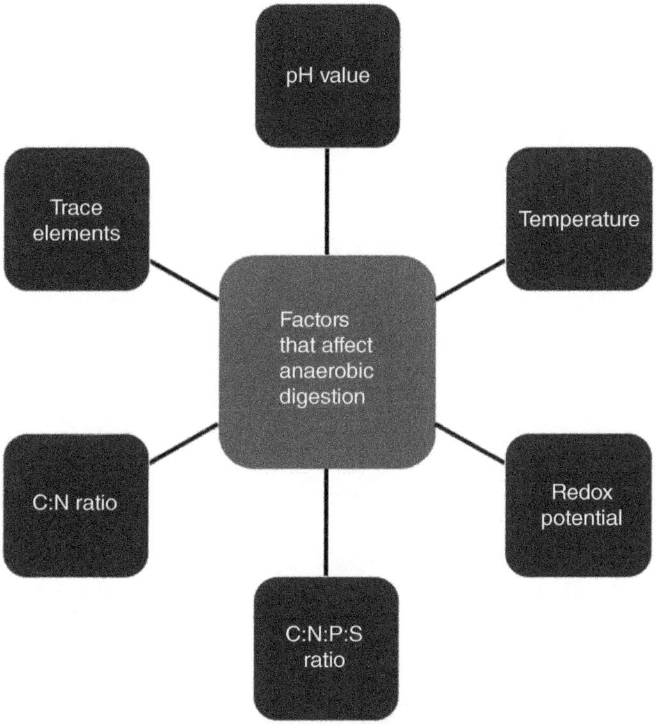

Figure 5.5 Factors involved in anaerobic digestion process.

5.3.5 Inhibitory Substances

It has been observed that in AD systems, some inorganic salts affect the process significantly; high concentrations of some substances may inhibit the function of microbial growth as some are necessary for small concentrations. Table 5.6 represents the optimum and inhibitory concentrations of inorganic salts and metal ions.

Heavy metals like lead, chromium, zinc, nickel, copper, and cadmium act as positive catalysts on the anaerobic decomposing digestion process when present in low concentrations, but they become fatal and toxic in more significant amounts. These substances are disruptive to the anaerobic decomposing digestion procedure. In particular, agricultural

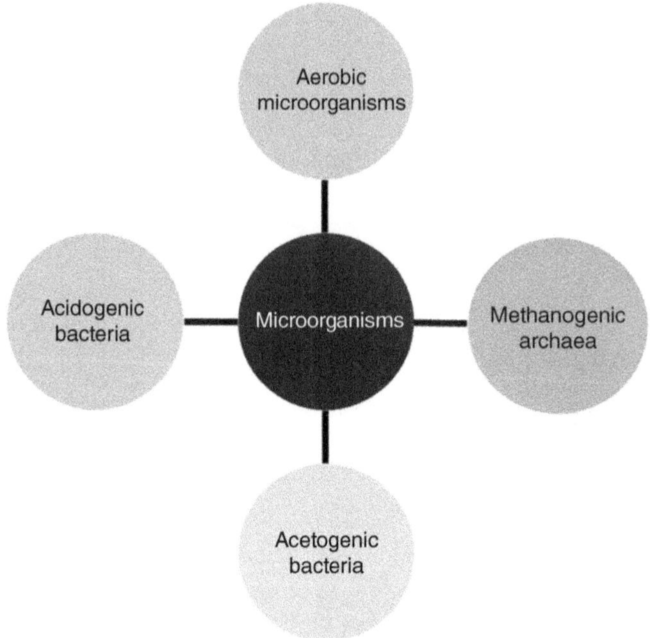

Figure 5.6 Major micro-organisms involved in anaerobic digestion process.

Table 5.6 Optimum and inhibitory concentrations of inorganic salts and metal ions.

Optimal/Inhibitory concentrations	Magnesium	Potassium	Calcium	Sodium
Optimal concentration (mg)	75–150	200–400	100–200	100–200
Moderate inhibition (mgL^{-1})	1000–1500	2500–4500	2500–4500	3500–5500
Inhibition (mg^{L-1})	3000	12 000	8000	16 000

wastes like in pig slurry, zinc causes inhibition starting from 400 mg/L. Antibiotic containing Zn as an additive **is present in pig fodder.** Table 5.7 shows the toxic and minimum inhibitory concentrations of heavy metals.

The presence of some organic substances along with the substratum can cause nonspecific inhibition. These substances can come from different sources and may be as varied as antibiotics, disinfectants, herbicides, pesticides, and surfactants having a specific chemical formula. It is difficult to figure out the inhibitory behaviour of such substances conclusively. Since such substances are present in the treated substratum, thorough research is required to find the inhibitory concentration of such chemicals on the micro-organisms.

5.3.6 Merits of Anaerobic Digestion

This process generates biogas. which is the form of renewable energy and net energy-producing process. The waste or feedstock is sanitized in the process when the temperature is set above the optimum temperature for a predefined interval of time, and it is

Table 5.7 Inhibitory and toxic concentrations of heavy metals.

Metal	Inhibition start (mgL^{-1})	Toxicity to adopted micro-organisms[3] (mgL^{-1})
Cd	70	600
Pb	340	340
Zn	400	600
Cr^{3+}	130	260
Cr^{6+}	110	420
Cu	40	170
Ni	10	30

pasteurization. The process is also responsible for reducing the odour of waste. If the fragrance before treatment and odour after treatment is compared, there is a huge difference; the smell of pretreated waste is much more pungent and obnoxious. The unprocessed trash is more likely to cause pollution than the treated waste if dumped on land. The process can be used as practical, valuable resource recovery and conserving nonrenewable energy resources. In the primary water treatment plant, the AD process produces lower sludge mass than the aerobic system under the same concentration of flow and contamination. The manure obtained from the anaerobic process is longer lasting compared to the untreated organic waste.

5.3.7 Challenges of Anaerobic Digestion

The investment cost of building large tanks and other processing vessels is high when AD is used commercially at the wastewater treatment plants. The odour problem can persist if the efficiency of the AD process is not taken **care of**. The conversion efficiency of the process is less than gasification because it does not efficiently convert the significant carbon content of biomass to biogas. The initial warm-up time is more than that of the aerobic system because of the slow growth of methane-producing organisms. In some applications, higher buffering chemicals are required to keep the pH range within 6.5–8. Nutrients are not removed efficiently in the anaerobic process compared to the aerobic process in wastewater treatment plants. Therefore a posttreatment is necessary for such discharges.

5.3.8 Economics of Anaerobic Digestion

Globally, there is a huge demand for energy because of rapid industrialization, urbanization, and increasing population growth (Mazur 1994; Zabel and Economics 2009; Stern 2011). To meet these growing energy demands, power plants are using fossil fuels, diminishing fossil fuels reserves. The burning of fossil fuels leads to harmful environmental effects, causing air pollution and impacting physiological health and well-being. Many health complications like bronchitis, respiratory diseases, asthma, allergic problems, and others arise from deteriorating air quality. To save humankind from this alarming situation, scientists are working throughout the globe on developing renewable and alternative sources for energy production. Agricultural waste to energy (WTE) generation is a widely adopted alternative technology to generate power from organic-rich waste streams via AD process or technology. Not only this new technology generates energy but also utilizes huge amount of agricultural wastes, saving the environment and nation from economic loss. Energy production mainly biogas by using AD has high initial set up cost but future recurring cost is very low

Newer technological development is the need of the hour to reduce the effective price, which generates eco-friendly sustainable energy (Koberg and Gedanken 2012; Kataki et al. 2017). Much scientific research provides suggestions on process challenges and economic potentials to help developers and investors make informed decisions before construction (Bioenergy Technologies Office 2017).

5.3.9 Market Opportunities for Anaerobic Digestion in India

In India, many recent developments have accelerated the demand for AD, which includes the following: the need for energy has increased, which is focusing attention on renewable energy sources to meet the demand. There have been concerns from an environmental perspective related to manure and agroindustrial waste management using the earlier method. Modified and or new policies are of the government that is supporting AD and developing renewable energy. Expanding AD market would open new green job growth **potential.**

According to detailed report by USEPA under 'Global Methane Initiative,' which states that India has the potential for anaerobic digestion of livestock and agro-industrial wastes **(The Global Methane Initiative** 2020**).** Government of India has taken various initiatives to help the staDrt-up industries regarding the biogas plant. **These initiatives have helped** developers, policymakers, and other interested stakeholders by bridging the gap and gives detailed information for generating energy using agricultural wastes. The report highlights the following areas in details:

1) **Uses of Biogas and Digestate:** They can be put to multiple uses. Biogas can be used in cooking, transportation, cooling, and heating. The digestate can be used as fertilizers, oil amendment, or as compost product.
2) **Current Biogas:** Policies and Incentives of Government of India: the objectives of India's policies are variable. The Government of India is aiming at improved manure management, also reducing the dependence on imports of oil and gas, as well as improved sanitation at the village level with the Swachh Bharat Mission.
3) **Biogas Potential from Agricultural Feedstock:** Many agricultural feedstocks in India are available, serving as an input into anaerobic digestion systems. It also estimates the potential for reducing methane emissions and production to be used as the source of energy. The various sectors that can provide the feedstock and benefit by using anaerobic digestion are distilleries, vegetable and fruit processing industries, sugarcane processing plants, milk processing and dairy farms, cornstarch, and tapioca production plants. The shared information would help all the stakeholders like policy makers, financiers, and industrialists focus on biogas development (The Global Methane Initiative 2020).

The technological options have been discussed, which give the details on current used anaerobic digester technologies and their key attributes, which can serve as baseline data and help the technical team develop alternative methods for improving based on feedstock, location, and region. The small-scale digester technologies are of use at the household or farm level. For commercial purposes, both medium- and large-scale systems are used based on the requirement and infrastructure. Successful business models and case studies have been discussed, which highlights the important aspects of successful AD systems, different aspects of revenue generation, potential owner, and operator models from India (The Global Methane Initiative 2020).

5.4 Way Forward

Untreated agricultural waste appears to be a financial and environmental burden but when this huge pool can be utilized effectively, they can generate bioenergy and bio-based products and thus could also resolve the energy and economic crisis of developing countries. The agricultural residue management network is very complex, and therefore economic and environmental implications of such a complex chain are challenging to assess. Moreover, differences also exist because of variations in seasons and regions.

5.4.1 Challenges

The first and foremost challenge is the harm to the environment due to inefficient residual management of agriculture waste and earlier agricultural practices. The components of the environment like atmosphere, lithosphere, and hydrosphere are all affected by direct and indirect usage of water, energy, land, and others) as well the practices yield a voluminous amount of waste that are unwanted and underutilized.

The second challenge is the upgradation of technology for generating biogas and biofertilizer from agricultural wastes. The most efficient and popular conversion technology used worldwide is the anaerobic digestion (AD) technique. AD plants are running and generating biogas in developed countries, but the yield has to be improved by technological upgradation. EBA. (2018). Statistical Report of the European Biogas Association (EBA). 7th edition, approximately 17 500 AD plants were operational in Europe, having a total installed capacity of 9.98 GWe. However, the total quantity of electricity generated from biogas is estimated at more than 65 TWh. Almost 500 plants of biomethane generation are operational in Europe, with a production capacity equal to 17 264 GWh.

Agricultural biomass can serve as building blocks for developing innovative bio-products and has the potential to meet sustainable development goals. The challenge is overcoming the bottlenecks for the conversion of agricultural residues into innovative materials.

If the building blocks of plastic can be replaced by biomaterials, the economic and environmental aspects can be addressed, but this requires in-depth knowledge and exploration of technology.

The application of the concept of circular economy to biomass residues is another concern that needs to be addressed. Frosch and Gallopoulos introduced the principles of circular economy in 1989.

Accordingly, the wastes generated by one unit or system can serve as raw material for another design or team to reduce the burden to the environment and the economy (Frosch and Gallopoulos 1989). The eco-industrial park can serve as a community for production, manufacturing, and service businesses and reap the collective benefit. It can manage issues related to the environment, energy, water, and materials. Kalundborg in Denmark is a very famous example of such eco-industrial park or industrial symbiosis (http://www.symbiosis.dk/en). Implementing such industrial symbiosis is still a great challenge, especially in developing economies of the world.

All the agricultural and livestock business stakeholders need to be connected, and knowledge needs to be shared – awareness created to manage the environmental and socioeconomic challenges at the local and regional levels.

5.4.2 Challenges of Existing Biogas Plants

The various critical challenges associated with biogas plants in emerging Asian countries (Pandyaswargo et al. 2019) have been highlighted in Table 5.8. In order to make the biogas plants sustainable, the challenges in terms of technicality, finance, and social aspects need to be addressed.

5.4.3 Initiatives Taken by Government of India to Prevent Stubble Burning

Strict measures need to be taken to stop the burning of stubble and synchronize the waste management system (Sidar 2018; Ravindra et al. 2019). Several government agencies have taken a lead role to control and regulate this grave situation. The Government of India has taken initiatives to educate the agricultural communities on proper and scientific agrarian waste management systems and sustainable development. The government officials and environmentalists are putting forth proposals and raising concerns in different forums so

Table 5.8 Summary of challenges of biogas plants in Asian countries.

Substrates	Technical challenges	Financial challenges	Social challenges
1) Municipal solid waste 2) Animal manure 3) Agricultural residue 4) Codigestion	5) Sustainable supply of substrate 6) Feasibility of environmental condition 7) Plant construction appropriateness 8) Sustainability of operation and maintenance 9) Continuous demand for biogas plant output	10) Support in terms of investment support 11) Financial support mechanism during operation 12) Competitive pricing with stable market demand of the output	13) Commitment of operators commitment 14) Existence of supporting government regulations 15) Other social factors like social taboos, local labor, or safety issues

that **the burning of crop residues can be curbed and alternate management for sustainability can be initiated.**

A number of laws have been enacted to check the burning of stubble

i) The Section 144 of the Civil Procedure Code (CPC) to ban burning of paddy
ii) The Air Prevention and Control of Pollution Act, 1981
iii) The Environment Protection Act, 1986
iv) The National Tribunal Act, 1995
v) The National Environment Appellate Authority Act, 1997.

National Green Tribunal (NGT) has taken strict measures against crop residue burning, mainly in Rajasthan, Uttar Pradesh, Haryana, and Punjab (Koberg and Gedanken 2012; Kumar et al. 2015; Lohan et al. 2018). The Government of India recently directed the National Thermal Power Corporation (NTPC) to add or mix agricultural crop residue pellets (approximately 10%) with coal for power generation to reduce agricultural waste (Bhuvaneshwari et al. 2019). The farmers benefited financially from this initiative; they earned approximately Rs. 5500 per ton of crop residue. **The farmers can take advantage of this initiative and** resolve the various problems associated with the burning of crop residue. The Government of India has taken measures to run the bio-composting plants. In August 2007, the State Plan Scheme of Additional Central Assistance (Rashtriya Krishi Vikas Yojana) was for the sustainable training development of farming as a part of the 11th Five Year Plan. In the first phase, various villages of Azamgarh and Marinath Bhanjam in the districts of eastern Uttar Pradesh were included in this scheme, where **eight demonstration and training projects were established. As per government records, around 456 farmers were trained** in bio-compost production and agrowaste bio-conversion. These training programmes have immensely helped farmers in reaping monetary benefits (Pratap Singh and Prabha 2017).

The Ministry of Agriculture of India has also developed a National Policy for Management of Crop Residue (NPMCR) (Pratap Singh and Prabha 2017). The main objectives of the NPMCR are as follows:

i) Promotion of technologies for optimum utilization and management of crop residue *in- situ,* prevention of loss of valuable soil nutrients, and diversification of uses of crop residue in industrial applications.
ii) Development and promotion of appropriate crop machinery in farming practices as, for example, modified grain recovery machines (harvesters having twin cutters). Also provide incentives and discounts in purchasing mechanized sowing machinery such as the happy seeder, turbo seeder, shredder, and baling machines.
iii) The National Remote Sensing Agency (NRSA) and Central Pollution Control Board would monitor crop residue management, by satellite-based remote sensing technology.
iv) Financial assistance to be provided through multidisciplinary approach and fund mobilization for innovative ideas and project proposals to accomplish above.

The use of remote sensing and aerial surveillance has helped the government regulatory body monitor and reinforce the measures effectively. **Fines have also been imposed on the farmers in Punjab, Rajasthan,** and Haryana, who were found violating the law and indulging in crop burning. There is no update available in the literature on the achievements of the other three objectives of the NPMCR, but identifying the exact location of crop burning has helped reduce stubble burning. The governmental agencies have successfully brought down the crop stubble burning in Haryana and Punjab by 25 and 38%, respectively, by vigilance through remote sensing.

5.5 Conclusion

One of the important ways to treat waste is AD, as it can not only handle waste but also utilize the agricultural biomass for energy production. Earlier, the process focused on sludge and manure stabilization, but presently, it is used for sustainable development. The process helps protect the environment, conserve natural resources, and serve as the continuous supply of renewable energy. In the present *era*, biogas production has added to the income of the farming sector. Specific research in the area of AD can help in the optimization and improvement of process stability.

One of the largest anthropogenic sources of methane, which is a potent greenhouse gas, is agricultural wastes. The biogas generated contains at least 50% methane and solid residue, a good source of nutrients used as organic fertilizer. An alternative pathway to managing large organic waste generated in livestock farming is AD, an excellent renewable energy source. If the methane gas is allowed to escape into the atmosphere, it will lead to global warming. Following the anaerobic process, this potential of methane gas is harnessed and serves as a continuous source of clean and green energy.

References

Abbasi, T. and Abbasi, S.A. (2011). *Renewable Energy Sources: Their Impact on Global Warming and Pollution*. PHI Learning Pvt. Ltd.

Abbasi, T., Premalatha, M., and Abbasi, S.A. (2011). The return to renewables: will it help in global warming control? *Renewable and Sustainable Energy Reviews 15* (1): 891–894.

Adejumo, I.O., Ologhobo, A.D., and Babalola, T.O. (2015). Effect of pre-planting seed dressers on serum enzymes of laying chickens. *Journal of the American Chemical Society 9*: 1–5.

Agricultural Waste usatanksales.com (2013). Archived from the original on 10 December 2013. Accessed 15 October 2013.

Aikawa, T. (2016). Recommendations for Woody Biomass Power Generation under FiT for the realization of a sustainable use of bioenergy. Renewable Energy Institute, Tokyo. http://www.renewable-ei.org/en/activities/reports_20161226 (accessed January 2021).

Angelonidi, E. and Smith, S.R. (2015). A comparison of wet and dry anaerobic digestion processes for the treatment of municipal solid waste and food waste. *Water Environment Journal 29* (4): 549–557.

Anon (2000). Behandling af bioaffald fra husholdninger i Nordøstsjælland (in Danish). Report by Rambol Ltd. for AFAV I/S, I/S Nordforbrænding and the communities of Fredensborg-Humlebæk and Elsinore.

Anukam, A., Mohammadi, A., Naqvi, M., and Granström, K. (2019). A review of the chemistry of anaerobic digestion: Methods of accelerating and optimizing process efficiency. *Processes 7* (8): 504.

Asumadu-Sarkodie, S. and Owusu, P.A. (2016). Feasibility of biomass heating system in Middle East Technical University, Northern Cyprus Campus. *Cogent Engineering 3* (1): 1134304.

Ayalon, O., Avnimelech, Y., and Shechter, M. (2000). Alternative MSW treatment options to reduce global greenhouse gases emissions-the Israeli example. *Waste Management & Research 18* (6): 538–544.

Bharathiraja, S., Suriya, J., Krishnan, M. et al. (2017). Production of enzymes from agricultural wastes and their potential industrial applications. In: *Advances in Food and Nutrition Research*, vol. 80 (ed. S.-K. Kim and F. Toldrá), 125–148. Academic Press.

Bhuvaneshwari, S., Hettiarachchi, H., and Meegoda, J.N. (2019). Crop residue burning in India: policy challenges and potential solutions. *International Journal of Environmental Research and Public Health 16* (5): 832.

Bioenergy Technologies Office (2017). Biofuels and bioproducts from wet and gaseous waste streams: Challenges and opportunities. Washington, DC, USA.

Borgens, M. (2019). Analysis of Toxic Pollutant Sources and Characteristics Contributing to Water Quality Impairments in the Willamette River Basin. Master of Environmental Management Project Reports. 55.

Chandra, R., Takeuchi, H., and Hasegawa, T. (2012). Methane production from lignocellulosic agricultural crop wastes: a review in context to second generation of biofuel production. *Renew Sust Energ Rev* 16 (3): 1462–1476.

Chiumenti, A., da Borso, F., and Limina, S. (2018). Dry anaerobic digestion of cow manure and agricultural products in a full-scale plant: efficiency and comparison with wet fermentation. *Waste Management 71*: 704–710.

Dien, B.V. and Vong, V.D. (2006). Analysis of pesticide compound residues in some water sources in the province of Gia Lai and DakLak. *Vietnam Food Administrator.*

Edenhofer, O., Pichs-Madruga, R., Sokona, Y. et al. (2014). Technical Summary. In: *Climate Change 2014: Mitigation of Climate Change. Contribution of Working Group III to the Fifth Assessment Report of the Intergovernmental Panel on Climate Change* (ed. O. Edenhofer, R. Pichs-Madruga, Y. Sokona, et al.). Cambridge, United Kingdom and New York, NY, USA: Cambridge University Press.

Eggleston, G. and Lima, I. (2015). Sustainability issues and opportunities in the sugar and sugar-bioproduct industries. *Sustainability 7* (9): 12209–12235.

Europe, G. I. (2018). European biomethane map. Infrastructure for biomethane production.

FAO (2012). World Agriculture Towards 2030/2050: The 2012 revision ESA E Working Paper No. 12-03. http://www.fao.org/economic/esa/esag/en (accessed January 2021).

Flannery, K.V. (1973). The origins of agriculture. *Annual Aeview of Anthropology 2* (1): 271–310. https://www.sciencedaily.com/terms/agriculture.htm (accessed January 2021).

Frosch, R.A. and Gallopoulos, N.E. (1989). Strategies for manufacturing. *Scientific American 261* (3): 144–153.

Goh, C.S., Junginger, M., Cocchi, M. et al. (2013). Wood pellet market and trade: a global perspective. *Biofuels, Bioproducts and Biorefining 7* (1): 24–42.

Han, Y., Sung, S., and Dague, R.R. (1997). Temperature-phased anaerobic digestion of wastewater sludges. *Water Science and Technology 36* (6–7): 367–374.

Holm-Nielsen, J.B., Al Seadi, T., and Oleskowicz-Popiel, P. (2009). The future of anaerobic digestion and biogas utilization. *Bioresource Technology 100* (22): 5478–5484.

Hossain, A.K. and Badr, O. (2007). Prospects of renewable energy utilisation for electricity generation in Bangladesh. *Renewable and Sustainable Energy Reviews 11* (8): 1617–1649.

Jain, N., Bhatia, A., and Pathak, H. (2014). Emission of air pollutants from crop residue burning in India. *Aerosol and Air Quality Research 14* (1): 422–430.

Jayathilakan, K., Sultana, K., Radhakrishna, K., and Bawa, A.S. (2012). Utilization of byproducts and waste materials from meat, poultry and fish processing industries: a review. *Journal of Food Science and Technology 49* (3): 278–293.

Jecu, L. (2000). Solid state fermentation of agricultural wastes for endoglucanase production. *Industrial Crops and Products 11* (1): 1–5.

Karak, T., Sonar, I., Nath, J.R. et al. (2015). Struvite for composting of agricultural wastes with termite mound: utilizing the unutilized. *Bioresource Technology 187*: 49–59.

Kataki, R., Chutia, R.S., Bordoloi, N.J. et al. (2017). Biohydrogen production scenario for Asian countries. In: *Biohydrogen Production: Sustainability of Current Technology and Future Perspective* (ed. A. Singh and D. Rathore), 207–235. New Delhi: Springer.

Kaygusuz, K. (2012). Energy for sustainable development: a case of developing countries. *Renewable and Sustainable Energy Reviews 16* (2): 1116–1126.

Kim, S. and Dale, B.E. (2004). Global potential bioethanol production from wasted crops and crop residues. *Biomass and Bioenergy 26* (4): 361–375.

Klich, M.A. (2007). Aspergillus flavus: the major producer of aflatoxin. *Molecular Plant Pathology 8* (6): 713–722.

Koberg, M. and Gedanken, A. (2012). Optimization of bio-diesel production from oils, cooking oils, microalgae, and castor and jatropha seeds: probing various heating sources and catalysts. *Energy & Environmental Science 5* (6): 7460–7469.

Kompogas (2011). "Green energy from organic waste." Retrieved 11th August 2011, from http://www.axpo-kompogas.ch/index.php?path=home&lang=en.

Kumar, P., Kumar, S., and Joshi, L. (2015). *Socioeconomic and Environmental Implications of Agricultural Residue Burning: A Case Study of Punjab, India*, 144. Springer Nature.

Kunatsa, T. and Xia, X. (2022). A review on anaerobic digestion with focus on the role of biomass co-digestion, modelling and optimisation on biogas production and enhancement. *Bioresource Technology 344*: 126311.

Logan, M. and Visvanathan, C. (2019). Management strategies for anaerobic digestate of organic fraction of municipal solid waste: current status and future prospects. *Waste Management & Research 37* (1_suppl): 27–39.

Lohan, S.K., Jat, H.S., Yadav, A.K. et al. (2018). Burning issues of paddy residue management in north-west states of India. *Renewable and Sustainable Energy Reviews 81*: 693–706.

Marinari, S., Masciandaro, G., Ceccanti, B., and Grego, S. (2000). Influence of organic and mineral fertilisers on soil biological and physical properties. *Bioresource Technology 72* (1): 9–17.

Mazur, A. (1994). How does population growth contribute to rising energy consumption in America? *Population and Environment 15* (5): 371–378.

Mentis, D., Howells, M., Rogner, H. et al. (2017). Lighting the World: the first application of an open source, spatial electrification tool (OnSSET) on Sub-Saharan Africa. *Environmental Research Letters 12* (8): 085003.

Møller, H.B., Sommer, S.G., and Ahring, B.K. (2002). Separation efficiency and particle size distribution in relation to manure type and storage conditions. *Bioresource Technology 85* (2): 189–196.

Nkoa, R. (2014). Agricultural benefits and environmental risks of soil fertilization with anaerobic digestates: a review. *Agronomy for Sustainable Development 34* (2): 473–492.

Obi, F.O., Ugwuishiwu, B.O., and Nwakaire, J.N. (2016). Agricultural waste concept, generation, utilization and management. *Nigerian Journal of Technology 35* (4): 957–964.

Oluseun, A.I., Durojaiye, O.A., Ajibade, A.I., and Idowu, O.S. (2014). Status of exposure of bio-systems to restricted aluminium phosphide pesticide in Kano State, Nigeria. *International Journal of Scientific Research in Knowledge 2* (7): 306.

Pandyaswargo, A.H., Jagath Dickella Gamaralalage, P., Liu, C. et al. (2019). Challenges and an implementation framework for sustainable municipal organic waste management using biogas technology in emerging Asian Countries. *Sustainability* 11 (22): 6331.

Pratap Singh, D. and Prabha, R. (2017). Bioconversion of agricultural wastes into high value biocompost: a route to livelihood generation for farmers. *Advances in Recycling and Waste Management 137*: 1–5.

Ramos, J.L., Valdivia, M., García-Lorente, F., and Segura, A. (2016). Benefits and perspectives on the use of biofuels. *Microbial Biotechnology 9* (4): 436–440.

Ramos, J.L., Udaondo, Z., Fernández, B. et al. (2016). First-and second-generation biochemicals from sugars: biosynthesis of itaconic acid. *Microbial Biotechnology 9* (1): 8.

Ravindra, K., Singh, T., and Mor, S. (2019). Emissions of air pollutants from primary crop residue burning in India and their mitigation strategies for cleaner emissions. *Journal of Cleaner Production 208*: 261–273.

Ravindran, R., Hassan, S.S., Williams, G.A., and Jaiswal, A.K. (2018). A review on bioconversion of agro-industrial wastes to industrially important enzymes. *Bioengineering 5* (4): 93.

Reardon, T., Echeverria, R., Berdegué, J. et al. (2019). Rapid transformation of food systems in developing regions: highlighting the role of agricultural research & innovations. *Agricultural Systems 172*: 47–59.

Riya, S., Suzuki, K., Meng, L. et al. (2018). The influence of the total solid content on the stability of dry-thermophilic anaerobic digestion of rice straw and pig manure. *Waste Management 76*: 350–356.

Scarlat, N., Fahl, F., Dallemand, J.F. et al. (2018). A spatial analysis of biogas potential from manure in Europe. *Renewable and Sustainable Energy Reviews 94*: 915–930.

Schieber, A., Stintzing, F.C., and Carle, R. (2001). By-products of plant food processing as a source of functional compounds – recent developments. *Trends in Food Science & Technology 12* (11): 401–413.

Schievano, A., D'Imporzano, G., Salati, S., and Adani, F. (2011). On-field study of anaerobic digestion full-scale plants (Part I): an on-field methodology to determine mass, carbon and nutrients balance. *Bioresource Technology 102* (17): 7737–7744.

Sidar, R.S. (2018). Crop residue burning in India: potential solutions. *Journal of Pharmacognosy and Phytochemistry 7* (5): 3451–3455.

Solomon, B.D., Barnes, J.R., and Halvorsen, K.E. (2007). Grain and cellulosic ethanol: history, economics, and energy policy. *Biomass and Bioenergy 31* (6): 416–425.

Spillages, M. (2011). Managing Fire water and major spillages: PPG18 https://www.spillcontrolcentre.co.uk/download/Managing_fire_water_and_major_spillages_%20Scottish_Environment_Protection_Agency_PPG_18.pdf.

Stagich, E.B. (2017). Fresh from the farm: regulating concentrated animal feeding operations for antibiotic abuse in Tennessee. *University of Memphis Law Review 48*: 281.

Stern, D.I. (2011). The role of energy in economic growth. *Annals of the New York Academy of Sciences 1219* (1): 26–51.

The Global Methane Initiative, US EPA (2020). Market Opportunities for Anaerobic Digestion in India. https://www.biocycle.net/market-opportunities-anaerobic-digestion-india/ (accessed 14 April 2020).

UNEP (2009). https://wedocs.unep.org/bitstream/handle/20.500.11822/7614/WasteAgriculturalBiomassEST_Compendium.pdf. (accessed January 2021).

Venglovsky, J., Martinez, J., and Placha, I. (2006). Hygienic and ecological risks connected with utilization of animal manures and biosolids in agriculture. *Livestock Science 102* (3): 197–203.

Wang, B., Dong, F., Chen, M. et al. (2016). Advances in recycling and utilization of agricultural wastes in China: based on environmental risk, crucial pathways, influencing factors, policy mechanism. *Procedia Environmental Sciences 31*: 12–17.

Wang, W., Lee, D.J., and Lei, Z. (2022). Integrating anaerobic digestion with microbial electrolysis cell for performance enhancement: a review. *Bioresource Technology 344*: 126321.

Weiland, P. (2004). Ist die gemeinsame Vergärung von Bioabfällen, landwirtschaftlichen Abfällen und Energiepflanzen möglich und sinnvoll. *EEG und Emissionshandel*, pp. 89–100.

Weiland, P. (2006). Biomass digestion in agriculture: a successful pathway for the energy production and waste treatment in Germany. *Engineering in Life Sciences 6* (3): 302–309.

Weiland, P. and Rieger, C. (2005). Ergebnisse des Biogas-Messprogramms. Report (in German), Fachagentur Nachwachsende Rohstoffe, Gülzow, Germany.

Yadav, S., Yadav, P.K., Yadav, D., and Yadav, K.D.S. (2009). Purification and characterization of pectin lyase produced by *Aspergillus terricola* and its application in retting of natural fibers. *Applied Biochemistry and Biotechnology 159* (1): 270–283.

Zabel, G. and Economics, E. (2009). Peak people: the interrelationship between population growth and energy resources. *Energy Bulletin 20*: 1–59. https://www.resilience.org/stories/2009-04-20/peak-people-interrelationship-between-population-growth-and-energy-resources/.

Zhou, S., Nikolausz, M., Zhang, J. et al. (2016). Variation of the microbial community in thermophilic anaerobic digestion of pig manure mixed with different ratios of rice straw. *Journal of Bioscience and Bioengineering 122* (3): 334–340.

6

Nothing is 'Waste' in Agriculture: From Nanotechnology and Bioprocesses Perspectives

Priyabrata Roy

Department of Molecular Biology and Biotechnology, University of Kalyani, Kalyani, West Bengal, India

CHAPTER OUTLINE

6.1 Introduction

For developing countries, agriculture contributes most to the economy, supplying food and allowing people to live better lives. Agriculture is presently experiencing a number of issues, including unexpected climate change, soil pollution from different damaging environmental contaminants such as fertilizers and pesticides, and, most importantly, rising

Agriculture Waste Management and Bioresource: The Circular Economy Perspective, First Edition.
Edited by Suruchi Singh, Pardeep Singh, Anu Sharma, and Moharana Choudhury.

food demands due to a growing worldwide population (Pouratashi and Iravani 2012). Nanotechnology, which is a cutting-edge innovation in today's agricultural techniques, is predicted to become a large influence in the future by contributing creative uses. This novel approach to agriculture based on nano-principles has enormous potential for addressing global concerns in the sectors of environment, sustainability, food safety and security, and global climate change. Nanotechnology is widely employed in modern agriculture, food technology, as well as the packing, dairy, and packaging, shipping, and quality management of agricultural products (Roy and Choudhury 2021). The present global population of 7.6 billion people is predicted to reach to 9.8 billion by 2050 (UNDESA 2015). This means that new food, water, and energy systems will be required to assure food security. Similarly, growing more food demands the use of natural resources (Glenn and Florescu 2016). As a result, scientific research will be required in the not-too-distant future to give new paradigms and approaches for solving exceedingly complex and diversified challenges.

Nanotechnology is making a significant contribution through a variety of utilization in agriculture and food sectors (Mishra et al. 2014; Rabbani et al. 2016) for effective nutrient management and operation, and nano-biosensors for sensing on-site water level, relative humidity, and nutrient availability in soil. As a result, we may conclude that nanotechnology will play a crucial role in the agricultural sector by revolutionizing the food demand and supply chain while ensuring ecological integrity, environment protection, and economic strength (Wang et al. 2016; Banotra et al. 2017). Modern agriculture is transforming into precision farming as technology advances, allowing farmers to get the most out of their resources. Other nano-based materials have distinct surface chemical, physical, and optical characteristics (Yaqoob et al. 2020), resulting in improved sensitivity, lower detection limits, and faster response times (Kumbhakar et al. 2014). Fertilizers are important for agricultural productivity, yet they deplete soil fertility by supplying the soil's mineral balance (Solanki et al. 2015).

6.2 Relevance and Opportunities to Use Nanotechnology

The scientific objective is to create and test NPs that allow for the limited release of less toxic compounds that might be linked with crop physiological needs. However, we are still at a point where research on nanomaterials (NMs) and their biological interactions yields contradictory results. This is also true for nano-based agricultural research. Agriculture's trajectory is uncertain due to its reliance on natural variables such as climatic, water, and soil condition. As a result, sensing, collecting, analysing, and storing trustworthy and precise data of all components are critical for meeting the problems of food quality and quantity while ensuring environmental sustainability (Servin et al. 2015). NPs interact with plants, causing physiological and structural changes that are influenced by the type of NPs (Zabrieski et al. 2015). Plant development and crop yield may be aided by the administration of NPs in safe doses. Plant–NP interactions could be a chemical phenomenon, involving inter-cellular transport, the formation of reactive oxygen species (ROS), lipid peroxidation, ion cell disturbance, and oxidative breakage (Vijaya et al. 2011; Suresh et al. 2013). Agriculture benefits from several technological developments such as hybrid lines, chemical fertilizers, and pesticides over time. However, agricultural experts are discovering that modernization is critical for agricultural expansion in order to address global food and nutritional security and achieve sustainable development goals. Nanotechnology applications in agriculture have just recently gained prominence, yet research on the subject began more than half a century ago (Mukhopadhyay 2014).

There are some important segments in the controlling of agricultural production on which further studies can create some significant improvements to meet the future demands.

6.3 Increase Productivity of Soil

Fertilizers are needed to preserve soil fertility and produce high-quality food and crops, especially with the emergence of high-yielding and technology-dependent cultivars. There are losses in biota and physical environment caused by traditional fertilizer using methods. As a result, relatively little concentration reaches the targeted location. However, excessive fertilizer and pesticide use pollutes the environment, degrades natural resources, creates pesticide resistance in pests and pathogens, hampers the soil microflora and nitrogen fixation, and accelerate pesticide bioaccumulation (Tilman et al. 2002). As a result, the best use of chemical/synthetic fertilizer based on the nutritional needs of the crop and the least amount of pollution are critical. This can be accomplished by using nanofertilizers. Nanofertilizers are NMs that can provide nutrients to plants, enhancing crop development and output without giving nutrients directly to the plants (Liu and Lal 2015). Nanoencapsulation of fertilizers provides high surface tension, binds more strongly, and possesses properties like solubility, stability, and specificity by providing desired concentration and reduce the toxicity for safe handling and disposal (Green and Beestman 2007). Nanofertilizers have previously been proven to be beneficial in a number of studies. When dissolved in water, nanofertilizer releases soluble ions of various nutrients which absorbs into the soil. Plants absorb such nutrient ions supplied by nanofertilizers because the soluble nutrient ions are similar to dissolved traditional fertilizers.

6.4 Precision Farming

Precision farming is described as using precise management practices at actual locations, at particular rates, and at perfect times to improve long-term site-specific multiple cropping performance and effectiveness using integrated relevant information and an improved agricultural system while reducing unforeseen ecological effects. The success of the project is dependent on an accurate assessment of localized crop environmental conditions, as well as the identification of input demands that vary by location. These nanotechnology-enabled monitoring systems, as well as tiny sensors, will have a significant impact on future precision agricultural methods. Appropriate sensing of location-specific differences in management practices, edaphoclimatic conditions as well as data acquisition, modification, storage, and exchange are therefore critical in dealing with this fluctuation and fine-tuning management operations accordingly. GPS and GIS are two technologies that have emerged from the electronics revolution over the previous several decades. A variety of sensors, monitors, and processors are used to teach precise movements of equipment, deliver at precise locations, apply fertilizers and pesticides, and evaluate all data in conjunction with agro-climatic data for precision crop environment monitoring. Nanosensors are essential in precision farming for monitoring the signalling events between crop and soil environment, determining nutrition status, and detecting infections in crops, among other things. Physical, chemical, and biological nanosensors are the three types of nanosensors. Wireless nanosensor systems are necessary for actual and continuous monitoring in precision farming. There are several components of the wireless nanosensor with diverse application areas. Under a small transmission range, nano-nodes conduct simple calculations with limited memory. Nanocommunication is a rapidly growing research subject focused on the advancement of novel communication strategies for nanodevices in agriculture (Dressler and Kargl 2012). Communication amongst nanodevices is a fundamental difficulty in the design and study of wireless nanosensor networks. However, developing gateways and organizing and managing network structures over the Internet are difficult tasks that are still being researched.

Some applications of nanosensors in precision agriculture are as follows:

- **Smart dust sensors** for remote monitoring of microclimates (ETC group 2004).
- **Agrinfortronics** (Opara 2002) to store data on the information on the communication technology (precise and consistence data on the crop management environment).
- **Ambient intelligence**, a more advanced form of intelligence targeted at creating smart settings that use artificial intelligence (AI) and sensors to anticipate and respond to individual requirements (Cook et al. 2009).

6.5 Restoration of Environmental Health

Pollutant degradation and sequestration are two types of environmental clean-up. The use of NMs for remediation, i.e. nanoremediation, would be more efficient and cost-effective. Nanoremediation (the use of NPs for the removal of pollutants) can be used to clean groundwater and freshwater, wastewater, soil, sediment, or other contaminants, as well as to control air pollution (Crane and Scott 2012). The reactive NPs are delivered to the contaminated site by groundwater. When NPs come into touch with pollutants, they can bind to them, immobilize them, and breakdown them into less harmful and mobile molecules by adsorption or complexation. The cost of drilling and packing a well is high. Many reactive NPs' performance has been assessed. For environmental remediation, nanosized zeolites, metals, CNTs, metal oxides, etc., have been employed (Pramanik et al. 2020). Air pollution can be controlled using NP-based filtration approaches. Nanofilters in automobiles and industrial smokestacks will be able to remove impurities and let them free into the atmosphere. Even at extremely low concentrations, nanosensors can detect harmful gas escape. There are different categories of NMs for environmental remediations as shown in Table 6.1.

Table 6.1 Different types of NMs for environmental remediations.

Nanomaterials	Applications	References
AgNPs	Water disinfectant.	Xiu et al. (2012) Pal et al. (2007)
TiO_2 NPs	Water soil disinfectant, removal of aromatic hydrocarbon, and pesticides.	da Silva et al. (2016) Gu et al. (2012)
Metal-dopped TiO_2	Removal of gaseous molecules.	Li et al. (2014)
Binary mixed oxide	Removal of dyes from water.	Sreeja and Shetty (2016) Rasalingam et al. (2014)
Fe-based NPs	Removal of heavy metal and organic solvents.	Han and Yan (2016) Mystrioti et al. (2016)
Bimetallic NPs	Removal of chlorinated and brominated contaminants from soil and water.	Zhang et al. (2010) Xie et al. (2014)
Silica-mediated NPs	Treatment of heavy metal and dyes in wastewater. Treatment of gases like, H_2S, CO_2, aldehydes, and ketones.	Nakanishi et al. (2015) Huang et al. (2003) Afkhami et al. (2010) Nomura and Jones (2013) Kuwahara et al. (2012)
Polymer-based nanomaterials	Removal of polynuclear aromatic hydrocarbon from soil, heavy metals from water, volatile organic compounds from gas, etc.	Tungittiplakorn et al. (2004) Guerra et al. (2018) Mittal et al. (2015) Ge et al. (2012)

6.6 Agricultural Application of Nanotechnology

Agriculture's trajectory is uncertain due to its reliance on natural variables and soil systems. As a result, sensing, collecting, analysing, and storing precise data of all biotic and abiotic factors are critical for eradicating the problems of food quality and quantity while ensuring environmental sustainability. Incorporating nanotechnology into agriculture aids in the achievement of better results. Farmers could possibly increase their agricultural production by providing robust and portable nanotechnology-based machines. Plant–NP interactions could be biologically important, involving cell wall transport, the formation of ROS, lipid peroxidation, ion cell disturbance, and oxidative breakage. Plant development and germination have been demonstrated to be greatly aided by NMs (Vijaya et al. 2011).

NMs not only help micro-organisms degrade waste and harmful materials more efficiently, but also help them degrade waste and toxic materials more efficiently. Toxins and hazardous compounds in agricultural soil and water are broken down or removed via bioremediation. Bioremediation (helpful bacteria), phytoremediation (plants), and mycoremediation (mycoremediation) are some of the other words that are commonly used (fungi and mushrooms).

Contaminants in soil and water could be removed by micro-organisms in an environmentally friendly and efficient manner through bioremediation (Dixit et al. 2015).

6.7 Nanofertilizers and Nanopesticides

Nanofertilizers enhance the accessibility and nutrient enrichment to different parts of a plant, resulting in better vegetative growth, quality, and yield. When employed at the right quantities, nanofertilizers have the ability to easily penetrate seeds and increase the nutrient supply to the growing seedling, resulting in increased root and shoot length (Tapan et al. 2010). Nanofertilizers may include nano-based different types of metals and metal oxides, as well as quantum dots (QDs), nanorods, etc. In the last decade, extensive research on the absorption, biological fate, and toxicity of many metal oxide NPs, including Al_2O_3, TiO_2, CeO_2, FeO, and ZnONPs, has been conducted for agricultural production (Zhang et al. 2016). Soil alkalinity and zinc deficiency have been reported as the key challenges limiting agricultural productivity (Sadeghzadeh 2013). Sustainable bioeconomy using different bioprocesses and renewable resources is continuing to reduce and replace toxic materials in ongoing applications and thus will play an important role in developing new technologies in future (Marzbani et al. 2015). The accumulation of knowledge in the domains of biological sciences and engineering brings up opportunities for increasing biomass productivity as well as highly efficient utilization of biomass and organic wastes.

In coming years, the need and utilization of NMs in agriculture and food industry will be understood. Insect pests are well known in agricultural fields and in farm produce; thus, NPs may play an important role in the management of pests and host-specific infections (Khot et al. 2012). A nanoencapsulated pesticide formulation with improved targeted application has been developed (Bhattacharyya et al. 2016). These benefits are mostly obtained by either preventing early degradation of the encapsulated active components or boosting their pest control activity for a prolonged period of time. The development of nanoencapsulated pesticides produced a reduction in pesticide dosage and human exposure, making crop protection more environmentally benign (Nuruzzaman et al. 2016). Distribution of chemical products to the targeted biological process is being developed using microencapsulation-like nanoencapsulation. Fertilizer manufacturers have recently promoted nanoscale pesticides to the market as 'microencapsulated pesticides', despite the fact that they are really nanoscale emulsions. This approach is generally used for organic NPs comprising active agrochemicals or targeted molecules (Gouin 2004). Nanotechnology-based specific applications have the significant engagements of using nanotechnology in agriculture (Prasad et al. 2017).

6.8 Nanotechnology for Food Industry

Aside from agriculture, nanotechnology offers various uses in the food business, from packaging and maintenance to smart distribution (Table 6.2). Consumers can customize and transform their food according to special taste and nutritional needs that enable nanotechnology into different manufacture and preservation strategies (Singh et al. 2017).

Table 6.2 Some applications of nanomaterials in food industry.

Types and area application	Use of the nanomaterials
Design of nano-based materials (composites, biocomposites, NPs, and emulsions)	Self-assembling, self-healing, and self-manipulation features are included in the novel product.
Nanosensors and nanobiosensors for food safety and food control	Sensor evaluation, detection of foodborne pathogen, detection of very low concentrations of contaminants, checking food quality and labelling of items, sensor
Food processing using nanotechnology, including filtration, enzymatic reactors, heat and mass transfer, manufacturing, and nanocapsules to change nutrient absorption	Process knowledge, increased high-temperature-resistant packaging, nanoceramic pan for shorter roasting time and oil consumption, higher usage of plant oil for reduced trans-fatty acid consumption, and insertion of nanocapsules into food for systemic bioavailability of required nutrients
Food packaging	Nanocomposites are employed as coatings, release devices, and unique packaging, altering the permeation behaviour of foils, improving mechanical and heat resistance behaviour, and boosting mechanical, thermal, chemical, and microbiological properties. Nanomycells are utilized in nutrition nanotherapy and regulated release of nutrients and bioactives and examined by nanoencapsulation and nanoscale enzymatic reactor to generate novel products and health promoters with fortified food items.

Source: Adapted and modified from Pramanik et al. (2020).

To improve the quality of the food, nanocomposites are being used in food industry as an antimicrobial coating agent (Grillo et al. 2016). They also have the possibility to improve plant tolerance to thermal and mechanical stress while lowering oxygen transport rates. When introduced to food, nanocapsules containing specialized nutrients such as vitamins, colourants, and flavour enhancers go dormant and release nutrients when needed by the body (Katouzian and Jafari 2016). Nanoscale biosensors can help in pathogen detection and diagnosis. While expanding nanoscale knowledge of food components, nanotechnology has the potential to give bioactive substances in foods to hosts (Martirosyan and Schneider 2014). It also helps with the creation of nanoscale filtration devices for manipulating food texture. Nanodevices interact with food to provide a pleasing surface that preserves the food's glaziers and colours, as well as magnetic nanoparticles for tag sensors (Savina et al. 2016). Recently, various nutrients, mostly vitamins, have been encapsulated and transported into the bloodstream with high efficiency through the digestive system. These NPs were added to some foods and beverages without altering their taste or appearance. In ice cream, NP emulsions are employed, and nanoemulsion spreads can improve the texture and consistency of the ice cream (Berekaa 2015). The nano- and microscale particles are being utilized in the sectors like food technology and preservations.

6.9 Microbial Techniques for Bio-Transforming Food and Agricultural Waste into Resources

Food waste from diverse sources has increased because of lifestyles and unplanned urbanization of the global population. Each year, nearly one-third of all food production is being wasted, according to the Food and Agriculture Organization (FAO). This is a major source of concern since not only are valuable resources being lost, but also their disposal in the environment is also a matter of concern. Due to the high organic content of food waste, traditional methods could pose a serious environmental and human health risk by emitting hazardous gases. As a result, using biological approaches to handle such waste provides a long-term solution for valorization. Cereals account for a large component of the human diet on a global scale. India is the world's largest producer of pulses, and therefore the processing produced a large number by-product (Parate and Talib 2015). Husk is reprocessed in a variety of methods to create high-end materials. As a by-product of processing, husk is produced by a variety of crops. Aside from animal feed, by-products from cereal crops are used in traditional ways to make several daily useful household items. Fruit and vegetable waste must be reduced in order to alleviate rising food demand and improve the efficiency throughout the supply chain (Matharu et al. 2016). Microbial treatment of biowaste has opened up new possibilities for value-added products from waste produce. Several high-end goods have been reported to be made from food waste. India is the world's largest milk producer that results in 3.739–11.217 million m^3 of waste per year for processed milk. A large number of nutrients, such as carbohydrates, lipids, and proteins, are found in wastewater. Milk proteins break down, releasing organic and inorganic nitrogen in the form of nucleic acids, urea, proteins, and other compounds (Kushwaha et al. 2011). This causes significant ecological effects and necessitates the use of microbially assisted waste management methods (Dias et al. 2014). Dairy products are a good substrate for *Saccharomyces cerevisiae* to make ethanol from fermentable sugars via enzymatic hydrolysis (Parashar et al. 2016). Asia produces 4.4 billion tonnes of solid trash each year. In India, more than 350 million tonnes of agro-industrial waste are generated each year from various sources (ICAR 2020). Agriculture wastes are economical and natural substitute for a variety of high-value products due to its chemical, physical, and biodegradability features. Microbes may easily feed on agricultural and food waste to produce colours, phytochemicals, antibiotics, and a variety of enzymes.

6.10 Bioconversion of Food Waste to Wealth

The current strategies of food and agricultural waste management failed to reduce the ecological and economic stress; thus, valorization is of interest. Waste could be transformed to different forms of value-added products.

6.10.1 Biosurfactants

Biosurfactants are biologically derived surface-active chemicals synthesized by a variety of micro-organisms, including bacteria, fungus, and yeast (Gaur et al. 2019). This rising need has prompted researchers to look at using low-cost waste materials as substrates, lowering production costs while also requiring waste management. As a result, waste from households and other sources could be used as a substrate for manufacturing biosurfactants, lowering production costs and reducing pollution.

6.10.2 Organic Fertilizers

Inorganic fertilizers contribute more in methane emissions, whereas organic fertilizers are a viable alternative for increasing agricultural yield while lowering methane emissions. Food and agricultural wastes have been using as animal feed and organic fertilizer, which has been composted and used as a substrate for mushroom production (Sharma et al. 2020). Using different methods, such as anaerobic digestion, composting, and chemical hydrolysis, organic fertilizers, can be made from food waste. Those materials can be directly administered to soil in an irrigation system and are thus more beneficial to plant growth than other fertilizers without causing environmental harm.

6.10.3 Bioplastics

In the past, plastics were made from petrochemicals using irreversible procedures (Tsang et al. 2019). In recent years, bioplastic has been found to be an alternative of that and more sustainable. Bioplastics can be made by micro-organisms that use a variety of waste items and are often preferred over synthetic ones because they reduce environmental impact while remaining environmentally benign. Substitution with bioplastics significantly reduces global warming concerns because petroleum-based synthetic plastic manufacture requires more energy (77 MJ/kg) than bioplastics (57 MJ/kg) (Gironi and Piemonte 2011). Considering the negative impact of landfilling of biowaste on groundwater and greenhouse gas emission, bioconversion of food and agricultural waste to make bioplastics is the best waste disposal approach.

6.11 Biowaste and its Products for Agricultural Sustainability

There has been a rapid use of synthetic fertilizers to increase agricultural output, which has resulted in a rapid deterioration of soil health, resulting in plant yield stagnation (Pathak et al. 2003). In recent years, academics have been paying increasing attention to composting and integrated vermicomposting. These processes can degrade organic waste materials, especially biowaste, resulting in nutrient-rich organic fertilizers, like vermicompost. Vermicompost has been shown to boost the growth and food production of the plants (Jayakumar et al. 2011; Najar et al. 2015; Ravindran et al. 2019). As a result, there is a lot of potential for using composting and integrated vermicomposting technologies to enrich nutrients and reduce pathogens in biowaste. Biogas can be made from a wide range of biowastes, and the resulting slurry can be used as an organic fertilizer or a nitrogen-rich supplement in composting and vermicomposting. Different composting processes can create organic fertilizer with improved agronomic qualities with the suitable bulking components, microbial agents, and nutrient supplements. Biochar with more functionalized surface ligands can be made from biowaste from agro-industries and could be used in heavy metal remediation. Biochar-integrated biowaste vermi-conversion is successful in nutrient management and produces vermicompost with less toxic chemical.

6.12 Post-Harvest Agriculture Waste Management Using NPs

Nanotechnology in agriculture has a significant possibility in post-harvest applications. The research in these fields is diverse, and it frequently necessitates in-depth knowledge of the application areas in question. Sustainability and a desire to reduce reliance on fossil-derived raw resources are two factors driving this interest. Nanocellulose, which may be made from different agri-based sources, is an excellent example in the context. Nanocellulose has been shown to be wise in different applications, including filters, aerogels, thickeners, flavour carriers, and stabilizers.

The majority of contemporary research on nanotechnology applications focuses on nanopesticides, nanofertilizers, nanosensors, and crop protection. Agricultural waste provides an abundant source of lignocellulose, which can be used to make useful NMs (Othman 2014; Shahabi-Ghahafarrokhi et al. 2015) and biochar. Cereal husks may be utilized to prepare value-added items, reducing waste disposal issues (Kaur et al. 2016). Electrospinning may turn cotton fibres and cellulose into high-value products (Ramakrishna et al. 2006). Cellulose is a common biopolymer and a crucial component of the paper business. It can be used to create nanocellulose and natural fibres, as well as to effectively apply them (Dai and Fan 2013). Rice husk, which is a rich source of nano and micro-silica, can be used to extract silica NPs, an inorganic element. Validamycin-based nano-silica is an effective way to distribute water-soluble pesticides in a controlled manner (Liu et al. 2006). Methyl methoxy silane-based nano-silica has been suggested for plant growth regulator and nano-fungicide (Huang et al. 2015). Biochar could be employed as a catalyst, a composite filler, a filtering agent, a water treatment agent, or a soil conditioner (Peterson et al. 2013). Carbon NPs were found in biochar, and they appeared to help wheat plants develop faster. Rice peel was also used to make biochar NPs through a hydrothermal carbonization technique. Biochar's structure has been revealed to have nanopores that could help with hydrogen storage (Cheng et al. 2016). Nanotechnology allows for better feed formulations, vaccinations, diagnostics, and treatments for livestock. The possibilities for post-harvest usage are diverse; several functions are specific and involve in-depth expertise and skill in the topics. However, it is still to be explored.

6.13 Sustainable Agriculture in India: Some Policy Recommendation

Transitioning to a cleaner and sustainable food production, existing agricultural policies must be aligned with UN Sustainable Development Goals (SDGs), especially for an edapho-climatically varied country like India. In 2015, the United Nations devised the SDGs, focusing two of them exclusively on the concerns of poverty, hunger, and malnutrition, as well as increasing agricultural productivity while minimizing damage to natural resources (www.unsdsn.org). Incorporating sustainability principles into traditional agricultural systems should aimed at enhancing agricultural productivity while reducing negative environmental impacts. Nevertheless, the mechanism for achieving this change is still up for question, as organic farming methods, while environmentally friendly, result in poorer yields. Meanwhile, farm intensification and mechanization boost output but have detrimental consequences for soil quality and agrobiodiversity (Eyhorn et al. 2019):

- Policy interventions for sustainable transition.
- Policy framework for sustainable agriculture.
- Coherence between sustainable developmental goal indicator and available options.
- Present role of effective policy framing for sustainable agriculture.
- Addressing the gaps between agriculture practices and government policies.
- Restructuring of policies based on scientific evidence.

6.14 Risk and Safety Aspects of Nanotechnology in Agriculture

Nano-based application in agriculture can be successful if standard techniques are followed in a more advanced scientific manner. Given that technique may enable for nanometre-scale assembly control, a variety of innovative plausible outcomes in improving precision techniques are possible. The use of nanotechnology to develop new approaches for modern agriculture allows for controlled administration of agrochemicals such as fertilizers, insecticides, herbicides, and plant growth regulators. Most advantage application would be in agriculture applications using nanoscale materials. Despite extensive research, a little improvement has been achieved in determining the risk and health problems associated with NM. The majority of NP applications in the agri-food industry lack a transparent and accurate risk governance framework across all sectors. Safety-by-design synthesis techniques have been the focus of the establishment of harmonized regulatory frameworks in the field of nanosafety (Kraegeloh et al. 2018). Studies have shown that cellulose nanocrystals from agricultural substances open up a completely new dimension for cellulosic NMs by allowing them to be used as a component of nanocomposite (Leistritz et al. 2007). Carbon nanotubes could help plants grow faster by increasing component absorption and supplement usage (Baruah and Dutta 2009). The advancement of nanofabrication and characterization techniques has enabled physicochemical and natural relationships between plant cell organelles and diseases. Perception on the plant pathogenic processes will lead to develop treatment mechanisms to control infection and protect the plant (Sangeetha et al. 2017). A detailed understanding of plant pathogenic interaction may lead to improved treatment procedures for infection control and plant security. Crop genome research is focusing on producing potentially enhanced plant types that are resilient to various stress circumstances such as droughts, salt, diseases, and others.

Various forms of NPs have been examined for its nano-ecotoxicity and risk assessment. This is due to the widespread usage of these NPs in a variety of industries, including agriculture. Although NPs in agriculture are still in their infancy, they have the potential to impact on the entire food system, from production through manufacturing, protection, hygiene, packaging, distribution, storage, and supply. Because of their great chance to reach the human circulatory transportation network and interface with essential organs, in vivo and in vitro toxicity testing have been conducted to ensure health and ecological safety (Mwaanga 2018). Growth promoters are commonly incorporated in nanofertilizers, allowing for delayed, targeted, and efficient release into the fields. Nanopesticides are often formulated in the same way as nanofertilizers, with active ingredients encapsulated in a nanoscale product. These nanopesticides have shown to increase stability and solubility which allow them to release the encapsulated products and effective pest management targeting (Bhattacharyya et al. 2016). Herbicides with target-specific smart delivery method that mix with soil effectively suppress the weeds' growth that has developed resistant to traditional herbicides. Nanoherbicides will be effective in weed management programmes if they are used strategically (Ditta 2012).

The most diversified class of NMs is inorganic-based NPs. Metals, metal oxides, and QDs are examples of NPs in this category. They offer a wide range of features which allow them to be used in a variety of disciplines such as medical, engineering, the environment, and agriculture. As previously noted, the use of large amounts of pesticides and fertilizers in agriculture causes major environmental contamination with negative consequences for humans, and these agrochemicals can sometimes influence the flavour and nutritional content of food crops. Nanotechnology plays a sophisticated and intelligent role in nanosystems that can give the necessary supplements to plants, as well as nanoencapsulated slow-discharge fertilizers and insecticides that can deposit the correct quantities at controlled rates. Organically derived NPs are also widely used in agriculture. Nanotechnology's prospective applications and benefits in agriculture are undeniable. Many more nanoproducts will be developed as a part of research in agriculture progresses to benefit the entire agricultural value chain. However, the current toxicity testing techniques may not be enough to provide useful information on the NPs for risk assessment design. Because the dose metric may introduce some uncertainty, risk evaluations for NPs must ensure that all elements associated with a given NP type that led to hazardous consequences are well recognized. This is especially essential because various NPs have varied properties, and some may even be biopersistent. The absence of proven procedures and methods for detecting the presence of NPs in the food matrix contributes to some uncertainty. Understanding the benefits and potential hazards of various forms of NPs in agricultural food necessitates their detection and final characterization. Risk assessment techniques for both aquatic and terrestrial creatures have substantially aided in the knowledge of the effects of NPs on organisms and, to a considerable extent, have supplied trustworthy information for the establishment of nanotechnology safety guidelines in general.

Future nanotechnologies in agriculture appear to be both intriguing and promising. However, a lack of sufficient information on key risk evaluation aspects still exists. The health and environmental risks posed by NMs were emphasized as being difficult to measure due to their various features and behaviour. The artificial usage and probable release of NMs have come before an assessment of the harm to the ecosystem, including humans.

6.15 Conclusion and Future Implications

Agrochemicals, fertilizers, and insecticides have undoubtedly contributed significantly to the growth and expansion of agricultural productivity. Despite only a slight increase in cultivated area, food production has expanded at an unprecedented rate during the last two decades. Despite the numerous advantages of agrochemicals in terms of increased agricultural production, their excessive use has resulted in significant degradation of terrestrial and aquatic environments. Nanotechnology and bioprocessing technologies have the potential to improve disease resistance, plant growth, waste management, and nutrient usage. Nanoencapsulated products demonstrate the ability to apply pesticides, insecticides, and herbicides more effectively and locally in an environmentally benign, greener manner. It has been effectively utilized in post-harvest to preserve the freshness, quality, and shelf life of stored products, as well as to prevent disease outbreaks in a relatively safe manner. NMs are a relatively new technology in agriculture, and more study is needed. Nanotechnology's usage in agriculture has social and ethical implications that must be explored. The toxicity of NMs must be assessed before they can be commercialized and used in the field. NMs have the potential to harm the ecosystem and human health. According to scientists and researchers, the potential concerns associated with NPs released into the ecosystem are somewhat hazy. From a public standpoint, the application of NPs in agriculture poses a number of unpredictability and ethical problems. The risk associated with human and environmental health of several types of NPs has been studied. The risk factor assessments of various NPs proposed for use in agriculture prevail as an unsettled subject.

As other businesses compete for the same agricultural-related bioresources, resource availability will continue to diminish. In agriculture, however, nothing goes to waste. Increased nutrient usage efficiency can be achieved by increasing the sustainable utilization of agricultural wastes through redistribution, recovery, value addition, and other methods. For it to succeed on the ground, public opinion and effective regulatory mechanisms are essential. Finally, micro-organisms play a critical role in optimizing resource utilization.

Acknowledgement

PR is thankful to the editors of the book for the invitation to contribute a book chapter.

Conflict of Interest

The author declares no conflict of interest.

References

Afkhami, A., Saber-Tehrani, M., and Bagheri, H. (2010). Simultaneous removal of heavy-metal ions in wastewater samples using nano-alumina modified with 2,4-dinitrophenylhydrazine. *J. Hazard. Mater.* 181 (1–3): 836–844.

Banotra, M., Kumar, A., Sharma, B.C. et al. (2017). Prospectus of use of nanotechnology in agriculture – a review. *Int. J. Curr. Microbiol. App. Sci.* 6 (12): 1541.

Baruah, S. and Dutta, J. (2009). Nanotechnology applications in pollution sensing and degradation in agriculture: a review. *Environ. Chem. Lett.* 7 (3): 191–204.

Berekaa, M.M. (2015). Nanotechnology in food industry; advances in food processing, packaging and food Safety. *Int. J. Curr. Microbiol. App. Sci.* 4: 345–357.

Bhattacharyya, A., Duraisamy, P., Govindarajan, M. et al. (2016). Nano-biofungicides: emerging trend in insect pest control. In: *Advances and Applications Through Fungal Nanobiotechnology*, Fungal Biology (ed. R. Prasad), 307–319. Cham: Springer.

Cheng, H.N., Klasson, K.T., Asakura, T., and Wu, Q. (2016). Nanotechnology in agriculture. In: *Nanotechnology: Delivering on the Promise Volume 2* (ed. H.N. Chen), 233–242. American Chemical Society.

Cook, D.J., Augusto, J.C., and Jakkula, V.R. (2009). Ambient intelligence: technologies, applications, and opportunities. *Pervasive Mob. Comput.* 5 (4): 277–298.

Crane, R.A. and Scott, T.B. (2012). Nanoscale zero-valent iron: future prospects for an emerging water treatment technology. *J. Hazard. Mater.* 211: 112–125.

Dai, D. and Fan, M. (2013). Green modification of natural fibres with nanocellulose. *RSC Adv.* 3: 4659–4665.

DESA (2015). United nations department of economic and social affairs, population division. World population prospects: The 2015 revision, key findings and advance tables. Online Edition UN DESA, New York.

Dias, T., Fragoso, R., and Duarte, E. (2014). Anaerobic co-digestion of dairy cattle manure and pear waste. *Bioresour. Technol. 164*: 420–423.

Ditta, A. (2012). How helpful is nanotechnology in agriculture? *Adv. Nat. Sci. Nanosci. Nanotechnol.* 3 (3): 033002.

Dixit, R., Malaviya, D., Pandiyan, K. et al. (2015). Bioremediation of heavy metals from soil and aquatic environment: an overview of principles and criteria of fundamental processes. *Sustain. For.* 7 (2): 2189–2212.

Dressler, F. and Kargl, F. (2012). Towards security in nano-communication: challenges and opportunities. *Nano Commun. Netw.* 3 (3): 151–160.

ETC group (2004). http://www.etcgroup.org/documents/ETC_DOTFarm2004.pdf. Down on the farm (accessed 10 January 2022).

Eyhorn, F., Muller, A., Reganold, J.P. et al. (2019). Sustainability in global agriculture driven by organic farming. *Nat. Sustain. 2* (4): 253–255.

Gaur, V.K., Regar, R.K., Dhiman, N. et al. (2019). Biosynthesis and characterization of sophorolipid biosurfactant by Candida spp.: application as food emulsifier and antibacterial agent. *Bioresour. Technol. 285*: 121314.

Ge, F., Li, M.-M., Ye, H., and Zhao, B.-X. (2012). Effective removal of heavy metal ions Cd^{2+}, Zn^{2+}, Pb^{2+}, Cu^{2+} from aqueous solution by polymer-modified magnetic NPs. *J. Hazard. Mater.* 211: 366–372.

Gironi, F. and Piemonte, V. (2011). Bioplastics and petroleum-based plastics: strengths and weaknesses. *Energy Sources, A: Recovery, Util. Environ. Eff. 33* (21): 1949–1959.

Glenn, J.C. and Florescu, E. (2016). The millennium project team, 2015–2016 state of the future. *J. Social.* 5 (3): 1–6.

Gouin, S. (2004). Microencapsulation: industrial appraisal of existing technologies and trends. *Trends Food Sci. Technol.* 15: 330–347.

Green, J.M. and Beestman, G.B. (2007). Recently patented and commercialized formulation and adjuvant technology. *Crop Prot.* 26 (3): 320–327.

Grillo, R., Abhilash, P.C., and Fraceto, L.F. (2016). Nanotechnology applied to bio-encapsulation of pesticides. *J. Nanosci. Nanotechnol.* 16 (1): 1231–1234.

Gu, J., Dong, D., Kong, L. et al. (2012). Photocatalytic degradation of phenanthrene on soil surfaces in the presence of nanometer anatase TiO_2 under UV-light. *J. Environ. Sci.* 24 (12): 2122–2126.

Guerra, F.D., Campbell, M.L., Whitehead, D.C., and Alexis, F. (2018). Capture of aldehyde VOCs using a series of amine-functionalized cellulose nanocrystals. *ChemistrySelect* 3: 5495–5501.

Han, Y. and Yan, W. (2016). Reductive dechlorination of trichloroethene by zero-valent iron NPs: reactivity enhancement through sulfidation treatment. *Environ. Sci. Technol.* 50 (23): 12992–13001.

Huang, H.Y., Yang, R.T., Chinn, D., and Munson, C.L. (2003). Amine-grafted MCM-48 and silica xerogel as superior sorbents for acidic gas removal from natural gas. *Ind. Eng. Chem. Res.* 42: 2427–2433.

Huang, S., Wang, L., Liu, L. et al. (2015). Nanotechnology in agriculture, livestock, and aquaculture in China. *A review. Agro. Sustain. Dev.* 35: 369–400.

ICAR (2020). *Creating Wealth from Agricultural Waste*, 172. New Delhi: Indian Council of Agricultural Research.

Jayakumar, M., Sivakami, T., Ambika, D., and Karmegam, N. (2011). Effect of turkey litter (*Meleagris gallopavo* L.) vermicompost on growth and yield characteristics of paddy, *Oryza sativa* (ADT-37). *Afr. J. Biotechnol. 10* (68): 15295–15304.

Katouzian, I. and Jafari, S.M. (2016). Nano-encapsulation as a promising approach for targeted delivery and controlled release of vitamins. *Trends Food Sci. Technol.* 53: 34–48.

Kaur, T., Singh, G.P., Kaur, G. et al. (2016). Synthesis of biogenic silicon/silica (Si/SiO_2) nanocomposites from rice husks and wheat bran through various microorganisms. *Mater. Res. Expr.* 3: 085026.

Khot, L.R., Sankaran, S., Maja, J.M. et al. (2012). Applications of nanomaterials in agricultural production and crop protection: a review. *Crop Prot.* 35: 64–70.

Kraegeloh, A., Suarez-Merino, B., Sluijters, T., and Micheletti, C. (2018). Implementation of safe-by-design for nanomaterial development and safe innovation: why we need a comprehensive approach. *Nano* 8: 239.

Kumbhakar, P., Ray, S.S., and Stepanov, A.L. (2014). Optical properties of NPs and nanocomposites. *J. Nanomater.* 2014: 181365.

Kushwaha, J.P., Srivastava, V.C., and Mall, I.D. (2011). An overview of various technologies for the treatment of dairy wastewaters. *Crit. Rev. Food Sci. Nutr. 51* (5): 442–452.

Kuwahara, Y., Kang, D.Y., Copeland, J.R. et al. (2012). Dramatic enhancement of CO_2 uptake by poly(ethyleneimine) using zirconosilicate supports. *J. Am. Chem. Soc.* 134: 10757–10760.

Leistritz, F.L., Hodur, N.M., Senechal, D.M. et al. (2007). Biorefineries using agricultural residue feedstock in the great plains (No. 1193-2016-94551).

Li, D., Cui, F., Zhao, Z. et al. (2014). The impact of titanium dioxide NPs on biological nitrogen removal from wastewater and bacterial community shifts in activated sludge. *Biodegradation* 25 (2): 167–177.

Liu, R. and Lal, R. (2015). Potentials of engineered nanoparticles as fertilizers for increasing agronomic productions. *Sci. Total Environ. 514*: 131–139.

Liu, F., Wen, L.X., Li, Z.Z. et al. (2006). Porous hollow silica NPs as controlled delivery system for water-soluble pesticide. *Mater. Res. Bull.* 41: 2268–2275.

Martirosyan, A. and Schneider, Y.J. (2014). Engineered nanomaterials in food: implications for food safety and consumer health. *Int. J. Environ. Res. Public Health* 11 (6): 5720–5750.

Marzbani, P., Afrouzi, Y.M., and Omidvar, A. (2015). The effect of nano-zinc oxide on particleboard decay resistance. *Maderas Cienc. Tecnol.* 17: 63–68.

Matharu, A.S., de Melo, E.M., and Houghton, J.A. (2016). Opportunity for high value-added chemicals from food supply chain wastes. *Bioresour. Technol. 215*: 123–130.

Mishra, S., Singh, A., Keswani, C., and Singh, H.B. (2014). Nanotechnology: exploring potential application in agriculture and its opportunities and constraints. *Biotech Today* 4 (1): 9–14.

Mittal, H., Maity, A., and Ray, S.S. (2015). Synthesis of co-polymer-grafted gum karaya and silica hybrid organic–inorganic hydrogel nanocomposite for the highly effective removal of methylene blue. *Chem. Eng. J.* 279: 166–179.

Mukhopadhyay, S.S. (2014). Nanotechnology in agriculture: prospects and constraints. *Nanotechnol. Sci. Appl.* 7: 63.

Mwaanga, P. (2018). Risks, uncertainties, and ethics of nanotechnology in agriculture. In: *New Visions in Plant Science* (ed. Ö. Çelik). IntechOpen 22(3).

Mystrioti, C., Xanthopoulou, T.D., Tsakiridis, P. et al. (2016). Comparative evaluation of five plant extracts and juices for nanoiron synthesis and application for hexavalent chromium reduction. *Sci. Total Environ.* 539: 105–113.

Najar, I.A., Khan, A.B., and Hai, A. (2015). Effect of macrophyte vermicompost on growth and productivity of brinjal (*Solanum melongena*) under field conditions. *Int. J. Recycl. Org. Waste Agric. 4* (2): 73–83.

Nakanishi, K., Tomita, M., and Kato, K. (2015). Synthesis of amino-functionalized mesoporous silica sheets and their application for metal ion capture. *J. Asian Ceram. Soc.* 3: 70–76.

Nomura, A. and Jones, C.W. (2013). Amine-functionalized porous silicas as adsorbents for aldehyde abatement. *ACS Appl. Mater. Interfaces* 5: 5569–5577.

Nuruzzaman, M.D., Rahman, M.M., Liu, Y., and Naidu, R. (2016). Nanoencapsulation, nano-guard for pesticides: a new window for safe application. *J. Agric. Food Chem.* 64 (7): 1447–1483.

Opara, L.U. (2002). Agricultural engineering education and research in knowledge-based economy. Aktualni zadaci mehanizacije poljoprivrede. Zbornik radova, 30. međunarodnog simpozija iz područja mehanizacije poljoprivrede, Opatija, Hrvatska, 12-15 ož 2002, 33-46.

Othman, S.H. (2014). Bio-nanocomposite materials for food packaging applications: types of biopolymer and nano-sized filler. *Agric. Agricul. Sci. Proc.* 2: 296–303.

Pal, S., Tak, Y.K., and Song, J.M. (2007). Does the antibacterial activity of silver NPs depend on the shape of the NP? A study of the gram-negative bacterium Escherichia coli. *Appl. Environ. Microbiol.* 73 (6): 1712–1720.

Parashar, A., Jin, Y., Mason, B. et al. (2016). Incorporation of whey permeate, a dairy effluent, in ethanol fermentation to provide a zero waste solution for the dairy industry. *J. Dairy Sci. 99* (3): 1859–1867.

Parate, V.R. and Talib, M.I. (2015). Characterization of tur dal (*Cajanus cajan*) husk carbon and its kinetics and isotherm study for removing Cu (II) ions. *IOSR J. Environ. Sei., Toxicol. Food Technol. 9*: 27–41.

Pathak, H., Ladha, J.K., Aggarwal, P.K. et al. (2003). Trends of climatic potential and on-farm yields of rice and wheat in the Indo-Gangetic Plains. *Field Crop Res. 80* (3): 223–234.

Peterson, S.C., Jackson, M.A., and Appell, M. (2013). Biochar: sustainable and versatile. In: *Advances in Applied Nanotechnology for Agriculture* (ed. B. Park and M. Appell), 193–205. American Chemical Society.

Pouratashi, M. and Iravani, H. (2012). Farmers' knowledge of integrated pest management and learning style preferences: implications for information delivery. *Int. J. Pest Manage.* 58: 347–353. https://doi.org/10.1080/09670874.2012.724468.

Pramanik, P., Krishnan, P., Maity, A. et al. (2020). Application of nanotechnology in agriculture. In: *Environmental Nanotechnology* 4, vol. 32 (ed. N. Dasgupta, S. Ranjan and E. Lichtfouse), 317–348. Cham: Springer.

Prasad, R., Bhattacharyya, A., and Nguyen, Q.D. (2017). Nanotechnology in sustainable agriculture: recent developments, challenges, and perspectives. *Front. Microbiol.* 8: 1014.

Rabbani, M.M., Ahmed, I., and Park, S.J. (2016). Application of nanotechnology to remediate contaminated soils. In: *Environmental Remediation Technologies for Metal-Contaminated Soils* (ed. H. Hasegawa, I. Rahman and M. Rahman), 219–229. Tokyo: Springer.

Ramakrishna, S., Fujihara, K., Teo, W.E. et al. (2006). Electrospun nanofibers: solving global issues. *Mater. Today* 9: 40–50.

Rasalingam, S., Peng, R., and Koodali, R.T. (2014). Removal of hazardous pollutants from wastewaters: applications of TiO_2-SiO_2 mixed oxide materials. *J. Nanomater.* 2014: 617405. 1–42.

Ravindran, B., Lee, S.R., Chang, S.W. et al. (2019). Positive effects of compost and vermicompost produced from tannery waste-animal fleshing on the growth and yield of commercial crop-tomato (*Lycopersicon esculentum* L.) plant. *J. Environ. Manag. 234*: 154–158.

Roy, P. and Choudhury, M. (2021). Environmental susceptibility and nanowaste. In: *Waste Recycling Technologies for Nanomaterials Manufacturing* (ed. M. ASH and A. GAM), 65–87. Cham: Springer.

Sadeghzadeh, B. (2013). A review of zinc nutrition and plant breeding. *J. Soil Sci. Plant Nutr.* 13 (4): 905–927.

Sangeetha, J., Thangadurai, D., Hospet, R. et al. (2017). Agricultural nanotechnology: concepts, benefits, and risks. In: *Nanotechnology* (ed. R. Prasad, M. Kumar and V. Kumar), 1–17. Singapore: Springer.

Savina, E., Karlsen, J.D., Frandsen, R.P. et al. (2016). Testing the effect of soak time on catch damage in a coastal gillnetter and the consequences on processed fish quality. *Food Control* 70: 310–317.

Servin, A., Elmer, W., Mukherjee, A. et al. (2015). A review of the use of engineered nanomaterials to suppress plant disease and enhance crop yield. *J. NP Res.* 17 (2): 1–21.

Shahabi-Ghahafarrokhi, I., Khodaiyan, F., Mousavi, M., and Yousefi, H. (2015). Preparation and characterization of nanocellulose from beer industrial residues using acid hydrolysis/ultrasound. *Fibers Polym.* 16: 529–536.

Sharma, P., Gaur, V.K., Kim, S.H., and Pandey, A. (2020). Microbial strategies for bio-transforming food waste into resources. *Bioresour. Technol. 299*: 122580.

da Silva, M.B., Abrantes, N., Nogueira, V. et al. (2016). TiO_2 NPs for the remediation of eutrophic shallow freshwater systems: efficiency and impacts on aquatic biota under a microcosm experiment. *Aquat. Toxicol.* 178: 58–71.

Singh, M.D., Gautam, C., Patidar, O.P., and Meena, H.M. (2017). Nano-fertilizers is a new way to increase nutrients use efficiency in crop production. *Int. J. Agric. Sci.* 9 (7): 3831–3833.

Solanki, P., Bhargava, A., Chhipa, H. et al. (2015). Nano-fertilizers and their smart delivery system. In: *Nanotechnologies in Food and Agriculture* (ed. M. Rai, C. Ribeiro, L. Mattoso and N. Duran), 81–101. Cham: Springer.

Sreeja, S. and Shetty, V. (2016). Microbial disinfection of water with endotoxin degradation by photocatalysis using Ag@ TiO_2 core shell NPs. *Environ. Sci. Pollut. Res.* 23 (18): 18154–18164.

Suresh, A.K., Pelletier, D.A., and Doktycz, M.J. (2013). Relating nanomaterial properties and microbial toxicity. *Nanoscale* 5 (2): 463–474.

Tapan, A., Biswas, A.K., and Kundu, S. (2010). Nano-fertiliser-a new dimension in agriculture. *Indian J. Fertil.* 6 (8): 22–24.

Tilman, D.J.D.P.M., Knops, J., Wedin, D., and Reich, P. (2002). Plant diversity and composition: effects on productivity and nutrient dynamics of experimental grasslands. In: *Biodiversity and Ecosystem Functioning: Synthesis and Perspectives* (ed. M. Loreau, S. Naeem and P. Inchausti), 21–35. Oxford, UK: Oxford University Press.

Tsang, Y.F., Kumar, V., Samadar, P. et al. (2019). Production of bioplastic through food waste valorization. *Environ. Int. 127*: 625–644.

Tungittiplakorn, W., Lion, L.W., Cohen, C., and Kim, J.Y. (2004). Engineered polymeric NPs for soil remediation. *Environ. Sci. Technol.* 38: 1605–1610.

Vijaya, S., Mrudula, T., Naga Deepth, C., and Sree, V. (2011). Novel applications of nanotechnology in life sciences. *J. Bioanal. Biomed.* 11: 1–4.

Wang, P., Lombi, E., Zhao, F.J., and Kopittke, P.M. (2016). Nanotechnology: a new opportunity in plant sciences. *Trends Plant Sci.* 21 (8): 699–712.

Xie, Y., Fang, Z., Cheng, W. et al. (2014). Remediation of polybrominated diphenyl ethers in soil using Ni/Fe bimetallic NPs: influencing factors, kinetics and mechanism. *Sci. Total Environ.* 485: 363–370.

Xiu, Z.M., Zhang, Q.B., Puppala, H.L. et al. (2012). Negligible particle-specific antibacterial activity of silver NPs. *Nano Lett.* 12 (8): 4271–4275.

Yaqoob, A.A., Parveen, T., Umar, K., and Mohamad Ibrahim, M.N. (2020). Role of nanomaterials in the treatment of wastewater: a review. *Water* 12 (2): 495.

Zabrieski, Z., Morrell, E., Hortin, J. et al. (2015). Pesticidal activity of metal oxide NPs on plant pathogenic isolates of Pythium. *Ecotoxicology* 24 (6): 1305–1314.

Zhang, Z., Shen, Q., Cissoko, N. et al. (2010). Catalytic dechlorination of 2,4-dichlorophenol by Pd/Fe bimetallic NPs in the presence of humic acid. *J. Hazard. Mater.* 182 (1-3): 252–258.

Zhang, Q., Han, L., Jing, H. et al. (2016). Facet control of gold nanorods. *ACS Nano* 10 (2): 2960–2974.

7

Agro-Wastes as Low-Cost Biosorbent for Dyes Removal from Wastewater

Navamallika Gogoi[1], Palas Samanta[2], and Papori Dahutia[3]

[1] *Department of Chemistry, Arunachal University of Studies, Namsai, Arunachal Pradesh, India*
[2] *Department of Environmental Science, Sukanta Mahavidyalaya, University of North Bengal, Dhupguri, West Bengal, India*
[3] *Department of Physics, Dibrugarh University, Dibrugarh, Assam, India*

CHAPTER OUTLINE

Agriculture Waste Management and Bioresource: The Circular Economy Perspective, First Edition.
Edited by Suruchi Singh, Pardeep Singh, Anu Sharma, and Moharana Choudhury.

7.1 Introduction

One of the most crucial political issues in many countries is environmental protection. India is an agribusiness country where 70% of its rural household still relies solely on farming to sustain it (Prasad et al. 2020). The undesirable or invalid materials produced by different agricultural activities are agricultural wastes. The agricultural wastes consist of various sources of high-quantity, repeatable, and biodegradable trash (Dai et al. 2018). Approximately 15% of overall trash generation has been made from agribusiness (Sadh et al. 2018). Globally, it is anticipated that 998 million tonnes of agricultural residues are generated each year. More than 600 tonnes, which resulted in a significant problem with disposal, have been made from farm waste in India (Prusty et al. 2016). In India, 350 million tonnes of agricultural wastes, with green fertilizer used in agriculture, can produce more than 18 000 MW power annually. The treatment of agricultural wastage before releasing into water plays a vital role for healthy biotic and aquatic life for both flora and fauna. Furthermore, agriculture accounts for 21% of all greenhouse gases (GHGs) emissions. Given that the world's population is expected to exceed 10 billion by 2050, agriculture must find a sustainable path. The developed countries get an opportunity to use agro-wastes in the construction industry due to high utility of natural resources and difficulty in the removal of agricultural wastes (Prusty et al. 2016).

Many sectors, such as plastics, rubber, printing, textiles, cosmetics, leather, and many more, employ dyes to colour their goods. Dye is a brightly coloured synthetic/natural substance used as diverse substances. Aqueous solutions are commonly used for dyes (Aragaw and Bogale 2021). In consequence, they have a considerable volume of colourful effluent to deal with. Over 105 tonnes of dyestuff are produced each year from commercial dye supplies with over 10 000 different dyes. About 2% of the dyes made each year are emitted in effluents from the dye-producing industries (Bharathi and Ramesh 2013). Approximately 90% of the entire dye used by the textile industry globally is used up in the production of fabrics. Dyes are released as an effluent flow by the textile industrial activities, and on average, one thousand tonnes of dyes are released each year by the entire textile industry (Bharathi and Ramesh 2013). Pigments-contaminated wastewater discharged into water system causes significant difficulties for the aquatic ecosystem, food chain, and environmental beauty. Water containing dyes will also reflect sunlight, causing bacterial development and photosynthesis to be hindered in aquatic plants. Once the dyes' complex aromatic structures are degraded, the difficulties get much worse because they no longer function effectively in the existence of thermal energy, sunlight, micro-organisms, and chemical oxidation (Ashrafi et al. 2015). Because of this, human health and water quality are at risk, which is a problem of significant concern. Affected enterprises/sectors are obligated to remediate dye-contaminated wastes before throwing away into aquatic ecosystem to preserve the necessity of colour removal. Thus, science bears the burden of working to find appropriate waste treatment techniques. The effects of dyes on organisms exposed to them can be acute and chronic, influenced by time and dye concentration. Irritant contact dermatitis, sensitive to skin, occupational asthma, and mutation can occur due to dyes. This has led to remediation of toxic potentiality of dye substances in order to increase the acceptability of water resources.

Most commonly, researchers use physical (adsorption and membrane treatment), chemical (ion exchange), and biological treatments (flocculation) to clean up these contaminants from aqueous phases (Rosales et al. 2017). Activated carbon adsorption, ion exchange, membrane technology, and coagulation are some traditional methods for removing dyes from wastewater effluents. Activated carbon adsorption has been reported to be most effective ways for removing dyes from wastewater (Rahimian and Zarinabadi 2020) due to wide adsorption capacity. However, even after being exhausted, it remains as an expensive adsorbent due to the increasing price of regeneration (Varghese et al. 2019). As a result, there is a need to look for novel and cost-effective ways to remove dyes. Utilizing agro-waste as cost-effective adsorbents for treating dyes more recently has emerged as a substantial global research focus. Agricultural waste has been studied extensively for its potential use as a treatment for wide range of contaminants including dyes. Actually, the practicality of employing these materials could benefit both the environment and the economy by eliminating waste disposal problem. For example, banana waste may be used to extract zinc ions from aqueous solutions (Houda and Moussa 2017). Cadmium ions were successfully removed from orange peels, according to Salmania et al. (2017). In addition, a considerable number of researchers dedicated themselves to the task of removing thick metal. Melon peel-packed columns were studied by Djelloul and Hamdaoui (2015) for treating methylene blue (MB). Shang et al. (2015) tested the heat treatment of pine cone as a means of removing dimethyl trisulfide from contaminated water to generate a low-cost adsorbent. In additional to this, literature review disclosed that a number of agricultural wastes have been utilize as adsorbents like rice husk, spent tea leaves, and wheat straw (Choi and Yu 2018; Bardhan et al. 2020; Siqueira et al. 2020). However, dye removal investigations in combination dye systems have been few in recent years. Accordingly, the present book chapter will address the feasibility of agricultural trashes as low-cost substance that can remove dyes from wastewater that are present separately or in a mixture.

7.2 Dyes Classification and Their Toxicity

Chemical structure, hue, application, and solution's particle charge are generally used to classify dyes. Dyes are often classified as synthetic and natural dyes. Mineral/animal dyes are the most general sources of natural dyes. But natural dyes were unable to meet the demands of the public as industrial activity advanced. As a result, synthetic dyes have been produced and have steadily supplanted natural colours in different industrial sectors. Synthetic dyes could be both ionic and non-ionic (NI) (Zhou et al. 2019). NI dyes are divided into cationic or anionic (reactive, acid, and direct) dyes, while ionic one is divided into disperse and vat dyes. Dyes categorized by chemical structure are relatively widespread, in addition to the aforementioned categorization (Zhou et al. 2019).

Colour is a key constituent of commercial food production, textile industry, etc. Although dyes provide vibrant hues, they are posing a major threat to human health owing to their discharge into water bodies as waste. The majority of dyes are made up of a mixture of organic molecules such as aromatics, trace heavy metals, and amines (Tan et al. 2015). Dyes are harmful due to their components, and their presence in water bodies interfere with aquatic life's photosynthesis process by blocking sunlight transmission. The majority of dyes have carcinogenic, mutagenic, or teratogenic properties towards our ecosystems, and these properties are quite concerning (Amirza et al. 2017). Dye compounds in wastewater, as far as we know, can cause human dysfunction including central nervous system (CNS) and genital system (Farhadi et al. 2021). All of dyes (acidic, cationic, dispersion, and direct) have the potential to cause benign and malignant tumours as well as bladder cancer. Allergic conjunctivitis, dermatitis, occupational asthma, and rhinitis are all possible side effects of reactive dyes (Tan et al. 2015). Furthermore, most dyes are strongly resistant and non-degradable, leading to long persistency in effluents (Zhou et al. 2019). Because of their detrimental effects and long-term presence in effluents, it is critical to highlight effective dye removal technologies from wastewater.

7.3 Conventional Dye Removal Technologies from Wastewaters

There are numerous techniques reported for the separation of pigments from effluents such as biological, physical, and chemical methods as shown in Figure 7.1.

7.3.1 Physical Methods

Due to its easy separation technique, effectiveness, and economic nature, physical dye elimination technology is commonly used in practice (Zhou et al. 2019). Membrane treatment, ion exchange, adsorption, and coagulation-flocculation are the different types of separation methods frequently used in physical dye separation process. Out of this, adsorption is one of the most commonly applied techniques in numerous fields (Farhadi et al. 2021). To carry out this process, mass transfer machinery is widely utilized. Additionally, minimal quantity of chemicals are applied in physical process in comparison with chemical/biological dyes separation method (Crini and Lichtfouse 2019). Biological approaches for colour removal, which utilize living organisms, tend to be more unpredictable than this technique (Katheresan et al. 2018). However, there are several drawbacks of physical process for treating wastewater; the treatment does not work with some colours, and the restoration is expensive (Shamraiz et al. 2016).

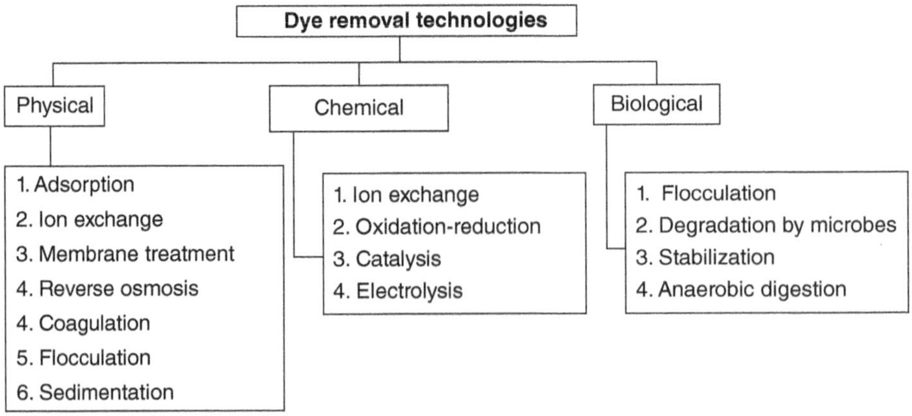

Figure 7.1 Conventional processes for dye removal from wastewaters.

7.3.2 Chemical Methods

Chemical dye sequestration methods are conventionally applied for removing organic contaminants (Ukiwe et al. 2013). Numerous and expensive chemicals required for coagulation and flocculation processes have been described by Nidheesh et al. (2013) to be extensive and costly (Nidheesh et al. 2013). The benefits of the more common chemical methods include their relative efficiency, and the presence of some drawbacks are large-scale consumption of chemicals and reagents, the creation of sludge, and the need for disposal of sludge.

There are various disadvantages in the chemical processes, including the consumption of significant amounts of chemicals and reagents, the requirement for electrical energy, the need for precipitation, sludge development, and its disposal. These chemicals may also lead to pollution via secondary effects. Hence, they cannot be used in large-scale applications due to their excessive electricity consumption to power reactors or equipment wherein chemical dye separation occurs (Aragaw and Bogale 2021; Farhadi et al. 2021).

7.3.3 Biological Methods

Biological colour separation procedures are distinguished based on their varying oxygen requirements and have therefore classed as aerobic and anaerobic. A particular enzyme is required to catalyse oxidation reactions for aerobic biological techniques (Al Prol 2019). Factors such as susceptibility of micro-organisms to seasonal variations, the poisonousness of particular chemicals, and the demand of enormous land expanses for bioreactor creation all impact the performance of biological treatment systems. Natural treatments are much more prevalent, simple, and environmentally friendly than physical and chemical procedures. The usual practice calls for several microbial species, such as bacteria, fungus, and yeast, to be applied to the colour elimination of synthetic dye under aerobic or anaerobic conditions. Nevertheless, in biological process, outer ecological factors like nutrition, temperature, and pH must be maintained to avoid adverse effects. The bioreactor also requires a specific quantity of land, which adds to the process's complexity (Ngulube et al. 2017). Because of this, the use of synthetic dyes is typically not as efficient as that of biological therapy, which is because majority of synthetic dyes have steady and composite chemical composition and are difficult to biological decadence. The concern is also raised about the creation and disposition of extra sludge. One of the fast-developing treatment procedures is adsorption. It is an inexpensive, simple method that has a high-quality output. It also does not produce any hazardous compounds.

7.4 Agricultural Wastes and Characterization

Hemicelluloses, proteins, lignin, simple sugars, lipids, water, starches, and hydrocarbons are the primary components of agricultural wastes, and they contain diverse functional groups namely carbonyl, carboxyl, hydroxyl, ether, and amino acids (Zhou et al. 2019). These groups link molecules/ions to the surfaces. These resources are originated by both natural and man-made activities and are an affordable adsorbent. In addition to this, agricultural wastes are not only ecologically ideal for restoration but also have a solid mechanical structure and chemical stability. Moreover, agricultural wastes offer several benefits, and the primary ones are inexpensive, sustainable, and abundantly available. Accordingly, wastewater treatment can be inexpensively and sustainably achieved by transforming agricultural waste products into cost-effective adsorbents (Aragaw and Bogale 2021). But the chemical composition and nature of agriculture wastes adversely affect adsorption mechanism (Rosales et al. 2017). So, for better adsorption and dye sequestration processes, the characterization of this kind of biosorbent is essential. Table 7.1 summarizes the analytical procedure utilized in different literature for determining adsorbent's surface characteristics and structural functionality.

Table 7.1 Characterization technique used in dye removal process.

Technology	Outcomes
Fourier transform infrared (FTIR) spectroscopy	Gives information about functional group carbon. It also provides information about the mechanism involved.
X-ray photoelectron spectroscopy (XPS)	Identifies adsorbent's chemical composition, empirical formula, and functional units.
Scanning electron microscope (SEM)	Provides information about adsorbent's particle size adsorbent, morphology, surface flaws, and porosity nature.
Brunauer–Emmett–Teller (BET)	Identifies surface area.
Energy-dispersive X-ray spectroscopy (EDX)	Analyses surface elemental composition of carbon.
Transmission electron microscopy (TEM)	Analyses nanoscale resolution and adsorbent's inner morph ology.
X-ray diffraction (XRD)	Analyses crystal structure of adsorbents.
Nuclear magnetic resonance (NMR)	Analyses chemical composition of compound as well as solubility and diffusion.
Vibrating sample magnetometer (VSM)	Identifies magnetic nature of adsorbent.
Boehm titration (BT)	Quantifies O-containing surface as well as pH of adsorbents.
Elemental analyzer (EA)	Identifies elemental composition of a sample.
Zeta potential (ZP) analyzer	Evaluates surface charge strength.
X-ray fluorescence (XRF)	Identifies elemental study of adsorbent.
Raman spectroscopy (RS)	Identifies adsorbent's monolayer and multilayer character.
Electron diffraction (ED)	Identifies nature of electron and crystal structure of solid in adsorbent.
N_2 adsorption-desorption (NAD)	Analyses surface area and porosity of adsorbent.
Thermo gravimetric analysis (TGA)	Characterizes sorption behaviour, material description, and kinetics of reaction rate.

7.5 Agricultural Wastes: An Excellent Dye Adsorbent

Agro-waste is a cost-effective, available, and ecologically acceptable source that works well as an adsorbent. Coconut shells, banana peel, squash coal, pomegranate peel, pistachio peel, orange peel, tea waste, coffee bean powder, rice hulls ash, garlic/pomelo peels, and bagasse peat are the most popular sources for removing dyes. Figure 7.2 represents use of agro-waste in dye removal process. Actually, the functional units present in this adsorbent result in higher number of active sites and improved pollutant adsorption (Farhadi et al. 2021). Various acids have been employed to stimulate biosorbents in order to improve binding sites, porosity, surface area, or aqueous solution chemistry. Phosphoric acid, for example, increases carbon output by stimulating bond breakage in agricultural waste (Bello et al. 2019). By grafting phosphorus functionality onto biomass and boosting acid activities involved in colour absorption, phosphoric acid increased dye biosorption (Benabbas et al. 2021).Furthermore, the activating agents such as KOH, H_2SO_4, and NaOH are commonly used in the manufacture of agricultural waste-based bio-adsorbents. Various investigations on the adsorption of dyes from wastewater effluents have been undertaken thus far. The efficiency and best circumstances for using agro-waste to separate a number of dyes are reviewed in this section. Table 7.2 shows the adsorption capacity as well as the best adsorption conditions for dyes removal using agricultural waste-derived adsorbent.

Dyes are typical contaminants found in aquatic habitats, and even at low quantities, they can be hazardous (Pathak et al. 2015). Modified agro-wastes enhanced adsorption power on an efficient and commercially suitable condition. In addition to this, chemical/physical treatment of agro-wastes, including acidic/basic solution treatment, esterification, and carbonization produce more active binding position on the surface, thereby sequestering dye at favourable circumstances. Amin (2009) investigated pomegranate peel-derived charcoal efficiency to remediate Direct Blue 106 from aqueous solution. They recorded that under ideal conditions (pH = 2 and 2.5 g/l adsorbent dosage) and 20 g/l dye starting level, the removal effectiveness was 72%. Furthermore, they opined that the adsorption mechanism resembled the Freundlich equilibrium adsorption (FEA) model and the pseudo-second-order kinetic (PSOK) model (Amin 2009). Hameed and Ahmad (2009) demonstrated garlic

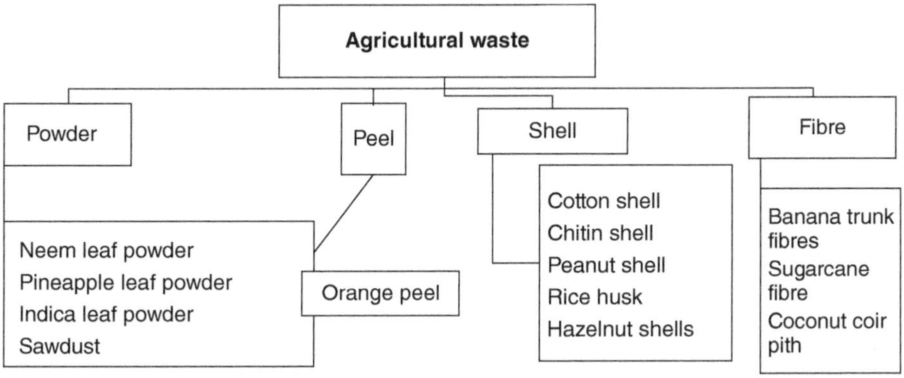

Figure 7.2 Different agro-waste used during dye removal process from wastewaters.

Table 7.2 Different agricultural wastes sorbents for removing dyes from wastewaters.

Agro-waste adsorbent	Dye	Dye conc. (mg/l)	Sorbent dose (g/l)	pH	Fitted adsorption model	Fitted kinetics model	Adsorption capacity (mg/g)	Removal efficiency (%)	References
Pine cone	Malachite green (MG)				Langmuir	PSOK	111.1	—	Kavci (2020)
Factionalized zea mays cob	Malachite green (MG)			8	Freundlich	PSOK	64.52	—	Ojediran et al. (2020)
Activated carbon (AC) beverage sludge	Allura Red AC	60	0.25	2	Freundlich	PSOK	287.1	90	Streit et al. (2019)
ACBS	CV	60	0.25	8	Sips	PSOK	640.7	98	Streit et al. (2019)
Corncobs	MB	150	0.3	5	Langmuir	PSOK	523.18	99.52	Sun et al. (2021)
Pineapple crown leaf	Methyl violet	20	6	5	—	—	—	>97	Astuti et al. (2019)
Lime peel	Malachite green	100	2	—	—	—	—	94.68	Ahmad et al. (2017)
Locust bean pod	Rhodamine B (RhB)	1000	0.1	2.87	Langmuir	PSOK	1111.1	95	Bello et al. (2019)
Grass waste	MB	—	0.8	10	Langmuir and Temkin	—	241.2	—	Jawad et al. (2020)
Dead leaves of oak trees	CV	50	0.1	7	—	—	—	—	Sulyman and Gierak (2020)
Rice hulls ash	Indigo Carmine	50	10	6.5	Freundlich	PSOK	65.9	96	Shen and Gondal (2017)
Powder (coffee bean)	Rhodamine 6G	15	0.05	2	Freundlich	PSOK	17.369	—	Argun et al. (2014)
Garlic peel	Direct Red 12B	50	0.3	2	Langmuir	PSOK	37.96	99	Amin (2009)

Pomegranate peel	Congo red	20	2.5	2	Langmuir	PSOK	42.59	72	Ghaedi et al. (2012)
Grape peel	MB	—	—	11	Langmuir	PFOK & PSOK	215.7	—	Ma et al. (2018)
Silica-coated soya waste	Methylene blue	—	—	7	Langmuir	PSOK	90	97.12%	Batool and Valiyaveettil (2021)
Sugarcane bagasse	Methylene blue	—	—	—	—	—	239	—	Mpatani et al. (2020)
Pumpkin peels treated with beetroot AC	Methylene blue	200	0.5	7	Langmuir	PSOK	198.15	—	Rashid et al. (2019)
Walnut shell	Methylene blue	—	—	—	—	—	84.5	—	Liu et al. (2019)
Terminalia catappa (TC) shells	Methylene blue	800		5	Freundlich	PSOK	88.62	90.56%	Hevira et al. (2021)
Piaçava fiber	—	—	—	10	—	—	427.3	75 %	Marques et al. (2019)
Modified orange peel	Methylene blue	—	—	6.2	Non-linear Sips and Langmuir	PSOK	307.63	—	Guediri et al. (2020)
Physically treated coffee husk	Methylene blue	37.78	0.740	7.6	Langmuir	PSOK	199.73	93.52%	Krishna Murthy and Gowrishankar (2020)
Chemically treated coffee husk	Methylene blue	84.49	0.406	7.9	Freundlich	PSOK	199.73	92.48%	Krishna Murthy and Gowrishankar (2020)
Biosorbent *Saccharum arundinaceum*	Methylene blue	—	—	10	Langmuir	PSOK	25.4	93%	Wazir et al. (2020)
Soyabean hulls	Methylene blue	—	—	7	Langmuir	PSOK	169.9	—	Cusioli et al. (2020)

(*Continued*)

Table 7.2 (Continued)

Agro-waste adsorbent	Dye	Dye conc. (mg/l)	Sorbent dose (g/l)	pH	Fitted adsorption model	Fitted kinetics model	Adsorption capacity (mg/g)	Removal efficiency (%)	References
Treated banana peel	Methylene blue	33.19	—	—	Langmuir	PFOK	991	—	Hashem et al. (2020)
Moringa oleifera seed husk and seed pulp	Acid Blue 9	—	—	—	Liu isotherm	—	329.5 and 694.2	—	Escobar et al. (2021)
Apple stem	Crystal violet	100	—	10	Langmuir	PSOK	153.85	96%	Takabi et al. (2021)
Water hyacinth (*Eichhornia crassipes*)	BF-4B red	—	—	—	Langmuir	PSOK	43.28	95%	Rigueto et al. (2020)
Biochar from pomelo peels	Methyl orange	—	—	—	Freundlich model	PSOK	141	—	—
Shaddock peel AC	Methyl orange	—	—	—	Freundlich	PSOK	94.6	54.25%	Tao et al. 2019
Banana peel powder	Congo red	—	—	—	Langmuir	PSOK	164.6	+90%	Munagapati et al. (2018)

peel efficiency for MB removal from aqueous solution using Freundlich, Langmuir, and Temkin isotherms. At 303, 313, and 323 K, the highest monolayer sorption activity was recorded as 82.64, 123.45, and 142.86 mg/g, respectively, during batch investigations. The inclusion of a polar functional group increased the adsorption capability of MB, according to the author (Hameed and Ahmad 2009). Lakshmi et al. (2009) noticed maximum adsorption of Indigo Carmine colour by rice husk ash at pH 5.4 and 10 g/ladsorbent dosage after eight hours. The authors also demonstrated that adsorption process was spontaneous as well as endothermic by analysing thermodynamic properties like enthalpy, entropy, and Gibbs free energy changes. Chang et al. (2009) used agricultural trash-derived fly ash to test the adsorption of Rhodamine B. Rhodamine B removal effectiveness by fly ash (80 g/l) was just 54%. The authors also demonstrated that removal effectiveness can be improved to 98% by combining Fenton oxidation with fly ash adsorption (Chang et al. 2009). Sureshkumar and Namasivayam (2008) employed micelle coconut coir pith to separate Rhodamine B and Direct Red 12B dyes. With a pH 5, 20 g/l dosage, and 20 mg/l starting dye concentrations, micelle coconut coir pith able to eliminate Rhodamine B and Direct Red 12B dyes by 14.9 and 76.3 mg/g, respectively (Sureshkumar and Namasivayam 2008). Dulman and Cucu-Man (2009) used beech wood sawdust to check the removal efficiency of dyes, viz. C.I. Direct Brown, C.I. Direct Brown 2, C.I. Direct Green 26, C.I. Basic Blue, C.I. C.I. Direct Blue 6, and Basic Blue 86 from aqueous solution. With increasing pH (above pH 10), the percentage clearance of brown dyes (both Brown 2 and Brown) reduced, with maximum remedial rate occurring at pH 3. As a result, the colour removal percentages of brown dyes (both Brown and Brown 2) decreased from 94.4 to 28.5 and 98.6 to 34.7%, respectively. At pH 4.43–7.06, the sorption of Basic Blue reaches a high of 97% (Dulman and Cucu-Man 2009). To separate two anionic/cationic dyes from wastewater, Vučurović et al. (2014) used maize stem and observed highest capabilities for anionic and cationic dye removal 167.01 and 160.84 mg/g, respectively. To remove Congo red, Ghaedi et al. (2012) used activated carbon (AC) derived from pomegranate peel and *Myrtus communis*. The maximal dye adsorption capabilities recorded for *Myrtus communis* and pomegranate are 10 and 19.231 mg/g, respectively, with improved adsorption performance with increasing pH value. The Langmuir isotherm model was followed by pomegranate peel; however, the Freundlich isotherm was followed by *Myrtus communis* (Ghaedi et al. 2012).

Pathania et al. (2017) developed *Ficus carica* bast-derived AC, which demonstrated a promising MB adsorption capacity under operating parameters like sorbent dose (5 g/l), 0.5 g/l of initial dye level, contact time of 1.3 hours, and temperature 30 °C at pH 7.8. The authors also demonstrated that isotherm models revealed that sorption is heterogeneous and occurs as a result of physicochemical interactions. Argun et al. (2014) used pomelo peel waste to remove Reactive Blue 114 (RB114) dye. They found that at 303 K temperature, pH 2, and starting dye level (1–200 mg/l), the dye's maximal remedial capacity was 16 mg/g (Argun et al. 2014). Asfaram et al. (2014) looked into how effective garlic peel is at removing Direct Red 12B from the solution. For adsorbent dosage of 4 and 50 g/l colour level, the desired and spectacular removal efficiency of 99% was reached in 25 minutes. The authors also conducted Langmuir equilibrium adsorption (LEA), endothermic adsorption procedure, and PSOK models. Rhodamine 6G and Rhodamine B were removed using coffee powder by Shen and Gondal (2017). Rhodamine 6G and Rhodamine B displayed highest adsorption capabilities of 17.369 and 5.255 mg/gl, respectively, at pH 2, 0.05 g/l of adsorbent dosage, and 15 mg/l of initial dye level. Additionally, for Rh 6G and RhB, the pseudo-first-order kinetic (PFOK), PSOK models, and FEA and LEA models were performed (Shen and Gondal 2017). Biochar made from Korean cabbage (KC), wood chip, and rice straw were developed by Sewu et al. (2017) as a substitute to AC to separate Crystal Violet (CV) and Congo red (Sewu et al. 2017); however, KC had greater absorption capacity than the others. Basic yellow dye separation from wastewater was recommended using AC from *Persea americana*, which removed 98% colour and recommended instead of commercial adsorbents (Regti et al. 2017). MB dye degradation using citrate-treated pomelo peel has been reported with enhanced adsorption power from 81.7 to 199.2 mg/g (Ren et al. 2018).

Watermelon peels were utilized to make modified watermelon biochar/adsorbent. Initially, biochar was treated with ozone, and then it was further modified with triethylenetetramine and ammonium hydroxide for removing Acid Yellow 11 (AY11) (El-Nemr et al. 2020a). At pH 1.0, 76.94–462.18 mg/g concentrations, and contact time of three hours at ambient temperature, 96% eradication of AY11 can be achieved. Furthermore, the LEA and FEA isotherms models were suited, and reaction followed the PSOK model. The principal mechanism influencing the AY11 dye adsorption onto the biochars was electrostatic interaction (El-Nemr et al. 2020a). In another work, El-Nemr et al. (2020b) developed pea (*Pisum sativum*) peels-derived biochar for Acid Orange 7 (AO7) dye removal following similar conditions like watermelon peels. At equilibrium, the improved biochar's adsorption capacity increased from 78.18 to 523.12 mg/g. Triethylenetetramine-modified biochar was shown to remove up to 98% of AO7, but unmodified biochar only removed 96% (El-Nemr et al. 2020b). Adsorbent made from potato peels was thought to be a potential adsorbent for separating the Cibacron Blue. The sorption capacity of potato peels is native<activated<calcined materials in that sequence (Bouhadjra et al. 2021). With a maximal adsorption capacity of 464 mg/g, adsorbent dose of 0.4 g/l, contact time of two hours, and 84.75% of remediation, mandarin (*Citrus reticulata*) peel was used to eliminate SO from aqueous solution. The putative adsorptive processes of SO removal were hydrogen bonds, electrostatic interactions, and μ-interactions (Januário et al. 2021). Furthermore, the desorption investigation revealed that even after four cycles, the adsorption ability was controlled at >50%. Maximum adsorption (129 mg/g) for MB colour sequestration using chemically treated (H_3PO_4) corn stalks and wheat straw was also demonstrated by Tang et al. (2019) and Batzias et al. (2009). Chemically treated swede rape straw was documented (Q_m = 432 mg/g) to be fine, inexpensive adsorbent for the treatment of MB dye (Feng et al. 2013). In another study, chemically treated corn steam (870 mg/g), camphor leave (741 mg/g), and *C. camphora* sawdust (787 mg/g) also displayed significant absorptivity (Tang et al. 2021).

7.6 Adsorption Kinetics and Isotherm

During adsorption process, the adsorbates bind to the adsorbents. Adsorption is the process of converting a liquid into a solid interface, and it is basically a surface phenomenon (Salleh et al. 2011). This process generally takes place based on the type of contact between the adsorbent's surface and the adsorbed molecule (physisorption and chemisorption). Physical adsorption is the process in which van der Waals forces and electrostatic interactions occur. They yield a reversible reaction. Electrostatic, H-bonding, and Π–Π stacking interactions have improved the dye's adsorption via the adsorption mechanism on the adsorbent surface. Covalent bonds appear to be the significant interaction between adsorbate and adsorbent in the chemical process. The majority of molecular diffusion from surface to substance occurs through a monolayer. While the production of multilayer was detected in the physical process, multilayer deposition on surface was seen in different processes. During physical process, the adsorption process followed free energy and entropy reduction and occur at the same time with optimum conditions. The various adsorption isotherm models show how the adsorbate (pollutant) concentration and equilibrium (q_e) adsorption ability influence equilibrium concentration of the adsorbate (C_e) at equilibrium. These equilibrium data disclose details on how contaminant concentrations interact with adsorbent surface.

Kinetics of adsorption system illustrates the rate and requisite time for the adsorption practice. The reaction rate mechanism and reaction rate are two key variables to be evaluated in an adsorption method. The adsorption mechanism can only be grasped by considering the adsorption rate and rate-limiting phases. In-depth research on kinetics factors is required to comprehend this process entirely. This assists in the design and implementation of strategies. Colour adsorption using agricultural waste has been explored by several authors using several kinetics models like PFOK, PSOK, Boyd plot, film diffusion, Elovich, Brouers-Weron-Sotolongo model (BWS), film diffusion, intraparticle diffusion (IPD), Avrami, and Freundlich. In the kinetic model, adsorption parameters corresponding to the best fit to the kinetic model to remove dyes from agricultural wastes are shown.

Models like adsorption isotherm give important information on maximal adsorption capacity and are therefore utilized in assessing adsorbent performance. The isotherm concept defines the equilibrium concentration connection between the liquid and solid phases at a specific temperature. Among the several techniques for analysing adsorption performance, isothermal adsorption models are the most important (Huang et al. 2018). The monolayer or multilayer surface phase is measured as a monolayer or multilayer (Huang et al. 2018) and to find out a model for designing reasons, it is vital to use isotherm models to analyse equilibrium adsorption data (Bharathi and Ramesh 2013). These isotherms are represented by the Langmuir, Freundlich, Brunauer–Emmett–Teller (BET), Temkin, Sips, and Temkin. Adsorption isotherms are generally applied to elucidate the performance of adsorption and the mechanism behind it.

7.6.1 Langmuir's Isotherm

According to the Langmuir equilibrium, single-layer sorption upon a surface with a fixed number of adsorbates where most of the sites are identical will be stable only if adsorbate does not migrate across the surface plane (Bharathi and Ramesh 2013). The Langmuir isotherm monolayer adsorbed model was constructed to provide an ideal framework for predicting monolayer adsorption, which implies that chemicals adsorbed onto the surface of adsorbent formed a single layer, and that each chemical has similar sorption energy (Zhou et al. 2011). Research investigators in the disciplines of adsorption are using this equilibrium data the most as it is well to be used at trace levels, easy to implement, and adjustable with computational models (Laskar and Hashisho 2020). The Langmuir isotherm's mathematical equation can be expressed as non-linear equation (Eq. 7.1):

$$Q_e = K_L Q_L C_e / 1 + K_L C_e \tag{7.1}$$

where Q_L represents maximal single later coverage capability (mg/g), K_L represents Langmuir constants (l/mg), C_e is equilibrium adsorbate level (mg/l), and Q_e represents equilibrium adsorbat equantity (mg/g).

7.6.2 Freundlich Isotherm

In the Freundlich isotherm model, the Langmuir equation is modified to include surface energy terms as functions of surface coverage (Bharathi and Ramesh 2013). The Freundlich isotherm is a well-known multilayered adsorption model. Adsorbates are thought to be non-linearly absorbed by water or soil particles (Zhou et al. 2014). In general, the Freundlich model is based on empirical theory and is used to describe the adsorption behaviour of single adsorbate in an aqueous solution at rigid temperature of organic components (Foo and Hameed 2010).

Freundlich equilibrium can be depicted by non-linear (Eq. 7.2) and linear (Eq. 7.3) equations as follows:

$$Q_e = K_F (C_e)^{1/n} \tag{7.2}$$

$$L_n(Q_e) = L_n(K_F) + 1/n L_n(C_e) \tag{7.3}$$

where C_e indicates residual concentration at equilibrium (mg/l), K_F is Freundlich constants (mg/g) $(l/g)^n$, n represents sorption intensity, and Q_e indicates equilibrium adsorption efficiency of adsorbent (mg/g).

7.6.3 Brunauer–Emmett–Teller (BET) Isotherm

This model takes into account adsorption energy to keep the initial layer of adsorbed species in place, as well as extra power to keep the next layer of adsorbate species in place. The model considers the saturation phenomena, which is when the solids are completely dissolved in the solvent and account solubility of the particles in the solvent. In its non-linear Eq. (7.4), the connection between sorption capacity and contents/levels at equilibrium can be expressed by isotherm Eq. (7.5) models as follows:

$$Q_e = \frac{Q_s C_{BET} C_e}{(C_s - C_e)[1 + (C_{BET} - 1)\left(\frac{C_e}{C_s}\right)} \tag{7.4}$$

$$\frac{C_e}{Q_e(C_s - C_e)} = \frac{1}{Q_s C_{BET}} + \frac{(C_{BET} - 1)\left(\frac{C_e}{C_s}\right)}{Q_s C_{BET}} \tag{7.5}$$

where C_e represents equilibrium concentration (mg/l), Cs is the monolayer saturation level (mg/l), Q_e indicates theoretical saturation capacity (mg/l), and Q_s represents the point at which BET adsorption isotherm occurs (mg/g).

Other isotherm models, such as Dubinin–Radushkevich model (DR), Temkin model, and Halsey and Harkins–Jura isotherms, are commonly employed in adsorption phenomena. These models were more descriptive of adsorption colours onto agricultural waste.

7.7 Factors Influencing Dye Removal Using Agro-Wastes

There are several characteristics, including temperature, pH, dye concentration before adsorption, and the dose of the adsorbent, and the adsorption performance of agricultural wastes were impacted. Some variables (pH, primary dye concentration, temperature, adsorbents quantity, etc.) affect the adsorption performance of agricultural wastes. Adsorption mechanisms are generally complex and difficult to understand, but increasing their efficiency can help broad utilization and the scientific community's understanding of adsorption.

7.7.1 Temperature

The temperature is an important indicator to predict the reaction behaviour of adsorption by predicting the method is exothermic or endothermic (Bharathi and Ramesh 2013). Increase in temperature increases the mobility of dye composition from liquid phase to adsorbent surface and their access to deeper layers. As adsorption capacity increased, it was thought that pore size enhancement, adsorbent surface stimulation, or the formation of binding sites might have occurred. In addition, temperature influences the adsorption capability on functional group of the adsorbents surface. Further, thermodynamic properties, such as $\Delta H°$, $\Delta G°$, and $\Delta S°$, govern the adsorption rate of adsorption. The endothermic and exothermic characteristics of the method are determined by the sum of $\Delta H°$ positive and negative values. By looking at the value and confirming the possible physical adsorption existence, the positive result showed that physical adsorption was possible. Enhanced lack of arrangement and unpredictability at the solid-solution boundary of the adsorbent are consistent by the overall increase in the +ve values of $\Delta S°$. Exothermic adsorption of MB dye experimented by Yagub et al. (2014) is due to the effect of reducing adsorptive force of MB dye and agro-waste (treated celery residue) surface with rise in temperature.

7.7.2 pH

pH is a crucial component, which can influence the surface charge and ionize the adsorbents. Initial pH influences a critical function during the adsorption process, since this determines the ionization yield, surface charge density, and adsorption power. For cationic dye, the dye separation capacity decreases at acidic condition, while for anionic dye it is reversal. High protonated surfaces promote the adsorption of acidic groups, thus leading to a rise in adsorption. Accordingly, when pH rises, the separation efficiency for anionic dyes reduces, and as the pH falls, the separation efficiency for basic dyes increases (Salleh et al. 2011). The maximum MB dye adsorption was recorded at pH 4 by banana stalk as a biosorbent (Hameed et al. 2008). In order to gain a thorough grasp of the operation and performance of the adsorption practice, it is essential to have the ability to control the pH to a value of zero charges (pHpzc). pHpzc is critical because it dictates the surface's ability to absorb and the kind of binding site that is accessible to enhance the understanding of adsorption (Mishra et al. 2015). The surfaces of adsorbents acquire a positive charge due to anionic dye adsorption at low pH values (below pHpzc). When the pH >pHpzc, the cationic dye is readily absorbed. However, acidic dye adsorption is observed when the solution has a positive charge.

7.7.3 Initial Dye Concentration

The adsorption sites become less available at higher dye concentrations due to relationship between dye quantity and adsorption sites. It was found that when the sorption sites on adsorbent surface get more saturated with dye, the proportion of the dye removal increases. The high driving force is provided by the initial dye concentration, making it easier for the dye to travel from liquid to solid phase and facilitates the dye–adsorbent interaction. For instance, MB separation increased from 93.48 to 98.28% by rising the initial concentration (5–15 mg/L) onto *Ficus palmata* leaves (Fiaz et al. 2019). Adsorption efficiency of adsorbent raises at higher initial concentration of analyte. It is basically based on the dye concentration and overall surface-active sites of adsorbent. Due to a rise in the ratio of adsorbent sites to dye molecules, a drop in dye concentration is required to maintain a constant colour removal (Kallel et al. 2016).

7.7.4 Amount of the Adsorbent

Amount of adsorbent is a significant factor that affects the amount of adsorbent–adsorbate solution. It is directly proportional to the surface area. The percentage of colour separation capacity increases with the increase in adsorbent dosage. Adsorption efficiency of different agro-wastes has been observed to decrease rising adsorbent amount. As for example, MB dye separation efficiency increases from 94 to 97% with the amount of modified corn stalk dosage increase (2 g/l) (Tang et al. 2019). Enhanced sorption percentage of MB dye from 55 to 97.1% with the increasing quantity of treated soya waste have also been reported by Batool and Valiyaveettil (2021). These proved that adsorbent dosage is very crucial to improve the quantity of active sites and total surface area of adsorbent.

7.8 Economic and Environmental Concern of Agricultural Wastes-Based Dye Removal

Although cost–benefit analysis (CBA) was not conducted, it has been revealed that agricultural waste-based adsorbents are less expensive than standard treatment procedures. To be cost-effective, the availability of a large number of adsorbents, ease processing, and activation requirements, as well as green chemistry concepts, are very critical (Bulgariu et al. 2019). Further, numerous works on agro-waste-related adsorption have been performed using simulated wastewater at laboratory scale, limiting the CBA analysis of agro-based adsorbents.

As a result, low-cost substances with adsorption capacity comparable to other sorbents are desperately needed. For such adsorption efficiency of dye, Basic Red 09 removal from aqueous solutions, a detailed CBA analysis of agro-waste-derived adsorbents (groundnut shell, rice husk, and coconut shell) has been conducted. Considering operating costs such as feedstock costs, maintenance charges, manufacturing charges, labour charges, transportation costs, and circulation charges, the 1 g adsorbent cost was calculated to be ₹0.97, ₹0.91, and ₹4.54 for rice husk, groundnut shell, and coconut shell, respectively (Praveen et al. 2021). Groundnut shell-based adsorbent showed highest sorption performance (46.3, 44 mg/g for rice husk, and 10 mg/g for coconut shell) and lowest cost of Basic Red 09 dye removal (€0.01) per unit gram. Rice husk and coconut shell cost €0.011 and €0.052 per Basic Red 09 gram removal, respectively (Praveen et al. 2021). The average price of adsorbent/biochar was €2.13 per kg worldwide, ranging from €0.072 in Philippines to €7.11 per kg in the United Kingdom (Akom et al. 2020), while the average price of blended adsorbent/biochar in India ranged from €0.064 to €10.83 per kg (Jirka and Tomlinson 2015). Bello et al. (2019) developed H_3PO_4-treated locust bean pod AC for RhB dye remediation and investigated adsorbent's CBA. They found that plant-derived AC was around six times less expensive than traditional AC. Deionized water and phosphoric acid are most expensive components (Bello et al. 2019). Although commercial adsorbents need more money compared with agro-waste-based adsorbents, agro-waste-based adsorbents need further CBA analysis to address the challenges of life cycle assessment (LCA) (Nishikawa et al. 2018; Aragaw and Bogale 2021). In addition to this, agriculture waste materials have unique properties that make LCA methods difficult to use, such as balancing agro-waste carbon and recycling, reuse, or energy/material recovery (Nishikawa et al. 2018; Aragaw and Bogale 2021). Furthermore, the safe disposal of spent (exhausted) agro-waste adsorbent should be looked in future studies, advancing the agricultural waste vapourization concept. Given that (waste)water dye treatment using agricultural wastes-derived adsorbents need multidisciplinary and sustainable long-term solutions to improve the water treatment.

7.9 Conclusions and Recommendations

During the past decades, dye production and its application have increased a lot, and they pose a risk to aquatic life as well as public health. In the last few years, many researchers have been focusing on agro-waste as low-costing biosorbent development because of their abundance availability as a resource, cost-effectiveness, fine adsorbent quality, and eco-friendly nature and have a plenty of possible utilizations in the near future. Researchers have established that agriculture-based biosorbent has offered affordable green filtration technology and offered promising results with maximum efficiency in wastewater management, especially contaminants remediation. A number of significant points were highlighted from the work which are described as follows:

- The majority of the adsorption processes using agro-wastes have been found to comply with PSOK as well as to follow monolayer or multilayer adsorption equilibrium isotherms.
- Maximum number of works in the literature indicate that agro-waste resources are efficient biosorbent.

- The importance of adsorption variables like pH, temperature, adsorbent quantity as well as primary dye level in evaluating sorption power can be inferred from the presented data. Most of the reports have shown fine removal capacity.

This paper endeavoured to review agricultural waste as eco-efficient biosorbent in treating dyes from polluted water. In addition, present review highlights the following future perspectives and challenges:

- Detailed perception is essential to understand adsorption mechanisms.
- Intensive study needed for the characterization of different agro-wastes biosorbent to promote extensive utilization of biosorbent.
- A combined characterization technique is required to describe the structural activity, morphology, and surface structure of biosorbent.
- Intensive work is needed for the utilization of inexpensive sorbents in dye separation for pilot- and commercial-level purposes.
- Additional research is essential to assess the industrial application of highly developed methods based on energy, cost, and environmental impacts.
- In-depth study is needed for modelling and regeneration of bio-adsorbents to enhanced efficiency and recovery of dyes.
- Detailed investigation is required for the application of selective and higher affinity biosorbent so that dyes can utilize positively with other species in aqueous media.

References

Ahmad, M.A., Afandi, N.S., and Bello, O.S. (2017). Optimization of process variables by response surface methodology for malachite green dye removal using lime peel activated carbon. *Appl. Water Sci.* 7: 717–727. 10.1007/s13201-015-0284-0.

Akom, M., Fanyin-Martin, A., Oti-Boateng, C. et al. (2020). Yield and cost of biochar produced by a locally fabricated reactor. *J. Agric. Environ. Sci.* 9: 1.

Al Prol, A.E. (2019). Study of environmental concerns of dyes and recent textile effluents treatment technology: a review. *Asian J. Fish. Aquat. Res.* 3 (2): 1–18.

Amin, N.K. (2009). Removal of direct blue-106 dye from aqueous solution using new activated carbons developed from pomegranate peel: adsorption equilibrium and kinetics. *J. Hazard. Mater.* 165: 52–62. 10.1016/j.jhazmat.2008.09.067.

Amirza, M.A.R., Adib, M.M.R., and Hamdan, R. (2017). Application of agricultural wastes activated carbon for dye removal – an overview. *MATEC Web Conf.* 103: 6013–6034. 10.1051/matecconf/201710306013.

Aragaw, T.A. and Bogale, F.M. (2021). Biomass-based adsorbents for removal of dyes from wastewater: a review. *Front. Environ. Sci.* 9: 764958. 10.3389/fenvs.2021.764958.

Argun, M.E., Güclü, D., and Karatas, M. (2014). Adsorption of reactive blue 114 dye by using a new adsorbent: pomelo peel. *J. Ind. Eng. Chem. 20*: 1079–1084. 10.1016/j.jiec.2013.06.045.

Asfaram, A., Fathi, M.R., Khodadoust, S., and Naraki, M. (2014). Removal of direct red 12B by garlic peel as a cheap adsorbent: kinetics, thermodynamic and equilibrium isotherms study of removal. *Spectrochim. Acta, Part A.* 127: 415–421. 10.1016/j.saa.2014.02.092.

Ashrafi, S.D., Kamani, H., Arezomand, H.S. et al. (2015). Optimization and modeling of process variables for adsorption of Basic Blue 41 on NaOH-modified rice husk using response surface methodology. *Desalin. Water Treat.* 57 (30): 14051–14059.

Astuti, W., Sulistyaningsih, T., Kusumastuti, E. et al. (2019). Thermal conversion of pineapple crown leaf waste to magnetized activated carbon for dye removal. *Bioresour. Tech.* 287: 121426. 10.1016/j.biortech.2019.121426.

Bardhan, M., Novera, T.M., Tabassum, M. et al. (2020). Adsorption of methylene blue onto betel nut husk-based activated carbon prepared by sodium hydroxide activation process. *Water Sci. Technol.* 82 (9): 1932–1949.

Batool, A. and Valiyaveettil, S. (2021). Chemical transformation of soya waste into stable adsorbent for enhanced removal of methylene blue and neutral red from water. *J. Environ. Chem. Eng.* 9: 104902.

Batzias, F., Sidiras, D., Schroeder, E., and Weber, C. (2009). Simulation of dye adsorption on hydrolyzed wheat straw in batch and fixed-bed systems. *Chem. Eng. J.* 148 (2–3): 459–472.

Bello, O.S., Adegoke, K.A., Sarumi, O.O., and Lameed, O.S. (2019). Functionalized locust bean pod (*Parkia biglobosa*) activated carbon for rhodamine B dye removal. *Heliyon* 5: e02323. 10.1016/j.heliyon.2019.e02323.

Benabbas, K., Zabat, N., and Hocini, I. (2021). Study of the chemical pretreatment of a nonconventional low-cost biosorbent (*Callitriche obtusangula*) for removing an anionic dye from aqueous solution. Euro-mediterr. *J. Environ. Integr.* 6: 1–17. 10.1007/s41207-021-00265-4.

Bharathi, K.S. and Ramesh, S.T. (2013). Removal of dyes using agricultural waste as low-cost adsorbents: a review. *Appl. Water Sci.* 3: 773–790.

Bouhadjra, K., Lemlikchi, W., Ferhati, A., and Mignard, S. (2021). Enhancing removal efficiency of anionic dye (*Cibacron blue*) using waste potato peels powder. *Sci. Rep.* 11: 1–10. 10.1038/s41598-020-79069-5.

Bulgariu, L., Escudero, L.B., Bello, O.S. et al. (2019). The utilization of leaf-based adsorbents for dyes removal: a review. *J. Mol. Liquids* 276: 728–747. 10.1016/j.molliq.2018.12.001.

Chang, S.H., Wang, K.S., Li, H.C. et al. (2009). Enhancement of Rhodamine B removal by low-cost fly ash sorption with Fenton preoxidation. *J. Hazard. Mater.* 172: 1131–1136. 10.1016/j.jhazmat.2009.07.106.

Choi, H.J. and Yu, S.W. (2018). Biosorption of methylene blue from aqueous solution by agricultural bioadsorbent corncob. *Environ. Eng. Res.* 24 (1): 99–106. 10.4491/eer.2018.107.

Crini, G. and Lichtfouse, E. (2019). Advantages and disadvantages of techniques used for wastewater treatment. *Environ. Chem. Lett.* 17 (1): 145–155.

Cusioli, L.F., Quesada, H.B., Baptista, A.T. et al. (2020). Soybean hulls as a low-cost biosorbent for removal of methylene blue contaminant. *Environ. Prog. Sustain. Energy.* 39 (2): 13328.

Dai, Y., Sun, Q., Wang, W. et al. (2018). Utilizations of agricultural waste as adsorbent for the removal of contaminants: a review. *Chemosphere* 211: 235–253. https://doi.org/10.1016/j.chemosphere.2018.06.179.

Djelloul, C. and Hamdaoui, O. (2015). Dynamic adsorption of methylene blue by melon peel in fixed-bed columns. *Desalin. Water Treat.* 56 (11): 2966–2975.

Dulman, V. and Cucu-Man, S.M. (2009). Sorption of some textile dyes by beech wood sawdust. *J. Hazard Mater.* 162: 1457–1464.

El-Nemr, M.A., Abdelmonem, N.M., Ismail, I.M.A. et al. (2020a). Removal of acid yellow 11 dye using a novel modified biochar derived from watermelon peels. *Dwt* 203: 403–431. 10.5004/dwt.2020.26207.

El-Nemr, M.A., Abdelmonem, N.M., Ismail, I.M.A. et al. (2020b). The efficient removal of the hazardous azo dye acid orange 7 from water using modified biochar from pea peels. *Dwt* 203: 327–355. 10.5004/dwt.2020.26190.

Escobar, O.D.S., Azevedo, C.F.D., Swarowsky, A. et al. (2021). Utilization of different parts of *Moringa oleifera* Lam. seeds as biosorbents to remove acid blue 9 synthetic dye. *J. Environ. Chem. Eng.* 9 (4): 105553. ISSN 2213-3437, https://doi.org/10.1016/j.jece.2021.105553.

Farhadi, A., Ameri, A., and Tamjidi, S. (2021). Application of agricultural wastes as a low-cost adsorbent for removal of heavy metals and dyes from wastewater: a review study. *Phys. Chem. Res.* 9 (2): 211–226. 10.22036/pcr.2021.256683.1852.

Feng, Y., Dionysiou, D.D., Wu, Y. et al. (2013). Adsorption of dyestuff from aqueous solutions through oxalic acid-modified swede rape straw: adsorption process and disposal methodology of depleted bioadsorbents. *Bioresour. Technol.* 138: 191–197.

Fiaz, R., Hafeez, M., and Mahmood, R. (2019). Ficcuspalmata leaves as a low-cost biosorbent for methylene blue: thermodynamic and kinetic studies. *Water Environ. Res.* 91 (8): 689–699.

Foo, K.Y. and Hameed, B.H. (2010). Insights into the modeling of adsorption isotherm systems. *Chem. Eng. J.* 156 (1): 2–10.

Ghaedi, M., Tavallali, H., Sharifi, M. et al. (2012). Preparation of low cost activated carbon from *Myrtus communis* and pomegranate and their efficient application for removal of Congo red from aqueous solution. *Spectrochim. Acta, Part A 86*: 107–114. 10.1016/j.saa.20.11.10.012.

Guediri, A., Bouguettoucha, A., Chebli, D. et al. (2020). Molecular dynamic simulation and DFT computational studies on the adsorption performances of methylene blue in aqueous solutions by orange peel-modified phosphoric acid. *J. Mol. Struct.* 1202: 127290. 10.1016/j.molstruc.2019.127290.

Hameed, B.H. and Ahmad, A.A. (2009). Batch adsorption of methylene blue from aqueous solution by garlic peel, an agricultural waste biomass. *J. Hazard. Mater.* 164: 870–875.

Hameed, B.H., Mahmoud, D.K., and Ahmad, A.L. (2008). Sorption equilibrium and kinetics of basic dye from aqueous solution using banana stalk waste. *J. Hazard. Mater.* 158: 499–506.

Hashem, A.H., Saied, E., Hasanin, M., and Sayed, H. (2020). Green and ecofriendly bio-removal of methylene blue dye from aqueous solution using biologically activated banana peel waste. *Sustain. Chem. Pharm.* 18: 100333.

Hevira, L., Rahmayeni, Y., Ighalo, J.O. et al. (2021). *Terminalia catappa* shell as low-cost biosorbent for the removal of methylene blue from aqueous solutions. *J. Ind. Eng. Chem.* 97 (3): 10.1016/j.jiec.2021.01.028.

Houda, B. and Moussa, B. (2017). Electrosorption removal of the zinc ions from aqueous solution on an artificialelectrode based in the banana wastes. *J. Electrochem. Sci. Technol.* 8 (1): 77–86.

Huang, W., Hu, Y., Li, Y. et al. (2018). Citric acidcrosslinked b-cyclodextrin for simultaneous removal of bisphenol A, methylene blue and copper: the roles of cavity and surface functional groups. *J. Taiwan Inst. Chem. Eng.* 82: 189–197.

Januário, E.F.D., Vidovix, T.B., Araújo, L.A.d. et al. (2021). Investigation of *Citrus reticulata* peels as an efficient and low-cost adsorbent for the removal of safranin orange dye. *Environ. Tech.* 1: 1–15. 10.1080/09593330.2021.1946601.

Jawad, A.H., Mohd Firdaus Hum, N.N., Abdulhameed, A.S., and Mohd Ishak, M.A. (2020). Mesoporous activated carbon from grass waste via H_3PO_4- activation for methylene blue dye removal: modelling, optimisation, and mechanism study. *Int. J. Environ. Anal. Chem.* 1–17. 10.1080/03067319.2020.1807529.

Jirka, S. & Tomlinson, T. (2015). State of the biochar industry 2014 – a survey of commercial activity in the biochar sector. In: *International Biochar Initiative* (IBI) (ed. I.B.I. (IBI)).

Kallel, F., Chaari, F., Bouaziz, F. et al. (2016). Sorption and desorption characteristics for the removal of a toxic dye, methylene blue from aqueous solution by a low cost agricultural by-product. *J. Mol. Liq.* 219: 279–288. 10.1016/j.molliq.2016.03.024.

Katheresan, V., Kansedo, J., and Lau, S.Y. (2018). Efficiency of various recent wastewater dye removal methods: a review. *J. Environ. Chem. Eng.* 6 (4): 4676–4697.

Kavci, E. (2020). Malachite green adsorption onto modified pine cone: isotherms, kinetics and thermodynamics mechanism. *Chem. Eng. Commun.* 208: 318–327. 10.1080/00986445.2020.1715961.

Krishna Murthy, T.P. & Gowrishankar, B.S. (2020). Process optimisation of methylene blue sequestration onto physical and chemical treated coffee husk based adsorbent. *SN Applied Sciences.* 2(5) http://doi.org/10.1007/s42452-020-2603-9

Lakshmi, U.R., Srivastava, V.C., Mall, I.D., and Lataye, D.H. (2009). Rice husk ash as an effective adsorbent: evaluation of adsorptive characteristics for Indigo Carmine dye. *J. Environ. Manage.* 90: 710–720. 10.1016/j.jenvman.2008.01.002.

Laskar, I.I. and Hashisho, Z. (2020). Insights into modeling adsorption equilibria of single and multicomponent systems of organic and water vapors. *Sep. Purif. Technol.* 241: 10.1016/j.seppur.2020.116681116681.

Liu, M., Li, X., Du, Y., and Han, R. (2019). Adsorption of methyl blue from solution using walnut shell and reuse in a secondary adsorption for Congo red. *Bioresour. Technol. Rep.* 5: 238–242.

Ma, L., Jiang, C., Lin, Z., and Zou, Z. (2018). Microwave-hydrothermal treated grape peel as an efficient biosorbent for methylene blue removal. *Int. J. Environ. Res. Public Health.* 15: 239.

Marques, B.S., Frantz, T.S., Sant'Anna Cadaval Junior, T.R. et al. (2019). Adsorption of a textile dye onto piacavafibers: kinetic, equilibrium, thermodynamics, and application in simulated effluents. *Environ. Sci. Pollut. Res. Int.* 26 (28): 28584–28592. 10.1007/s11356-018-3587-5.

Mishra, Y., Sowmya, V., and Shanthakumar, S. (2015). Adsorption studies of basic dyes onto Teak (*Tectona grandis*) leaf powder. *J. Urban Environ. Eng.* 9: 102–108.

Mpatani, F.M., Aryee, A.A., and Kani, A.N. (2020). Removal of methylene blue from aqueous medium by citrate modified bagasse: kinetic, equilibrium and thermodynamic study. *Bioresour. Technol. Rep.* 11: 100463. 10.1016/j.biteb.2020.100463.

Munagapati, V., Yarramuthi, V., Kim, Y. et al. (2018). Removal of anionic dyes (reactive Black 5 and Congo red) from aqueous solutions using banana peel powder as an adsorbent. *Ecotoxicol. Environ. Saf.* 148: 601–607.

Ngulube, T., Gumbo, J.R., Masindi, V., and Maity, A. (2017). An update on synthetic dyes adsorption onto clay based minerals: a state-of-art review. *J. Environ. Manag.* 191: 35–57. 10.1016/j.jenvman.2016.12.031.

Nidheesh, P.V., Gandhimathi, R., and Ramesh, S.T. (2013). Degradation of dyes from aqueous solution by Fenton processes: a review. *Environ. Sci. Pollut. Res.* 20: 2099–2132. 10.1007/s11356-012-1385-z.

Nishikawa, E., da Silva, M.G.C., and Vieira, M.G.A. (2018). Cadmium biosorption by alginate extraction waste and process overview in life cycle assessment context. *J. Clean.Prod.* 178 (2018): 166–175.

Ojediran, J.O., Dada, A.O., Aniyi, S.O., and David, R.O. (2020). Functionalized Zea Mays Cob (FZMC) as low-cost agrowaste for effective adsorption of malachite green dyes data set. *Chem. Data Coll.* 30: 100563.

Pathak, P.D., Mandavgane, S.A., and Kulkarni, B.D. (2015). Fruit peel waste as a novel low-cost bio adsorbent. *Rev. Chem. Eng.* 3: 361–381.

Pathania, D., Sharma, S., and Singh, P. (2017). Removal of methylene blue by adsorption onto activated carbon developed from *Ficus carica* bast. *Arab. J Chem.* 10: S1445–S1451. 10.1016/j.arabjc.2013.04.021.

Prasad, M., Ranjan, R., Ali, A. et al. (2020). Efficient transformation of agricultural waste in India. In: *Contaminants in Agriculture* (ed. M. Naeem, A. Ansari and S. Gill), 271–287. Springer Nature Switzerland AG. https://doi.org/10.1007/978-3-030-41552-5_13.

Praveen, S., Gokulan, R., Pushpa, T.B., and Jegan, J. (2021). Techno-economic feasibility of biochar as biosorbent for basic dye sequestration. *J. Indian Chem. Soc.* 98 (8): 100107.

Prusty, J.K., Patro, S.K., and Basarkar, S.S. (2016). Concrete using agro-waste as fine aggregate for sustainable built environment – a review. *Int. J. Sustain. Built. Environ.* 5: 312–333.

Rahimian, R. and Zarinabadi, S. (2020). A review of studies on the removal of methylene blue dye from industrial wastewater using activated carbon adsorbents made from almond bark. *Prog. Chem. Biochem. Res. J. Homepage* 3: 251–268. 10.33945/SAMI/PCBR.2020.3.8.

Rashid, J., Tehreem, F., Rehman, A., and Kumar, R. (2019). Synthesis using natural functionalization of activated carbon from pumpkin peels for decolourization of aqueous methylene blue. *Sci Total Environ.* 2019 (671): 369–376. 10.1016/j.scitotenv.2019.03.363.

Regti, A., Laamari, M.R., Stiriba, S.E., and El Haddad, M. (2017). Potential use of activated carbon derived from *Persea* species under alkaline conditions for removing cationic dye from wastewaters. *J. Assoc. Arab Universities Basic Appl. Sci.* 24: 10–18. 10.1016/j.jaubas.2017.01.003.

Ren, Y., Cui, C., and Wang, P. (2018). Pomelo peel modified with citrate as a sustainable adsorbent for removal of methylene blue from aqueous solution. *Molecules* 23: 1342.

Rigueto, C.V.T., Piccin, J.S., Dettmer, A. et al. (2020). Water hyacinth (Eichhornia crassipes) roots, an amazon natural waste, as an alternative biosorbent to uptake a reactive textile dye from aqueous solutions. *Ecological Engineering.* 150: 105817.

Rosales, E., Meijide, J., Pazos, M., and Sanromán, M.A. (2017). Challenges and recent advances in biochar as low-cost biosorbent: from batch assays to continuous-flow systems. *Bioresour. Technol.* 246: 176–192.

Sadh, P.K., Duhan, S., and Duhan, J.S. (2018). Agro-industrial wastes and their utilization using solid state fermentation: a review. *Bioresour. Bioprocess.* 5: 1. 10.1186/s40643-017-0187-z.

Salleh, M.A.M., Mahmoud, D.K., Abdul Karim, W.A.W., and Idris, A. (2011). Cationic and anionic dye adsorption by agricultural solid wastes: a comprehensive review. *Desalination* 280: 1–13.

Salmania, M.H., Miria, M., Ehrampousha, M.H. et al. (2017). Comparing cadmium removal efficiency of a magnetized biochar based on orange peel with those of conventional orange peel and unmodified biochar. *Desalin. Water Treat.* 82: 157–169.

Sewu, D.D., Boakye, P., and Woo, S.H. (2017). Highly efficient adsorption of cationic dye by biochar produced with Korean cabbage waste. *Bioresour. Tech.* 224: 206–213. 10.1016/j.biortech.2016.11.009.

Shamraiz, U., Hussain, R.A., Badshah, A. et al. (2016). Functional metal sulfides and selenides for the removal of hazardous dyes from Water. *J. Photochem. Photobiol. B* 159: 33–41.

Shang, J.G., He, W., and Fan, C.X. (2015). Adsorption of dimethyl trisulfide from aqueous solution on a low-costadsorbent: thermally activated pinecone. *Chin. J. Oceanol. Limn.* 33 (1): 169–175.

Shen, K. and Gondal, M.A. (2017). Removal of hazardous Rhodamine dye from water by adsorption onto exhausted coffee ground. *J. Saudi Chem. Soc. 21*: S120–S127. 10.1016/j.jscs.20.13.11.005.

Siqueira, T.C.A., Zanette da Silva, I., Rubio, A.J. et al. (2020). Sugarcane bagasse as an efficient biosorbent for methylene blue removal: Kinetics, isotherms and thermodynamics. *Int. J. Environ. Res. Public Health.* 17: 1–13. 10.3390/ijerph17020526.

Streit, A.F.M., Côrtes, L.N., Druzian, S.P. et al. (2019). Development of high quality activated carbon from biological sludge and its application for dyes removal from aqueous solutions. *Sci. Total Environ.* 660: 277–287. 10.1016/j.scitotenv.2019.01.027.

Sulyman, M. and Gierak, A. (2020). Green environmental approach for adsorption of hazardous dye from water using tree and sea plant leaves (Dead L.), green-environmental-approach-for-adsorption-of-hazardous-dyefrom-water-using-tree-and-sea-plant-leaves-dead-l. *Act. Scie. Agri.* 4: 1–10. 10.31080/asag.2020.04.

Sun, Z., Qu, K., Cheng, Y. et al. (2021). Corncob-derived activated carbon for efficiently adsorption dye in sewage. *ES Food Agrofor.* 4: 61–73. 10.30919/esfaf473.

Sureshkumar, M.V. and Namasivayam, C. (2008). Adsorption behavior of Direct Red 12B and Rhodamine B from water onto surfactant-modified coconut coir pith. *Colloids Surf.* 317: 277–283. 10.1016/j.colsurfa.2007.10.026.

Takabi, S.H., Shirani, M., and Semnani, A. (2021). Apple stem as a high performance cellulose based biosorbent for low cost and eco-friendly adsorption of crystal violet from aqueous solutions using experimental design: mechanism, kinetic and thermodynamics. *Environ. Technol. Innov.* 24: 101947. https://doi.org/10.1016/j.eti.2021.101947.

Tan, K.B., Vakili, M., Hord, B.A. et al. (2015). Adsorption of dyes by nanomaterials: recent developments and adsorption mechanisms. *Sep. Purif. Technol.* 150: 229–242.

Tang, Y., Zhao, Y., Lin, T. et al. (2019). Adsorption performance and mechanism of methylene blue by H_3PO_4- modified corn stalks. *J. Environ. Chem. Eng.* 7: 103398.

Tang, Y., Lin, T., Jiang, C. et al. (2021). Renewable adsorbents from carboxylate-modified agro-forestry residues for efficient removal of methylene blue dye. *J. Phys. Chem. Solids.* 149: 109811.

Tao, X., Wu, Y., and Cha, L. (2019). Shaddock peels-based activated carbon as cost-saving adsorbents for efficient removal of Cr (VI) and methyl orange. *Environ. Sci. Pollut. Control Ser.* 26 (19): 19828–19842.

Ukiwe, L.N., Ibeneme, S.I., Duru, C.E. et al. (2013). Chemical and electro-coagulation techniques in coagulation-floccculation in water and wastewater treatment – a review. *J. Adv. Chem.* 9: 1988–1999. https://doi.org/10.24297/jac.v9i3.1006.

Varghese, A.G., Paul, S.A., and Latha, M.S. (2019). Remediation of heavy metals and dyes from wastewater using cellulose-based adsorbents. *Environ. Chem. Lett.* 17: 867–877. 10.1007/s10311-018-00843-z.

Vučurović, V.M., Razmovski, R.N., Miljić, U.D., and Puškaš, V.S. (2014). Removal of cationic and anionic azo dyes from aqueous solutions by adsorption on maize stem tissue. *J. Taiwan Inst. Chem. Eng.* 45: 1700–1708. 10.1016/j.jtice.2013.12.020.

Wazir, A.H., Waseem, I., Qureshi, I., and Manan, A. (2020). *Saccharum arundinaceum* leaves as a versatile biosorbent for removal of methylene blue dye from wastewater. *Environ. Eng. Sci.* 37 (11): 737–745. 10.1089/ees.2020.0075.

Yagub, M.T., Sen, T.K., Afroze, S., and Ang, H.M. (2014). Dye and its removal from aqueous solution 28 by adsorption: a review. *Adv. Colloid Interface Sci.* 209: 172–184. 10.1016/j.cis.2014.04.002.

Zhou, L., Jin, J., Liu, Z. et al. (2011). Adsorption ofacid dyes from aqueous solutions by the ethylenediamine-modified magnetic chitosan nanoparticles. *J. Hazard. Mater.* 185: 1045–1052.

Zhou, Z., Lin, S., Yue, T., and Lee, T. (2014). Adsorption of food dyes from aqueous solution by glutaraldehyde cross-linked magnetic chitosan nanoparticles. *J. Food Eng.* 126: 133–141.

Zhou, Y., Lu, J., Zhou, Y., and Liu, Y. (2019). Recent advances for dyes removal using novel adsorbents: a review. *Environ. Pollut.* 252: 352–365. 10.1016/j.envpol.2019.05.072.

8

Agricultural Waste as Source of Organic Fertilizer and Energy

Ajay Neearaj, Asha Humbal, R.Y. Hiranmai, and Bhawana Pathak

School of Environment and Sustainable Development, Central University of Gujarat, Gandhinagar, Gujarat, India

CHAPTER OUTLINE

8.1 Introduction

Agro-based products or by-products are being more widely used regularly worldwide because they are eco-friendly and safe, and can be made sustainably. Agricultural waste, also known as 'agricultural waste product', is being produced faster than organic, which is used in a less expensive alternative to petroleum-derived energy (Chitara et al. 2021). Agricultural wastes include waste from producing and processing basic agricultural items, for example, crops, vegetables, fruits, dairy products, meat, and animals (Obi et al. 2016). The majority of developed countries face the challenge of rising agricultural production without degrading the climate (Conrad et al. 2018). According to the Organization for Economic Cooperation and Development (OECD), agricultural waste is produced by

Agriculture Waste Management and Bioresource: The Circular Economy Perspective, First Edition.
Edited by Suruchi Singh, Pardeep Singh, Anu Sharma, and Moharana Choudhury.

various agricultural activities, including manure and crop residues. Fertilizer run-off from fields, farms, poultry houses, pesticides enters the water, air, soils, and silt drainage from fields. In India, approximately 350 million tonnes of organic wastes are generated annually from agricultural sources (Pappu et al. 2007). One such activity is proper agro-waste management, especially when it comes to agricultural and food processing wastes (Prasad et al. 2020). They are non-product outputs of agricultural product harvesting and management that can produce content that can support man but whose economic costs are less than the expense of selection, shipping, and processing for beneficial use (Obi et al. 2016). It is challenging to develop new ideas for effective waste management because waste breakdown necessitates special procedures involving time, resources, and money to recognize the issue at the beginning of waste generation. It is essential to consider 'waste' as a valuable 'resource' that can be recycled into several valuable items. Reducing the wastage into a resource that can be put to primary use can be regarded as a wealth-generating process.

Consequently, the word 'Waste to Wealth' was coined (Obi et al. 2016). The idea of waste-to-wealth has been used to solve the environmental crisis by shifting the conventional image of waste as an end commodity to be disposed of, and raising environmental consciousness and converting it into potential value had been a significant source of contention around the world. Given the volume of waste generated, creative waste conversion technologies can generate massive amounts of micro-entrepreneurship opportunities. The scope for waste-to-wealth business in India is enormous. Growing opportunities for this sector may have a variety of benefits. Agricultural and its allied activity play an important part in the livelihood and India's economy. It employs approximately 50% of the population and contributes about 17% to the total gross domestic product (GDP) (Van Arendonk 2015). So, by proper processing, agricultural waste can be converted into high-quality agricultural organic fertilizer, allowing for systemic optimization of agricultural fertilizer, improved farmland fertility, and the development of agriculture's long-term sustainability. India creates an average of 500 million tonnes (Mt) of agricultural waste every year, as per the Ministry of New and Renewable Energy (MNRE) (National Policy for Management of Crop Residues (NPMCR) 2014). According to the research, the bulk of agricultural waste is utilized for feed, fuel, and other home and industrial reasons. However, there is still a 140 million tonnes excess, of which 92 million tonnes is burn up every year (Krajick and Lee 2016). The comparison of agricultural waste produce by different countries in metric tonnes per year is shown in Table 8.1 (FAO 1996). The amount of agricultural wastes burned in India is also significantly higher than the amount produced in other countries in the region. However, at each stage of agricultural processing, consumption and solid agricultural wastes are generated in large quantities.

Table 8.1 In comparison with other countries in the area, India generates a lot of agricultural waste generation (MT/Y).

Country	Agricultural waste generated (million tonnes/year)
India	500
Bangladesh	72
Indonesia	55
Myanmar	19

8.2 Agricultural Waste

Agricultural solid wastes are primarily produced as a result of agricultural practices. Moreover, agricultural wastes also generate due to food chain and other aspects like farming (Pappu et al. 2007). Agricultural waste products can be produced at any point and step of the agricultural food chain. Cultivation is the main source of agricultural waste. However, each and every step of farming processes from land clearance to harvest. The following items are included in the general classification of agricultural solid wastes as shown in Figure 8.1.

Agricultural activities generate a large amount of agricultural residues. These residues further divide into two main types of agriculture residues: field residues and process residues (Sadh et al. 2018). After crop harvesting, residues left on the field are called field residues. Field residues include a mixture of leaf, stem, stalk, and seed pods, while a processing residues remain after the crop has been processed into another valuable resource. These include molasses, husks, straw, bagasse, leaves, seeds, stems, roots, stalks, shells, pulp, peels, stubble, etc., and are used for animal feed, soil improvement, fertilizers, manufacturing, and a variety of other purposes (Bhuvaneshwari et al. 2019). However, large amounts of field residues are generated, but they are mostly underutilized.

Process residues generated through food processing industries such as juice, chips, meat, confectionery, and fruit produce vast amounts of organic compound and associated effluents every year (Ramteke et al. 2022). Population growth has increased the demand for food and its use of it. Consequently, different food and beverage industries have remarkably increased in most countries in that region to meet the food needs of the residents (Kearney 2010). In the food industry, wastes contain a high biological oxygen demand (BOD), chemical oxygen demand (COD), and other suspended solids contents. In most cases, this solid waste is left untreated, resulting in harmful environmental impacts and adverse health impacts (Digman and Kim 2008). The oil extraction process also results in a tremendous amount of residues, referred to as oil cakes, which are especially common in the oil industry (Gullon et al. 2020). Industrial residues can cause the environmental pollution as they contain high concentration of suspended solids, dissolved solids, oil, and fat. In addition, during the cultivation activity, tropic climates are suitable for growing crops but at the same time, they are conducive to the development of insects and weeds (Obi et al. 2016). This creates a high demand for pesticides to control insects and stop the spread

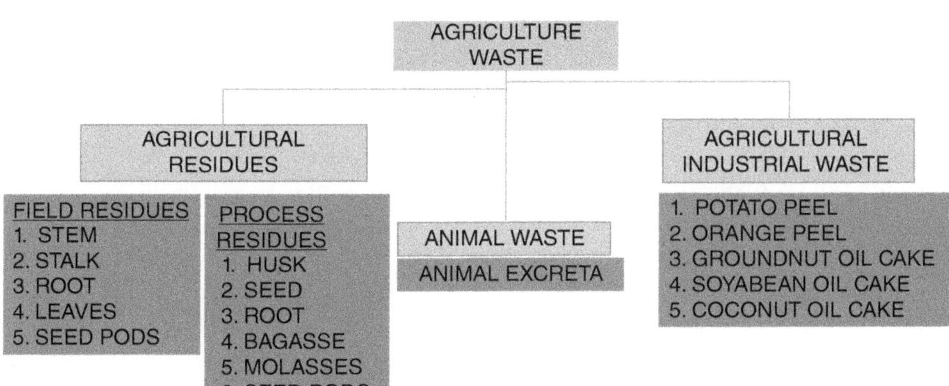

Figure 8.1 Different types of agriculture wastes and their sources.

of disease, but farmers often abuse pesticides. Most pesticide bottles and packages are dumped in fields or ponds after their use. Approximately 1.8% of chemical residues in their packaging remain, according to estimates made by the Plant Protection Department (PPD) (Obi et al. 2016). Due to their potentially lasting and toxic chemicals, these wastes can potentially cause many unexpected effects on the environment, such as food poisoning, poor food hygiene, and contaminated farmland. Furthermore, agricultural wastes also generated by livestock production include manure and organic materials in slaughterhouses and wastewater such as urine, cage wash water, and wastes generated from bathing and cleaning animals in slaughterhouses (Zakaria 2018). Moreover, solid wastes are also produced during horticultural activity, such as production and maintaining of garden plants and beautification flowers and plants they are used, and pruning and lawn cuttings are examples of such wastes (De Corato 2020).

8.3 The Potential Risk of Agriculture Waste on Human and Environment

Inappropriate controlling of agricultural waste can cause harmful effects to human health and environment by polluting air, water, and soil because of indiscriminately burning produced wastes (Adejumo et al. 2015). Agricultural solid wastes have been dumped and burned indiscriminately, resulting in wastes, which are a threat to human beings life, and other environmental issues that require global concentration; even though these wastes are rich in nutrients necessary for long-term agricultural output, they can be recycled to increase the soil fertility (Wang et al. 2015).

The burning of agriculture residues or manure generates many toxic gases, dust, and smoke, which can directly affect the air at the local and regional levels. Specifically, in New Delhi, during the winter, air quality becomes worsened due to post-harvest burning of crop residues. According to the new study, excessive fertilizer and ammonia gas are released from livestock waste in the air, which combines with pollutants such as nitrogen dioxide and sulphates, creating fine particulate matter (PM) (Krajick and Lee 2016). This fine PM can penetrate deep into the lungs and cause heart or pulmonary disease. In 2015, the amount of ammonia released into the atmosphere by agricultural activities accounted for 83% of agricultural air pollution. Moreover, according to the USEPA, agriculture activities are one of the major sources for the emission of greenhouse gases (USEPA 2015; Aruwajoye et al. 2020). Methane and nitrous oxide are the main gases released by agricultural activities. Methane is majorly produced from enteric fermentation and manure management, while the source of nitrogen dioxide includes soil management activity such as fertilizer (FAO 1996). Agriculture is considered as the biggest user of freshwater services, consuming an estimated percentage of all surface water supplies. As a result, it can cause major pollution to the surface water and groundwater resources (FAO 1996; Pedrero et al. 2010) Pollution caused by improper agricultural production, which includes the use of fertilizers and pesticides of various sorts (herbicides, insecticides, and fungicides) as well as other agricultural mismanagements, such as tillage of inappropriate soils, directly influences the water quality (FAO 2013). Pesticides may have a major impact on aquatic life and water quality. Surface run-off or spray drift could transport chemical residues to lakes, streams, and ponds, and they will have hazardous effects (for example, fish kills) and chronic poisoning. Pesticides stored in poor conditions are harmful to health and the environment, especially whenever they are stored in cities or near water bodies (Heaton et al. 2013). Farmers with a lack of pesticide managing knowledge are more likely to use older, more harmful, and environmentally persistent pesticides. This may be a specific concern in some developing countries. The problem is due to the lack of strict regulations or the lack of implementation of current regulations (Eddleston et al. 2002). Due to the run-off, farmers who used fertilizers to enhance agricultural production, particularly (N) nitrogen and (P) phosphorus, may end up in water bodies and groundwater. They could have a significant impact on river and groundwater quality, as well as on the eutrophication of wetlands and aquatic bodies (Pedrero et al. 2010). Excessive nutrient concentrations enhance algae growth, resulting in unstable marine environments with phytoplankton blooms, excess organic matter production, increased oxygen utilization, primary oxygen loss, and the loss of benthic species that live on or near the seafloor. Irrigation in agriculture also has consequences; for example, salt run-off induces salinization of surface waters, while fertilizer run-off and chemical run-off cause aquatic pollution and bioaccumulation in edible fish species. Also, animal waste is the major source of ammonia nitrogen (NH3-N) and total phosphorus (TP), which can cause eutrophication (USEPA 2014). Numerous studies found that high concentrations of nitrate in shallow freshwater aquifers are directly linked to agricultural land at different scales, ranging from controlled plot experiments to basin-sized inventories. Many studies found that the nitrate concentration in forested-pasture-grassland regions observed 2 mg/l NOa-N, and near highly cultivated and fertile (typically >5 mg/l to >100 mg/l NOs-N) are higher. Several studies have discovered connection between nitrogen fertilization rates or histories and nitrate seepage into groundwater. Furthermore, since cyazofamid is a fungicide with a short half-life in indifferently textured agricultural soils, it poses no danger to groundwater (Singh and Tandon 2015). Apart from that, the increased use of pesticides has been one of the main causes of soil contamination. Seventy percent apply pesticides for high production in agriculture, polluting farmland soil with pesticide residues (Zhang et al. 2018). Pesticides have a wide range of effects on soil quality and agricultural productivity (Jin et al. 2015). Increasing herbicide concentrations beyond their prescribed field application rates could change micro-organism development and activity, disrupting nutrient cycling in soil (Das et al. 2015). In groundnut fields, insecticides (acetamiprid and carbofuran) induced enzyme activity (arylamidase and myrosinase). However, high concentrations of these pesticides negatively affect enzyme activity in soils (Jaffer Mohiddin et al. 2015). Organophosphorus pesticides (OPs) are one of the most commonly found pesticides in polluted soils, and they pose a potential hazard to wildlife and human health (Jaramillo Colorado et al. 2016). Organochlorine pesticides (OCs), which are known as persistent organic contaminants (POPs) and banned for use due to their longevity in the atmosphere, were also present in soils and air (Jaramillo Colorado et al. 2016).

New technology and training are needed to utilize wastes best to enable farmers to create valuable products from wastes generated on their farms. Farmers must realize that there is no waste in agriculture and that anything can be turned into something useful. The disposal/utilization of agricultural wastes has recently become a big concern. Agricultural wastes may be used to produce electricity, biofuel, biofertilizer, compost, biogas, paper, and other products. Sustainable resource usage is the best substitute for reducing global environmental problems exacerbated by waste disposal. Power generation from agro-wastes is a new eco-friendly resource for sustainability, since it produces fewer emissions than traditional energy supplies (coal), and it will eventually meet the world's energy demand (Shehrawat et al. 2015).

8.4 Traditional Technologies Used for Recycling of Agricultural Waste

8.4.1 Composting

India has a long tradition of composting. There is limited literature about the benefits of composting in India's agriculture industry (Bhuvaneshwari et al. 2019), although backyard composting and MSW composting are popular. Agriculture is the first and most effective method of recycling organic wastes. By definition, composting involves decomposing organic material in aerobic conditions using micro-organisms and worms, e.g. worms or earthworms. Crop residue is high in organic content, similar to food waste and animal manure, making it an ideal raw material for composting (Misra et al. 2003). A compost pile is the result of a controlled process of microbial decay or decomposition of organic material (Hettiarachchi et al. 2018). By composting agricultural waste, a reduction in agricultural waste is possible and its environmental impact is minimized (Mohee 2007). In addition to the soil conditioner properties, this compost is rich in humus. Plants can access essential nutrients from the soil due to the microbial community.

Composting is mediated by a variety of micro-organisms that function in an aerobic environment. The decomposition of organic biomass is assisted by a variety of organisms, including bacteria, fungi, actinomycetes, protozoa, and algae (Torres et al. 2014). In aerobic conditions, smaller molecules are formed when the organic matter of animal or plant origin decomposes. Soil organic matter can be recycled and crops can grow in a humus-rich compost that is more stable and hygienic. Furthermore, it helps to establish the organic matter in soil and add specific nutrients to the soil and act as the organic fertilizer, which helps to reduce the use of chemical fertilizer. Compost can increase the soil water retention capacity by making soil porous (Sequi 1996). In addition, compost can help strengthen roots and resist pests and decay. Therefore, composting of agricultural waste is considered as low-cost, beneficial traditional method for recycling agricultural waste (Manderson 2009). Agricultural chemicals such as fertilizer and insecticides are completely replaced by compost in improving the soil's physicochemical and biological properties. A cultivated soil, with compost, has increased yield potential as well as increased resistance to droughts, diseases, and toxicity. As organic matter decomposes, it occurs in two stages: degradation and maturation. In the early stages of decomposition, organic substances such as sugars, amino acids, and organic acids are degraded. Carbon dioxide and energy are released by aerobic micro-organisms (Lei et al. 2010). During the thermophilic phase, temperatures are high, pH is high, and humidity is high, requiring several weeks to months for microbes to activate (Singh et al. 2021). Further microbial population reduction occurs as a result of breakdown of compound organic compounds in the second phase. When the temperature is lowered to 40–45 °C, the thermophilic phase switches to a mesophilic phase. The final stage of the process results in a temperature that is comfortable for the body, and thus it is no longer biologically active. Eventually, a dark brown to black substance similar to soil is produced. Numerous factors affect the aerobic composting process, including oxygen availability, moisture content, nitrogen availability, temperature, pH, lignin content, and nutrients. Composting is dependent on temperature; higher temperatures in the thermophilic range assist in disinfecting organic waste and killing pathogens (Sonesson et al. 2000). A study by Singh and Prabha looked at biocomposting in Uttar Pradesh, India (Pratap Singh and Prabha 2017). This study examined wheat straw, rice straw, vegetable crops, and garden plant leaves as the entire biomass. Overall, there were 45% solids, 26.7% organic matter, 15.3% carbon, and 1.36% nitrogen in the compost, which indicates a high carbon and organic matter content. Rice and wheat crops were significantly improved in agronomical properties after they were tested (Brady et al. 2008).

8.4.2 Vermicomposting

Vermicomposting is a well-known technique in India and a long-standing practice. It presents an attractive, efficient, and environmentally friendly method for treating and managing solid wastes generated from various sources, including industrial, agricultural, and household wastes. Vermicomposting is a solid waste management process in which solid waste converts into valuable organic fertilizer with the help of earthworms and microbial communities (Edwards et al. 2004). Degradation of the biowaste or biosolid occurs in aerobic conditions in the environment. Here, the symbiotic interaction between earthworms and micro-organisms promotes the degradation of the solid waste to convert into organic fertilizer (Gupta and Garg 2008). The main product of earthworm composting is the vermicompost or vermicast, which has a good fertilizer value. Also, vermicompost contains high organic content, suitable for organic farming and sustainable agriculture practices (Lim et al. 2012). However, to optimize the biodegradation processs, selection of suitable earthworm species is essential. The selection of the earthworm species is based on high consumption of organic waste, high productivity rate, and high tolerance capacity to an environmental factor. Thus, some earthworm species like *Epigeic* earthworm are the most suitable in the vermicomposting process. 'Epigeic' earthworms have effective biodegradation and nutrient release, are tolerant to disturbances, and aid in early litter decomposition. Epigeic earthworms include *Lumbricus rubellus, Eisenia fetida, Eiseniella tetraedra, Lumbricus castaneus, Lumbricus festivus, Bimastus minusculus, Bimastus eiseni, Dendrobaena veneta, Dendrodrilus rubidus,* and *Dendrobaena octaedra* (Maheshwari and Sudha 2013). The soil loosens and becomes porous as a result of earthworm activity. Aeration, water absorption, drainage, and root penetration are all improved by porosity. Earthworms and related bacteria assist to sustain soil ecosystems by forming soil aggregates. Earthworms help to break down cellulose, produce soil, and accumulate humus. The physical, chemical, and biological aspects of soil are all negatively impacted by earthworms' biological activities.

Earthworms consume organic waste and are remarkable in that they only take a tiny portion of it for the development and excrete the remainder of it partially digested (Jambhekar 1992). This is because earthworm intestines contain a variety of microorganisms, hydrolytic enzymes, and hormones that contribute to the fast breakdown of partially digested material, resulting in the transformation of complex organic waste into vermicompost in as short as 1–2 months (Nagavallemma et al. 2004). The breakdown and formation of vermicompost by earthworms occur in phases. The organic matter is ground into a fine powder by the earthworm's gizzard. Cellulase, amylase, lipase, protease, urease, and chitinase are hydrolytic enzymes (Munnoli et al. 2010). The fermentation processes

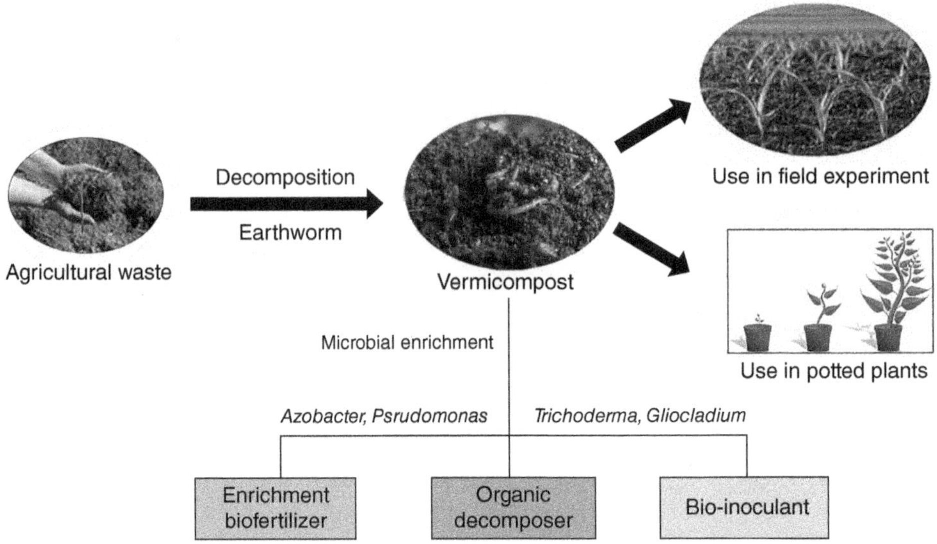

Figure 8.2 Recycling of agricultural waste using vermicomposting method. A. *Source:* Fotografie-Schmidt/Adobe Stock B. *Source:* kaliantye/Adobe Stock C. *Source:* Smileus / Adobe Stock.

of micro-organisms and other fermenting substances allow these compounds to break down in the stomach, resulting in 'casts'. This eventually becomes 'vermicompost' (Edwards et al. 2004). During vermicomposting, about half of the organic compounds in the feedstock are converted into carbon dioxide, while the remaining organic compounds are formed into more stable molecules. A more significant decrease in CO_2 was observed in vermicompost when compared to compost (Song et al. 2014). Vermicompost, as an organic manure, provides a better alternative to chemical fertilizers than chemical fertilizers, because it improves soil structure and yields, as well as promotes microbial biodiversity. The following wastes are managed via vermicomposting: food waste, paper, cardboard, manure, and biosolids, and vermicompost can even improve soils (Parthasarathi et al. 2016). Further, vermicomposting may lead to job creation. Both agriculture and horticulture can use vermicompost. Additionally, vermicomposting can benefit both agricultural and horticultural output. Vermicomposting does not dispose of materials or burn them but rather recycles them. This is why vermicomposting focuses on resource conservation and sustainable utilization, as shown in Figure 8.2 (Gupta et al. 2019).

8.5 Advanced Technology of Agricultural Waste for the Energy Production

8.5.1 Waste to Energy

In today's world, energy is one of the critical issues for discussing sustainable development. Currently, fossil fuels are predominantly used to meet energy demands, depleting at an alarming rate (Toor et al. 2020). Therefore, it is necessary that rather than depending on fossil fuel, there is a need to move towards the supply of affordable and clean energy or renewable energy, which do not cause any negative impact needs for long-term sustainability. Renewable sources of energy include solar, wind, tides, waves, and bioenergy, respectively. Energy from waste is also viewed as a sustainable energy source. By waste to energy technology, we can produce helpful energy forms such as biogas, bioethanol, biodiesel, and biohydrogen. The utilization of waste into energy serves as a platform for bioeconomy. Bioeconomy refers to all activities, services, processes, and sectors dependent on biological resources (Leal Filho 2018). Bioeconomy focuses on resources being converted into value-added goods and utilizing waste streams produced in the process sustainably as well. Several alternative techniques are available for utilizing a huge amount of agro-waste to exploit the inherent energy present in agricultural waste and protect the environment (Kumar et al. 2019).

8.5.2 Combustion

Biomass is widely burned using combustion in air as a practical and straightforward method. It is possible to utilize the chemical energy found in waste to generate heat, mechanical energy, and electricity. Many conversion devices such as steam turbines, boilers, and furnaces could be used to accomplish this process. An electric generating system is powered by a boiler that uses biomass burning to produce high-pressure steam that powers a turbine. The conversion efficiency of combustion technology generally ranges between 20 and 40% (Kumar et al. 2015).

8.5.3 Gasification

Gasification is the process through which agricultural waste is converted into synthetic gas. The gasification method can gasify and transform agricultural crop waste generated on a massive scale into syngas (Memon et al. 2016). Methane, carbon dioxide and hydrogen are the main components of the syngas. As a result of controlled oxidation of biomass, gasification necessitates high temperatures ranging from 800 to 1100 °C. It can be used to create mechanical or electrical power directly or as a fuel for internal combustion engines. It can be used to create mechanical or electrical power directly or as a fuel for internal combustion engines. Gasification can produce syngas with a calorific value of 4–5 MJ/m^3 composed of 40–70% H_2, 15–25% CO, and 1–2% CO_2 (Ravindranath and Balachandra 2009).

8.5.4 Pyrolysis

Pyrolysis is the process of converting biomass into biochar, bio-oil, and non-condensable gases, such as CO, CO_2, H_2, CH_4, and nitrogen. A higher temperature of 300–880 °C is needed when no oxygen is present. Pyrolysis products' composition is influenced by the temperature and residence time of the process. Bio-oil is produced through the pyrolysis of agricultural wastes and can be used to replace fossil fuels in a variety of applications, such as boilers and diesel engines for power generation, as well as an important source of chemicals (Kim et al. 2010).

8.5.5 Fermentation

The fermentation method is mainly used to generate bioethanol, from agricultural waste. Agricultural waste contains cellulose (grass and wood), starch (wheat, grains, and cone), and sugar (molasses and sugarcane). In this process, enzymes convert starch into sugar, which is then turned to ethanol by the yeast (Kumar et al. 2015). Mainly, sugarcane is used in India to produce bioethanol. The 1000 kg of dry maize can produce 450 l of bioethanol via fermentation. The solid waste generated through this method can be utilized as animal feed, while bagasse from sugarcane can be utilized as boiler fuel or for biomass gasification.

8.5.6 Anaerobic Digestion

In an anaerobic digestion process, bacteria consume organic matter to break it down in the absence of oxygen, for example, animal dung, liver stock, agricultural residues, food waste, and biosolids. Anaerobic digestion is facilitated via a microbial process that comprises four various phases: hydrolysis, acidogenesis, acetogenesis, and methanogenesis. The anaerobic digestion method is a step in a cascade agro-waste system that produces renewable energy such as biogas and a digested material known as biofertilizer. The main component of biogas includes methane and CO_2, with a minor amount of moisture, hydrogen sulphide (H_2S), and other gases (Ghosh et al. 2020). This biogas is used for heat and generating electricity. In addition, by-products of anaerobic digestion rich in humus material can be used as a soil enhancer which can provide the optimum condition, requiring nutrients such as nitrogen, potassium, and phosphorous with a pH of approximately 7, and the temperature should also be in control. Lignocellulosic part of the waste mainly present in agricultural waste is not fully utilized for energy recovery in anaerobic digestion due to recalcitrance of lignin, a significant portion of the organic material stays in the digested material (Antoniou et al. 2019). In addition, anaerobic digestion required the pre-treatment of agricultural waste, which is a complex and the primary obstacle to using agri-waste in anaerobic digestion. However, this technique has various advantages which make it a viable alternative for agricultural waste valorization; for example, it is easy to operate, is low cost, and has no requirement for electricity for biogas production. The main advantage of this method includes less emission of greenhouse gases and other pollutants compared to other techniques like combustion, pyrolysis, and gasification. Compared to combustion, gasification, and pyrolysis, the key advantage of this process is that it emits lower greenhouse gases (GHGs) and other air pollutants, resulting in net environmental benefits (Figure 8.3).

8.5.7 Transesterification Process

The transesterification process produces biodiesel from biomass. NaOH and KOH or alcohol, in general methanol, are used as catalysts in this process. Biodiesel consists of a long-chain alkyl ester (methyl, propyl, or ethyl) which can be used to power diesel engines or as a vehicle fuel. Various feedstocks can be converted to biodiesel through the esterification and transesterification processes, including animal fats, vegetable oil, non-edible oils,

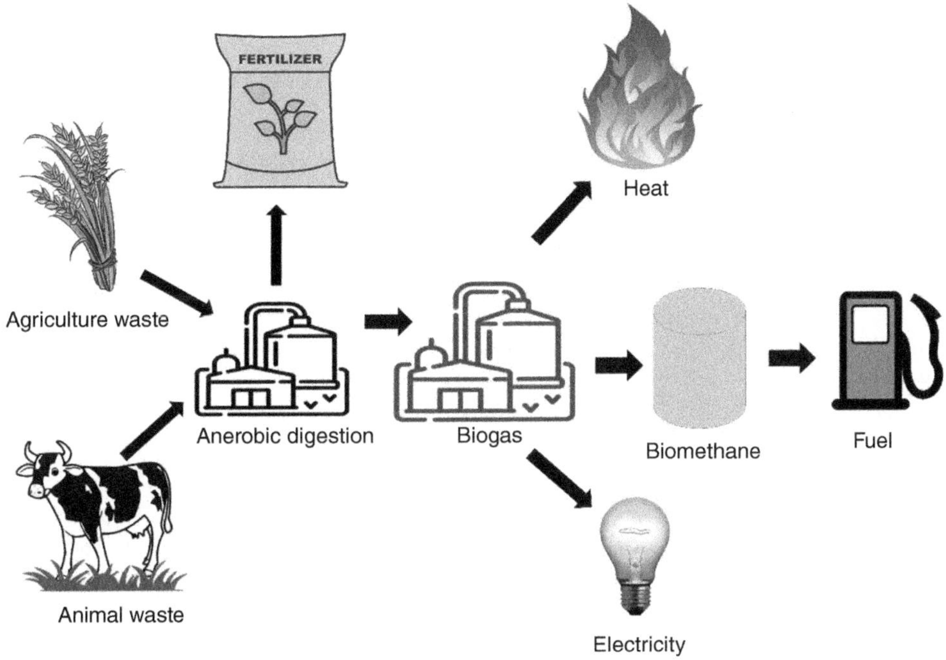

Figure 8.3 Anaerobic digestion of agricultural waste recycling.

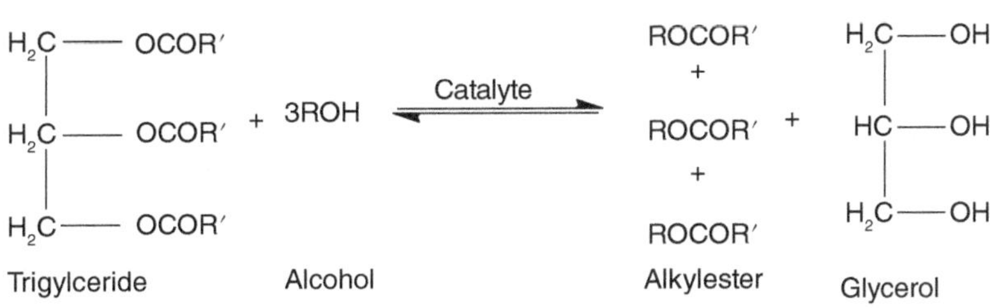

Figure 8.4 Transesterification process for the production of biofuel.

waste cooking oil, and so on (Borugadda and Goud 2012). Because these oils cannot be consumed for human consumption, they make better biodiesel feedstocks. Biodiesel is often manufactured from oilseeds such as Pongamia and Jatropha. In cosmetics, pharmaceuticals, and other industries, glycerol is a by-product of the transesterification reaction (Figure 8.4).

8.6 Sustainable Use of Agriculture Waste for Biofuel Production

Several plants, such as Pongamia and Jatropha, are used to produce biofuels. But production of this plant enhances the issue of food versus fuel, and therefore, this method is not considered as a long-term biofuel production. On the other hand, the rapid growth of the population requires high crop production, which generates a large amount of agricultural waste. Generated agricultural waste became a problem for the environment due to improper management practices to recycle agricultural waste. In addition, millions of tonnes of agro-waste are generated due to various agriculture activities and lack of cost-effective technology. Farmers are forced to burn this waste, and indirectly, it causes environmental or air pollution (Li et al. 2008).

Agriculture waste generally consists of complex lignocellulosic material, e.g. crop residue, wheat straw, rice straw, sawdust, bagasse, wood chips, and other lignocellulosic materials. Lignocellulosic material is the polymer of combined cellulose, hemicellulose, and lignin, a complex polymer. On the other hand, cellulose is a long linear polymer composed of glucose sugar linked together by beta1, 4-glycosidic connections, while hemicellulose is composed of D-xylose and D-arabinoside heteropolymers. Lignocellulosic material is protected from micro-attack, since lignin is secured to two carbohydrate polymers (Peiji et al. 1997). Using this complicated lignocellulosic biomass for biofuel production requires three steps. The first is to preprocess it for delignification, which breaks down the cellulose and hemicellulose. Xylose, glucose, mannose, galactose, and other sugars are obtained by hydrolysis of cellulose and hemicellulose (Panpatte and Jhala 2019). A grown from various agricultural wastes such as wheat straw, rice straw, corn straw, bagasse, and crop residue, lignocellulosic biomass can produce 442 billion litres of bioethanol each year (Pullammanappallil et al. 1998). The burning of agricultural waste generates greenhouse gases; crop residues can be used to generate energy, reducing emissions.

Furthermore, it helps to reduce the release of pollution from applying organic residue to land by reducing the release of pollutants to the air, twater, and soil (Champagne 2007). As a result, these crop residues also support soil health by improving soil organic matter, improving soil texture, enhancing nutrient availability, maintaining soil pH levels, increasing soil water retention, reducing soil erosion, and improving soil fertility (Reijnders 2008). Furthermore, bagasse is considered as a significant agricultural waste, since it is the residue of sugarcane juice. An acre of sugarcane yields approximately 280 kg of bagasse (Canilha et al. 2012). There are around 19–24% of lignin in bagasse, 27–32% of hemicellulose, 32–44% of cellulosic fractions, and 4.5% of ash, with a minor proportion of minerals, waxes, and well other constituents that are suitable for producing biofuel (Jacobsen and Wyman 2002). The study reported that corn stover and grain can be significantly used for biofuel production (Graham et al. 2007). Approximately 35 million litres of bioethanol produce from full utilization of corn waste, which could produce 25 million litres of gasoline (Pullammanappallil et al. 1998). Rice is consumed by humans in roughly 88% of cases, used for animal feed in 2.6% of cases, and discarded in 4.8% of cases. Biochemical composition of rice straw includes about 32–47% of cellulose, 19–27% of lignin, and 19% of ashes as well as it having different types of sugars, e.g. 41–43, 15–20, 3–5, 2, and 0.4% of glucose, xylose, arabinose, mannose, and galactose, respectively (Roberto et al. 2003). Rice straw would contain about 205 billion litres of bioethanol per year, accounting for about 5% of global ethanol intake (Belal 2013). Wheat straw is also an essential source of bioethanol. Also, wheat straw contains a key source for bioethanol production. It contains 33–40% of cellulose, 20–25% of hemicellulose, and 15–20% of lignin to manufacture cellulosic ethanol, and wood waste is a viable natural resource. The largest biomass available on the earth is wood waste created by forest operations (Dan et al. 2015). Without deforestation,

Table 8.2 Some study of the energy production from agricultural waste.

Type of waste	Type of energy production	References
Lignocellulosic waste	Natural gas	(Jaffar et al. 2020)
Lignocellulosic biomass	Diesel production	(Wong et al. 2016)
Crop residue	Hydrogen production	(Zhang et al. 2017)
Poultry litter	Fuel	(Staroń et al. 2017)
Agriculture and livestock wastes	Biogas	(Alatzas et al. 2019)
Agri-food and agricultural wastes	Biogas	(Tamburini et al. 2020)
Agricultural waste	Biogas	(Karellas et al. 2010)
Agricultural waste	Hydrogen production	(Guo et al. 2010)
Agricultural waste	Biobutanol production	(Cheng et al. 2012)
Agricultural waste	Biogas	(Moustakas et al. 2020)
Wheat straw, rice straw, and sugarcane biomass	Methane	(Bhuvaneshwari et al. 2019)
Wood, bark, and forest residue	Ethanol production	(Hu et al. 2008)
Agriculture biomass	Hydrogen production	(Li et al. 2017)
Mixed starch-based agriculture waste	Bioethanol and biohydrogen production	(Aruwajoye et al. 2020)

biomass from farm wood waste, plantation, and tree grown outside the woodland may be used to generate large amounts of ethanol. Generally, wood waste is categorized into two types: softwood and hardwood. Type of wood waste influences ethanol production as it contains different forms, e.g. hardwood includes more xylan and less mannan so that it is more resistant to the recalcitration (Zhu et al. 2010). In addition, construction and demolition waste comprise around 20–30% of wood material which can be an efficient source for ethanol production (Table 8.2) (Harris et al. 2019).

8.7 Conclusion

An essential source of income generation is agro-waste. In addition to lowering environmental pollution, increasing food security, exploring fiscal expansion, and improving crop production, creating wealth from agro-waste is an effective tool for lowering the environmental pollution. Agriculture, animal processing, and aquaculture all contribute to producing these leftovers. For a healthy ecosystem, it is essential to keep in mind that current research does not support effective waste creation, disposal, care, conversion, and usage. By managing waste properly, we can develop our agricultural industry and give many people to produce biofuels. Agriculture should be made aware of the problem of agricultural waste, the consequences of indiscriminate disposal of agricultural solid waste, and the benefits of effective agricultural waste management.

References

Adejumo, I.O., Ologhobo, A.D., and Babalola, T.O. (2015). Effect of pre-planting seed dressers on serum enzymes of laying chickens. *Journal of the American Chemical Society 9*: 1–5. 10.9734/ACSJ/2015/19687.

Alatzas, S., Moustakas, K., Malamis, D., and Vakalis, S. (2019). Biomass potential from agricultural waste for energetic utilization in Greece. *Energies 12* (6): 1095.

Antoniou, N., Monlau, F., Sambusiti, C. et al. (2019). Contribution to circular economy options of mixed agricultural wastes management: coupling anaerobic digestion with gasification for enhanced energy and material recovery. *Journal of Cleaner Production 209*: 505–514.

Aruwajoye, G.S., Kassim, A., Saha, A.K., and Gueguim Kana, E.B. (2020). Prospects for the improvement of bioethanol and biohydrogen production from mixed starch-based agricultural wastes. *Energies 13* (24): 6609. 10.3390/en13246609.

Belal, E.B. (2013). Bioethanol production from rice straw residues. *Brazilian Journal of Microbiology 44* (1): 225–234.

Bhuvaneshwari, S., Hettiarachchi, H., and Meegoda, J.N. (2019). Crop residue burning in India: policy challenges and potential solutions. *International Journal of Environmental Research and Public Health 16* (5): 832.

Borugadda, V.B. and Goud, V.V. (2012). Biodiesel production from renewable feedstocks: status and opportunities. *Renewable and Sustainable Energy Reviews 16* (7): 4763–4784.

Brady, N.C., Weil, R.R., and Weil, R.R. (2008). *The Nature and Properties of Soils*, vol. 13, 662–710. Upper Saddle River, NJ: Prentice Hall.

Canilha, L., Chandel, A.K., Suzane dos Santos Milessi, T. et al. (2012). Bioconversion of sugarcane biomass into ethanol: an overview about composition, pretreatment methods, detoxification of hydrolysates, enzymatic saccharification, and ethanol fermentation. *Journal of Biomedicine and Biotechnology* 2012: 989572.

Champagne, P. (2007). Feasibility of producing bio-ethanol from waste residues: a Canadian perspective: feasibility of producing bio-ethanol from waste residues in Canada. *Resources, Conservation and Recycling 50* (3): 211–230.

Cheng, C.L., Che, P.Y., Chen, B.Y. et al. (2012). Biobutanol production from agricultural waste by an acclimated mixed bacterial microflora. *Applied Energy 100*: 3–9.

Chitara, M.K., Chauhan, S., and Singh, R.P. (2021). Utilization of agricultural waste in agri-based biogas production and sustainable development. In: *Agri-Based Bioeconomy: Reintegrating Trans-disciplinary Research and Sustainable Development Goals* (ed. C. Keswani), 263–272. CRC Press.

Conrad, Z., Niles, M.T., Neher, D.A. et al. (2018). Relationship between food waste, diet quality, and environmental sustainability. *PLoS One 13* (4): e0195405.

Dan, M., Senila, L., Roman, M. et al. (2015). From wood wastes to hydrogen – preparation and catalytic steam reforming of crude bio-ethanol obtained from fir wood. *Renewable Energy 74*: 27–36.

Das, A.C., Das, R., and Bhowmick, S. (2015). Non-symbiotic N_2-fixation and phosphate-solubility in gangetic alluvial soil as influenced by pre-emergence herbicide residues. *Chemosphere 135*: 202–207.

De Corato, U. (2020). Agricultural waste recycling in horticultural intensive farming systems by on-farm composting and compost-based tea application improves soil quality and plant health: A review under the perspective of a circular economy. *Science of the Total Environment 738*: 139840.

Digman, B. and Kim, D.S. (2008). Alternative energy from food processing wastes. *Environmental Progress 27* (4): 524–537.

Eddleston, M., Karalliedde, L., Buckley, N. et al. (2002). Pesticide poisoning in the developing world – a minimum pesticides list. *The Lancet 360* (9340): 1163–1167.

Edwards, C.A., Dominguez, J., and Aranconl, N.Q. (2004). The influence ofvermicomposts on plant growth and pest incidence. In: *Soil Zoology for Sustainable Development in the 2 I st Century*, vol. 18 (ed. S.H.S. Hanna and W.Z.A. Mikhail), 397–420. Cairo.

FAO (1996). *Control of water pollution from agriculture.* Illustrate., vol. 55. Food & Agriculture Org., 1996.

FAO (2013). AQUASTAT database [online]. http://fao.org/nr/water/aquastat/ data/query/ index.html?lang=en (accessed 25 October 2013).

Ghosh, P., Kumar, M., Kapoor, R. et al. (2020). Enhanced biogas production from municipal solid waste via co-digestion with sewage sludge and metabolic pathway analysis. *Bioresource Technology 296*: 122275.

Graham, R.L., Nelson, R., Sheehan, J. et al. (2007). Current and potential US corn stover supplies. *Agronomy Journal 99* (1): 1–11.

Gullon, P., Gullon, B., Astray, G. et al. (2020). Valorization of byproducts from olive oil industry and added-value applications for innovative functional foods. *Food Research International 137*: 109683.

Guo, X.M., Trably, E., Latrille, E. et al. (2010). Hydrogen production from agricultural waste by dark fermentation: a review. *International Journal of Hydrogen Energy 35*: 10660–10673.

Gupta, R. and Garg, V.K. (2008). Stabilization of primary sewage sludge during vermicomposting. *Journal of Hazardous Materials 153* (3): 1023–1030.

Gupta, C., Prakash, D., Gupta, S., and Nazareno, M.A. (2019). Role of vermicomposting in agricultural waste management. In: *Sustainable Green Technologies for Environmental Management* (ed. S. Shah, V. Venkatramanan and R. Prasad), 283–295. Singapore: Springer.

Harris, P.A., Taylor, R., Minor, B.L. et al. (2019). The REDCap consortium: building an international community of software platform partners. *Journal of Biomedical Informatics 95*: 103208.

Heaton, E.A., Schulte, L.A., Berti, M. et al. (2013). Managing a second-generation crop portfolio through sustainable intensification: examples from the USA and the EU. *Biofuels, Bioproducts and Biorefining 7* (6): 702–714.

Hettiarachchi, H., Meegoda, J.N., and Ryu, S. (2018). Organic waste buyback as a viable method to enhance sustainable municipal solid waste management in developing countries. *International Journal of Environmental Research and Public Health 15* (11): 2483.

Hu, G., Heitmann, J.A., and Rojas, O.J. (2008). Feedstock pretreatment strategies for producing ethanol from wood, bark, and forest residues. *BioResources 3* (1): 270–294.

Jacobsen, S.E. and Wyman, C.E. (2002). Xylose monomer and oligomer yields for uncatalyzed hydrolysis of sugarcane bagasse hemicellulose at varying solids concentration. *Industrial & Engineering Chemistry Research 41* (6): 1454–1461.

Jaffar, M.M., Nahil, M.A., and Williams, P.T. (2020). Pyrolysis-catalytic hydrogenation of cellulose-hemicellulose-lignin and biomass agricultural wastes for synthetic natural gas production. *Journal of Analytical and Applied Pyrolysis 145*: 104753.

Jaffer Mohiddin, G., Srinivasulu, M., Maddela, N.R. et al. (2015). Influence of the insecticides acetamiprid and carbofuran on arylamidase and myrosinase activities in the tropical black and red clay soils. *Environmental Monitoring and Assessment 187* (6): 1–9.

Jambhekar, H.A. (1992). Use of earthworm as a potential source of decompose organic wastes. In *Proceeding of the National Seminar on Organic Farming, Mahatama Phule Krishi Vidyapeeth, Pune* (pp. 52–53).

Jaramillo Colorado, B.E., Bermúdez Tobón, A., and Tirado Ballestas, I. (2016). Bacterias degradadoras de pesticidas organofosforados presentes en suelos contaminados. *Revista Ciencias Técnicas Agropecuarias 25* (3): 13–22.

Jin, S., Bluemling, B., and Mol, A.P. (2015). Information, trust and pesticide overuse: interactions between retailers and cotton farmers in China. *NJAS-Wageningen Journal of Life Sciences 72*: 23–32.

Karellas, S., Boukis, I., and Kontopoulos, G. (2010). Development of an investment decision tool for biogas production from agricultural waste. *Renewable and Sustainable Energy Reviews 14* (4): 1273–1282.

Kearney, J. (2010). Food consumption trends and drivers. *Philosophical Transactions of the Royal Society, B: Biological Sciences 365* (1554): 2793–2807.

Kim, S.J., Jung, S.H., and Kim, J.S. (2010). Fast pyrolysis of palm kernel shells: influence of operation parameters on the bio-oil yield and the yield of phenol and phenolic compounds. *Bioresource Technology 101* (23): 9294–9300.

Krajick K. and Lee, K. (2016). A Major Source of Air Pollution: Farms. https://www.earth.columbia.edu/articles/view/3281 (accessed 16 May 2016).

Kumar, A., Kumar, N., Baredar, P., and Shukla, A. (2015). A review on biomass energy resources, potential, conversion and policy in India. *Renewable and Sustainable Energy Reviews 45*: 530–539.

Kumar, S.S., Kumar, V., Malyan, S.K. et al. (2019). Microbial fuel cells (MFCs) for bioelectrochemical treatment of different wastewater streams. *Fuel 254*: 115526.

Leal Filho, W. (2018). Bioeconomy meets the circular economy: the RESYNTEX and FORCE projects. In: *Towards a Sustainable Bioeconomy: Principles, Challenges and Perspectives* (ed. W. Leal Filho, D.M. Pociovălişteanu, P.R.B. de Brito and I.B. de Lima), 567–575. Cham: Springer.

Lei, Z., Chen, J., Zhang, Z., and Sugiura, N. (2010). Methane production from rice straw with acclimated anaerobic sludge: effect of phosphate supplementation. *Bioresource Technology 101* (12): 4343–4348.

Li, L., Wang, Y., Zhang, Q. et al. (2008). Wheat straw burning and its associated impacts on Beijing air quality. *Science in China Series D: Earth Sciences 51* (3): 403–414.

Li, B., Yang, H., Wei, L. et al. (2017). Hydrogen production from agricultural biomass wastes gasification in a fluidized bed with calcium oxide enhancing. *International Journal of Hydrogen Energy 42* (8): 4832–4839. 10.1016/j.ijhydene.2017.01.138.

Lim, S.L., Wu, T.Y., Sim, E.Y.S. et al. (2012). Biotransformation of rice husk into organic fertilizer through vermicomposting. *Ecological Engineering 41*: 60–64.

Maheshwari, N.U. and Sudha, S. (2013). Enumeration and detection of phosphate solubilizing bacteria from the gut of earthworm varieties. *Journal of Chemical and Pharmaceutical Research 5*: 264–267.

Manderson, G.J. (2009). Composting agricultural and industrial wastes. *BIOTECHNOLOGY-Volume VIII: Fundamentals in Biotechnology 8*: 93.

Memon, A.A., Shaikh, S.A., Shah, F.A. et al. (2016). Production of synthetic gas from agriculture and forest wastes using fixed bed gasifier. *Pakistan Journal of Agriculture, Agricultural Engineering and Veterinary Sciences 32* (2): 229–236.

Misra, R.V., Roy, R.N., and Hiraoka, H. (2003). *On-farm composting methods.* Rome, Italy: UN-FAO. *L. Water Discuss. Pap.*, vol. 2, no. 9, p. 51.

Mohee, R.O.M.E.E.L.A. (2007). Waste management opportunities for rural communities. Composting as an effective waste management strategy for farm households and others. *AGSF Working Document (FAO).*

Moustakas, K., Parmaxidou, P., and Vakalis, S. (2020). Anaerobic digestion for energy production from agricultural biomass waste in Greece: capacity assessment for the region of Thessaly. *Energy 191*: 116556.

Munnoli, P.M., Da Silva, J.A.T., and Saroj, B. (2010). Dynamics of the soil-earthworm-plant relationship: a review. *Dynamic Soil, Dynamic Plant 4* (1): 1–21.

Nagavallemma, K.P., Wani, S.P., Lacroix, S. et al. (2004). Vermicomposting: recycling wastes into valuable organic fertilizer. Global Theme on Agroecosystems Report no. 8.

National Policy for Management of Crop Residues (NPMCR) (2014). Government of India Ministry of Agriculture Department of Agriculture & Cooperation (Natural Resource Management Division), Krishi Bhawan, New Delhi.

Obi, F.O., Ugwuishiwu, B.O., and Nwakaire, J.N. (2016). Agri wastes. *Nigerian Journal of Technology* 35 (4): 957–964.

Obi, F.O., Ugwuishiwu, B.O., and Nwakaire, J.N. (2016). Agricultural waste concept, generation, utilization and management. *Nigerian Journal of Technology 35* (4): 957–964.

Panpatte, D.G. and Jhala, Y.K. (2019). Agricultural waste: a suitable source for biofuel production. In: *Prospects of Renewable Bioprocessing in Future Energy Systems* (ed. A.A. Rastegari, A.N. Yadav and A. Gupta), 337–355. Cham: Springer International Publishing.

Pappu, A., Saxena, M., and Asolekar, S.R. (2007). Solid wastes generation in India and their recycling potential in building materials. *Building and Environment 42* (6): 2311–2320.

Parthasarathi, K., Balamurugan, M., Prashija, K.V. et al. (2016). Potential of Perionyx excavatus (Perrier) in lignocellulosic solid waste management and quality vermifertilizer production for soil health. *International Journal of Recycling of Organic Waste in Agriculture 5* (1): 65–86.

Pedrero, F., Kalavrouziotis, I., Alarcón, J.J. et al. (2010). Use of treated municipal wastewater in irrigated agriculture – review of some practices in Spain and Greece. *Agricultural Water Management 97* (9): 1233–1241.

Peiji, G., Yinbo, Q., Xin, Z. et al. (1997). Screening microbial strain for improving the nutritional value of wheat and corn straws as animal feed. *Enzyme and Microbial Technology 20* (8): 581–584.

Prasad, M., Ranjan, R., Ali, A. et al. (2020). Efficient transformation of agricultural waste in India. In: *Contaminants in Agriculture* (ed. M. Naeem, A.A. Ansari and S.S. Gill), 271–287. Cham: Springer. 10.1007/978-3-030-41552-5_13.

Pratap Singh, D. and Prabha, R. (2017). Bioconversion of agricultural wastes into high value biocompost: a route to livelihood generation for farmers. *Advances in Recycling and Waste Management* 2 (3): 137.

Pullammanappallil, P.C., Svoronos, S.A., Chynoweth, D.P., and Lyberatos, G. (1998). Expert system for control of anaerobic digesters. *Biotechnology and Bioengineering 58* (1): 13–22.

Ramteke, P., Srikanth, P., Sivakumar, D., and Nouri, J. (2022). Agricultural waste management generated by agro-based industries using biotechnology tools. *Global Journal of Environmental Science and Management* 8 (2): 281–296.

Ravindranath, N.H. and Balachandra, P. (2009). Sustainable bioenergy for India: technical, economic and policy analysis. *Energy 34* (8): 1003–1013.

Reijnders, L. (2008). Ethanol production from crop residues and soil organic carbon. *Resources, Conservation and Recycling 52* (4): 653–658.

Roberto, I.C., Mussatto, S.I., and Rodrigues, R.C. (2003). Dilute-acid hydrolysis for optimization of xylose recovery from rice straw in a semi-pilot reactor. *Industrial Crops and Products 17* (3): 171–176.

Sadh, P.K., Duhan, S., and Duhan, J.S. (2018). Agro-industrial wastes and their utilization using solid state fermentation: a review. *Bioresources and Bioprocessing 5* (1): 1–15.

Sequi, P. (1996). The role of composting in sustainable agriculture. In: *The Science of Composting* (ed. M. de Bertoldi, P. Sequi, B. Lemmes and T. Papi), 23–29. Dordrecht: Springer.

Shehrawat, P.S., Nitu, S., and Parmila, D. (2015). Agricultural waste awareness and utilization for healthy environment and sustainable livelihood. *Scientific Papers Series-Management, Economic Engineering in Agriculture and Rural Development 15* (2): 371–376.

Singh, N. and Tandon, S. (2015). Dissipation kinetics and leaching of cyazofamid fungicide in texturally different agricultural soils. *International journal of Environmental Science and Technology 12* (8): 2475–2484.

Singh, G., Kaur, K., and Meetei, T.T. (2021). Effect of in stubble burning on physico chemical properties of soil, yield and environmental qualities. *The Pharma Innovation Journal* 10 (5): 298–305.

Sonesson, U., Björklund, A., Carlsson, M., and Dalemo, M. (2000). Environmental and economic analysis of management systems for biodegradable waste. *Resources, Conservation and Recycling 28* (1–2): 29–53.

Song, X., Liu, M., Wu, D. et al. (2014). Heavy metal and nutrient changes during vermicomposting animal manure spiked with mushroom residues. *Waste Management 34* (11): 1977–1983.

Staroń, P., Kowalski, Z., Staroń, A., and Banach, M. (2017). Thermal treatment of waste from the meat industry in high scale rotary kiln. *International journal of Environmental Science and Technology 14* (6): 1157–1168.

Tamburini, E., Gaglio, M., Castaldelli, G., and Fano, E.A. (2020). Biogas from agri-food and agricultural waste can appreciate agro-ecosystem services: the case study of Emilia Romagna region. *Sustainability 12* (20): 8392.

Toor, M., Kumar, S.S., Malyan, S.K. et al. (2020). An overview on bioethanol production from lignocellulosic feedstocks. *Chemosphere 242*: 125080.

Torres, A., Romero, J., Macan, A. et al. (2014). Near critical and supercritical impregnation and kinetic release of thymol in LLDPE films used for food packaging. *The Journal of Supercritical Fluids 85*: 41–48.

US Environmental Protection Agency (2015). *Connectivity of Streams and Wetlands to Downstream Waters: A Review and Synthesis of the Scientific Evidence*. EPA/600/R-14/475F. Washington: US Environmental Protection Agency.

USEPA (2014). *Surface Water Standards and Guidance*. Washington, DC: Water Quality Standards Handbook.

Van Arendonk, A. (2015). The development of the share of agriculture in GDP and employment. A case study of China, Indonesia, the Netherlands and the United States. Master's Thesis, Wageningen University, Wageningen, The Netherlands.

Wang, H.Y., Shen, L.I.U., Zhai, L.M. et al. (2015). Preparation and utilization of phosphate biofertilizers using agricultural waste. *Journal of Integrative Agriculture 14* (1): 158–167. 10.1016/S2095-3119(14)60760-7.

Wong, A., Zhang, H., and Kumar, A. (2016). Life cycle water footprint of hydrogenation-derived renewable diesel production from lignocellulosic biomass. *Water Research 102*: 330–345.

Zakaria, Z.A. (ed.) (2018). *Sustainable Technologies for the Management of Agricultural Wastes*. Singapore: Springer.

Zhang, Q., Wang, Y., Zhang, Z. et al. (2017). Photo-fermentative hydrogen production from crop residue: a mini review. *Bioresource Technology 229*: 222–230.

Zhang, P., Sun, H., Ren, C. et al. (2018). Sorption mechanisms of neonicotinoids on biochars and the impact of deashing treatments on biochar structure and neonicotinoids sorption. *Environmental Pollution 234*: 812–820.

Zhu, J.Y., Zhu, W., OBryan, P. et al. (2010). Ethanol production from SPORL-pretreated lodgepole pine: preliminary evaluation of mass balance and process energy efficiency. *Applied Microbiology and Biotechnology 86* (5): 1355–1365.

9

Production of Bioethanol Using Agricultural Waste: An Overview

Sangita Agarwal[1], Tathagata Deb[1], Anu Sharma[2], and Ravindra Pratap Singh[3]

[1] *Department of Applied Science, RCC Institute of Information Technology, Kolkata, West Bengal, India*
[2] *Govt. SPMR College of Commerce, Constituent College of Cluster University of Jammu, Jammu, Jammu and Kashmir (UT), India*
[3] *Central Public Works Department, New Delhi, India.*

CHAPTER OUTLINE

9.1 Introduction

Conventional fuels like coal, natural gas, and petroleum are obtained when year after year organisms decompose and the dead remnants of plants and animals accumulate, and this process takes a lot of time. Currently, the distribution of petroleum is abruptly declining day by day and this is the main reason for the inadequate source of precursors for oil production, but the demand for petroleum continues to increase with each passing day. This imbalance in the supply and demand of fossil fuels has led to an environmental crisis. Atmospheric gases like carbon dioxide, sulphur dioxide, nitrogen dioxide, and methane are evovled from burning of these fossil fuels and are labelled as greenhouse gases,

Agriculture Waste Management and Bioresource: The Circular Economy Perspective, First Edition.
Edited by Suruchi Singh, Pardeep Singh, Anu Sharma, and Moharana Choudhury.

increase the average temperature of the Earth leading to global warming (Forster et al. 2007). Due to this huge requirement of fossil fuels, we are prompted to use petroleum blended with bioethanol (Yüksel and Yüksel 2004; World Bioenergy Association 2017). The most commonly used alcohol is ethanol. The production of ethanol involves a process of fermentation using micro-organisms, and this type of ethanol is popularly known as bioethanol. This may be considered as another new version or new form of fuel that can help in reducing the emission of CO_2. Ordinary ethanol and bioethanol are almost the same. The fundamental difference is basically the precursors and the methodology by which they are being manufactured. Bioethanol has been an ideal substitute for the common fuels because of its many advantages. First, bioethanol possesses a high value of octane number of 108.6 with reference to gasoline that helps in curtailing the chances of eruption during the combustion. Gasoline–ethanol when mixed in the ratio of 60 : 40 increases the efficiency of gasoline. Second, the probability of emission of carbon monoxide is also lowered due to a much safer and cleaner combustion reaction, and this is a beneficial in reducing air pollution. Third, the end product of the combustion of bioethanol is less exothermic as compared to other fuels and thus increases the life expectancy of the engine. Fourth, having high thermodynamic efficiency, it provides more power in comparison to other fuels. Finally, bioethanol in the form of a fuel can help the farmers economically especially by selling the sugar-related products such as sugarcane and sugarbeet. Thereby to fulfil the needs of this fuel crisis, large-scale industries have started producing bioethanol, with raw materials like sugarcane, corn, cassava, and others as required, and this literally proved beneficial for the farmers as their surplus output would not get wasted and thus the price will also remain consistent. Meanwhile, job opportunities will also crop up gradually. Another alternative forerunner to produce ethanol can be agricultural waste, and this is because of its availability, cost-effectiveness besides its eco-friendly nature. Organic and agricultural waste can be used for conversion into ethanol since they are chemically enriched. Ethanol production from agricultural waste has many positive aspects that cannot be ignored such as it reduces the waste disposal issue and if waste is generated also, it can be reprocessed microbiologically leading to the production of ethanol. Second, we can also reduce the high cost during the ethanol manufacturing. High price of the raw materials is the chief reason for the great cost of production of ethanol. Hence, another alternative raw material for ethanol production has to be found out so that its production cost goes down. This alternative raw material could be the above-discussed agricultural waste as this can be very easily procured. Sugar, starch, and cellulose, respectively, are the three groups of raw materials required for the making of bioethanol (Bušić et al. 2018). Ethanol can also be directly manufactured from sugar that is derived from natural sources like beet sugar, cane sugar, molasses, and fruits, while starch derived from potatoes, plant, corn, and cassava first needs to be hydrolysed into sugar on a primary level. On the other hand, cellulose is obtained from agricultural waste, wood, wastes from pulp, and paper mills that needs acid hydrolysis using common mineral acids as found in the chemical laboratories (Lin and Tanaka 2006). There are three principal steps involved in the production of bioethanol transformation of polysaccharides into simple sugars, fermentation followed by distillation. In modern times, production of ethanol from agricultural waste has become viable due to its easy availability and also does not cause any environmental harm (Edeh 2020). The most commonly used micro-organism for the natural synthesis or biosynthesis of bioethanol is *Saccharomyces cerevisiae* (Faustine and Rustini 2021).

9.2 Origin of Bioethanol

In the primitive times, bioethanol was produced following the conventional fermentation process using sugar-containing feedstock. During the 12–14th century, the pure ethanol production ushered due to the distillation process. The most common uses of ethanol were in formulating medicinal drugs and making colour pigments used in paints. During the twelfth century, for the first time, ethanol was produced using feedstock-containing starch. During nineteenth century, large-scale industrial production of bioethanol started blooming as its production became economically viable (Roehr 2000). A German researcher named Nicholas Otto in 1860 used ethanol as a fuel for the first time in internal combustion engine. In 1908, Henry Ford invented an internal combustion engine that ran on ethanol and later on modified the engine with inbuilt carburettors using blend of ethanol and gasoline as fuel (Solomon et al. 2007). Ford was first person who quoted ethanol as 'the future fuel'. Later in the year 1925, Ford opined that bioethanol could be extracted from vegetables, fruits, sawdust, and weeds. The automobile industry employed fossil fuels due to their cheaper production cost and high heat of combustion values too. Hydrated ethanol can also replace bioethanol that can be used in engines besides gasoline. On mixing anhydrous ethanol with gasoline, the value of octane number increases (Turner et al. 2011).

9.2.1 Characteristics of Ethanol

Bioethanol is a transparent liquid substance having a distinctive sweet smell. Also, ethanol possesses a sweet flavour in diluted state, whereas in its concentrated form it gives a typical burning sensation. Ethanol (C_2H_5OH) has a density of 0.789g/ml at 20 °C with a boiling points of 78.5 °C. Ethanol possesses an octane number of 99 that is quite high and also it has a low ignition temperature that enables it to catch fire easily. Due its polar nature and ability to exhibit dipole–dipole interaction, it has miscibility in both nonpolar as well as polar solvents. Since ethanol is an organic solvent, it finds immense application in explosives, paints, and varnishes as well as in perfumery industry. Ethanol is dehydrated to form ether using concentrated sulphuric acid and can be oxidized into acetaldehyde using mild oxidants or to acetic acid using strong chemical oxidants. Tinctures are alcoholic solutions containing non-volatile substances while spirits are those containing volatile substances (Kwon et al. 2012).

9.3 Production Policy of Bioethanol in some Countries with Focus on India

Brazil has been one of the world's leading countries in terms of export and import of bioethanol. During the 1970s, when there was a big hollow in the market of oil and petrochemical industry, the Brazilian government introduced The National Ethanol Fuel Program to improve bioethanol production in order to make up for the scarcity of oil and related substances and also with the notion of reducing the price level of sugar (Tan et al. 2015).The program as launched went through three significant stages; between 1970 and 1990, the government gave top priority with regard to safe handling aqueous ethanol in dare oil crisis. Thereafter oil prices got optimized during 1990–2000. Brazil was able to produce successfully anhydrous ethanol when 20% ethanol (v/v) is mixed with gasoline, and the quality of ethanol improved in subsequent years. From 2000 till date, major quantity (almost 98%) of ethanol is used in transport and communication industries. Bioethanol blended fuels gained importance in the following years. (Ayadi et al. 2016). In Colombia, the government enunciated the biofuel policies clearly depicting that it is compulsory to blend 10% bioethanol in cities having a population of more than 500000 (United States Department of Agriculture 2009). The order came in practice in 2005, and this attempt proved to be worthy enough as statistical data says. The Government of Colombia aims to improve the content of ethanol up to 25% from 15% in the coming years. Colombia government has curtailed the prices of sugarcane, which has been an important precursor used in order to synthesize bioethanol on a macro-scale and also the VAT duty was exempted. All these policies aimed towards large-scale production of bioethanol that proved to be advantageous. (Torres-Jimenez et al. 2011).

9.3.1 India and Policy on Bioethanol

As a developing country, India is trying its level best to switch over from the traditional fossil fuels and replace by bioethanol. Currently, in India, molasses is mostly used for the production of ethanol. Proposal of combining 5% ethanol with gasoline was greatly acknowledged by twenty two States and four Union Territories of India, and bill was passed and finally the oil marketing companies started widening its horizon in this new area. The sugar industry also waved a positive sign in order to procure bioethanol straightway from sugarcane juice so that bare necessity of bioethanol can be manipulated at the earliest (Ministry of New and Renewable Energy, Government of India 2009). Beside usage of the conventional fuels, productivity of bioethanol is on its rise. But it has been suggested that sugar supply and requirement of alcohol should not be hampered due to this new horizon that has been ushered (Ministry of New and Renewable Energy, Government of India 2009).

The National Policy on Biofuels (NPB) was embraced by the Union Cabinet of India in the year 2018, which considered the second-generation feedstock for the production of biofuel. The second-generation feedstock includes nonfood biomass like leaves and stalks, surplus grains, and molasses, The newly proposed 2025 ethanol roadmap mainly focuses on food-based feedstock, thus indicating a paradigm shift from the colloquial deviation from the Ethanol Blended Program (EBP). The price of ethanol is increasing day by day, which is affecting the policy and hence the 2025 India's roadmap of bioethanol can be reached and achieve a balance between the supply as well as meet the blending targets. Statistical analysis on projected ethanol production from domestic grains finds a massive fourfold increase by 2025, and that ethanol produced from sugarcane would also increase. As per USDA, Foreign Agricultural Service report analysis reveals that in 2020 there had been a hike in coarse grains production only by 7% increase but much higher percentage production rate is required to meet the target goals. India produces ethanol mainly based on coarse grains and thus provides a maximum share to the economic market (The ICCT 2021).

9.4 Distribution of Bioethanol Worldwide

Globally, two leading bioethanol producers are Brazil and the US. Brazil produces bioethanol from sugarcane as the major crop biomass, while United States uses corn grain for bioethanol production. Bioethanol can be produced from primary sources like sugar and starch-containing crops as well as from other sources like agricultural biomass containing lignocellulose such as bagasse, corn stover, banana and orange peels, crop straw, forestry wastes, and wastepaper (Foyle et al. 2007). The worldwide circulation of ethanol production in various countries from 2007 till 2020 is shown in Figure 9.1 (Renewable Fuels Association-ethanolrfa 2012). As is evident from the graph that bioethanol production has steadily increased from 2012 worldwide except in the 2020, the year when COVID-19 pandemic struck the world. Out of the total world production of 26.059 billion gallons in 2020, the share of United States has been 13.9 billion gallons. Combined production of US and Brazil is about 84% share of the world's ethanol production.

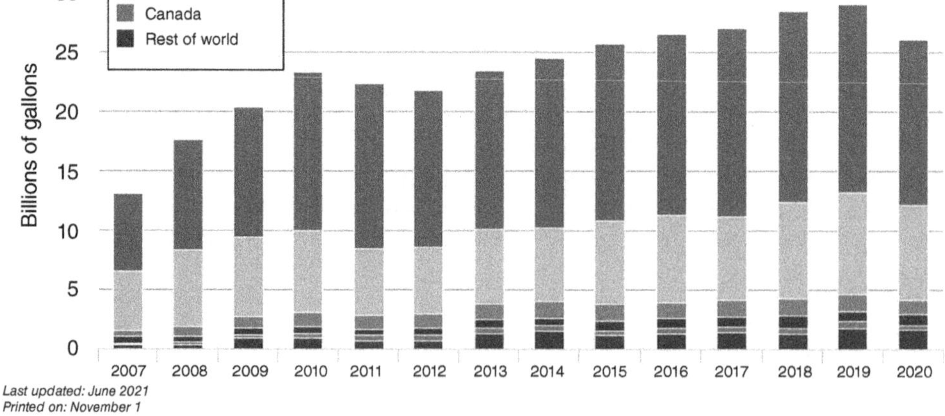

Figure 9.1 Bioethanol production globally by country or region from 2007 to 2020. *Source:* Maitah, Mansoor & Procházka, Petr & Smutka, Lubos & Maitah, Kamil & Honig, Vladimír. (2019). Analysis of the Impact of Ethanol Production on Agricultural Product Prices in Brazil. Sugar Tech. 21. 10.1007/s12355-019-00709-w.

9.5 Comparison of Ethanol and Gasoline

Gasoline finds application as fuel in transport and communication industry and is obtained as a by-product during fractional distillation of petroleum. Due to the presence of long-chain hydrocarbon moiety, it is hydrophobic in nature, insoluble in water, usually floats in water, and has a flash point of approximately −45 °F, with variable octane rating. Its specific gravity is 0.72–0.76 and having boiling point varying between 37 and 204 °C. Gasoline has an internal combustion point between 280 and 456 °C. A person when gets exposed to gasoline for a longer time develops respiratory problems (Ameri et al. 2008). Bioethanol comprises 35% oxygen. By far, ethanol is considered as a clean fuel as it can decrease the possibility of emission of particulate and nitrogen oxides (NO_X) as emitted due to the combustion of fossil fuels. Pure ethanol is a polar solvent, hydrophilic, and has 55 °F flash point. It is heavier than air. Gasoline is more toxic compared to bioethanol and when uniformly mixed with gasoline, it can function as a clean fuel. Rampant use of fossil fuels has increased the generation of greenhouse gases and so shifting to an alternative fuel can improve the efficacy of combustion process, increase the oxygen content required for combustion (Wang et al. 1999), as well reduce greenhouse emission. When ethanol is mixed with gasoline, 10% bioethanol gets converted to 90% gasoline (also designated as 'E10') and commonly called 'gasohol'.

According to the analyses based on the availability of feedstock, bioethanol can be classified into two types. The first one includes ethanol, derived from sugar and starch-like sugarcane (Figure 9.2), rice, and wheat, etc. and is referred to as first-generation bioethanol. These substrates are chiefly used as consumables by human being and hence used up soon. Henceforth, a scarcity of food supply already prevails with respect to the production of bioethanol. Therefore, there is a quest to switch over to some substitute sources that will not affect the food security and at the same time do not cause any harm to the environment. Further investigations and extensive research revealed that second-generation bioethanol has many advantages over first-generation bioethanol (Sun and Cheng 2002). This second-generation bioethanol reduces the possibility of emission of greenhouse gases thereby maintaining the ambient air quality. Each year on an annual basis almost 1×10^{10} metric ton lignocelluloses are being produced throughout the globe and has revealed that this second type of feedstock does happen to be the most abundant and renewable biomass. Hence, second-generation bioethanol ensures huge output, which acts as a good alternative to fossil fuels, at the same time food production remains unperturbed (Goh et al. 2010).

Figure 9.2 Representative diagrams of sugar-containing plant crops (a) sugarcane. *Source:* M.studio / Adobe Stock. (b) sweet sorghum. *Source:* Howard F. Schwartz / Colorado State University / CC BY 3.0 US. (c) sugarbeet commonly used in bioethanol production. *Source:* rsooll/Adobe Stock.

9.6 First-Generation Bioethanol

Bioethanol that is being synthesized from sources like sugar beet, corn, wheat, and sugarcane, is termed as first-generation bioethanol. The first-generation bioethanol does not have any complicated structural pattern, which is an added advantage in terms of its treatment and other industrial proceedings, say, for example, it does not require any pretreatment methods before we formally start with the fermentation process. It is a relatively cheaper method with respect to bioethanol production (Sun and Cheng 2002)

9.6.1 Sources of First-Generation Bioethanol Production

The production of biofuels can be executed in a befitting manner provided the source from where it is being extracted is relatively cheap source, so new technologies are being ushered. Latest research reveals that during recent times, crops like corn, sugarcane, wheat, and soyabeans are labelled as renewable energy sources and are used for bioethanol production. Usage of renewable plant substrates is apparently beneficial, while utilization of other types of biomass may lead to a series of issues, including scarcity of food stuff and deterioration of soil fertility (Balat 2009). Bioethanol feedstock can be differentiated under three categories namely: first category that includes feedstock comprising sucrose like sugar cane, sweet sorghum, sugar beet, fruits, etc., and the second category includes starchy substances like wheat, rice, potatoes, and corn, while the last category encompasses the lingo cellulosic biomass say for instance, grasses, stem of tall trees, and wood.

9.6.1.1 Sugar-Containing Plant Crops

9.6.1.1.1 Sugarcane

Sugarcane (*Saccharum* spp.) is generally harvested in tropical and subtropical climatic zones across the globe. Sugarcane production is very much essential due to the fact that it is a rich source for worldwide sugar production, at the same time its increasing demand as bio energy crop has made it a very essential commodity. Most of bioethanol produced is procured from sugarcane (Figure 9.2) and remaining bioethanol from various other sources that include sugar beet, sorghum, wheat, rice, etc. (Dufey 2006). Brazil occupies the topmost position in the global production market with respect to sugarcane production. Other parts of the world, like Europe and US mainly depend on wheat, starch, and barley (Linde et al. 2008) for bioethanol production.

9.6.1.1.2 Sweet Sorghum

Sweet sorghum is a perennial plant that belongs to Poaceae family (Ratnavathi et al. 2010). Sweet sorghum has some environmental benefits in terms of low carbon emissions. For proper growth of sorghum temperate, subtropical and tropical climates are the most suitable one. Every constituent of the plant has commercial utility, starting from grain that is used as food, stalks for fuel, the leaves for forage, and the fibrous substances containing cellulose that is consumed as either animal feed or mulch. The time duration as required for the plant to grow big and become a full-sized tree is relatively of short duration, almost 3–5 months in comparison to that of sugarcane. Saline soil does not affect its growth rate.

The juice as extracted from sorghum stalks leads to huge ethanol production (Figure 9.2). The sweet sorghum cultivars stock has a high content of carbohydrates (15–23%). The fermentable carbohydrates thus obtained are sub classed into three major components such as sucrose (70%), glucose (20%), and fructose (10%) (Prasad et al. 2007a).

9.6.1.1.3 Sugarbeet

In countries like Europe, bioethanol is produced from the feedstock constituting beet molasses, which is a prime source of sucrose. Sugarbeet crop (Figure 9.2) produces a larger amount of bioethanol as compared to other common crops that are being cultivated annually. The advantage of this crop is the time required for its maturation is very short, the yield is ambient, and the crop can adapt itself to various climatic changes or natural disasters, using minimum quantity of water as well as chemical fertilizers (Balat 2009).

9.6.1.2 Starch-Containing Plant Crops

When the composition of biofuels like corn, wheat, and rice (Figure 9.3) was analysed, it revealed that the starch content in such crops is very high (Balat et al. 2008). In the beginning of the process, starch-containing materials are pretreated with acid and/or enzyme, which lead to the production of biofuels having high sugar concentration. Starch is a polysaccharide and consists of linear chains of D-glucose as its monomeric units. Through acid hydrolysis, starch is decomposed into fermentable sugars. When starch is used as the precursor for extraction of bioethanol, the polysaccharide chains degrade and liberate glucose syrup. This syrupy substance when react with yeast cells gets easily transformed into bioethanol. In the list of various important sources for bioethanol production, starch is considered as the most important and naturally available source. The second most important source that adds to the feedstock is wheat (Cardona and Sanchez 2007). US is leading nation based upon corn-based bioethanol in comparison to other countries.

9.6.2 Second Generation of Bioethanol

Bioethanol created from agriculture waste is now considered as one of the best practices of production and is called as second-generation bioethanol. This process is eco-friendly and minimizes waste generation. One of the advantages is that the agricultural residues

Figure 9.3 Representative diagrams of starch-containing plant crops (a) wheat plant. *Source:* Don Graham / Flickr / CC BY-SA 2.0. (b) corn plant. *Source:* Don Graham / Flickr / CC BY-SA 2.0. (c) rice plant commonly used in bioethanol production. *Source:* Toshiyuki IMAI / Flickr / CC BY-SA 2.0.

or by-products can be regenerated through various physical and chemical means and present in abundance (Sun and Cheng 2002). Second-generation biofuels are effective as they decrease the generation of greenhouse gases, possesses high thermodynamic efficiency and can mitigate the energy crisis (Taherzadeh and Karimi 2008). Regions where harvesting of energy crops is challenging, agricultural biomass can be the best outcome. Agricultural wastes like wheat straw, barley husks, corn cobs, paper pulp, sugar cane bagasse as well as fruit extracts such as banana peel, orange peels and pineapple peels are illustrations of second-generation feedstocks. The easy availability of agricultural residues has made them a very demanding entity in the global market (Koltermann et al. 2014). Polysaccharides like cellulose, hemicelluloses and lignin are now-a-days on the verge of replacing the first-generation ethanol producing substrates in the field of agriculture and bio technology.

9.6.2.1 Composition of Agricultural Waste

Agricultural debris comprises of the following: cellulose occupies 35–50% of the total dry weight, hemicellulose consisting of 20–35%, and lignin having 10–25%, the structures have been elucidated in Figure 9.4. Cellulose and hemicelluloses are resistant to degradation to the simpler forms because of special structural modifications when they are linked with lignin via hydrogen and covalent bonds (Verma et al. 2011).

9.6.2.1.1 Cellulose

Cellulose $(C_6H_{10}O_5)_n$ is one of the primary constituents of agricultural extract, which is commonly present in cell wall of plants. It is a polysaccharide molecule that is linear and crystalline polymer. It consists of 500–15 000 repeating units of D-glucose linked end to end through the β-1,4-glycosidic bonds (Figure 9.4). Cellulose has a crystalline plus rigid structure hence strong chemical treatment is necessary to extract glucose (Gray et al. 2006) from it. Cellulose chains are joined with one another by intramolecular hydrogen bonds and weak Van der Waals' forces. The presence of microfibrils within cellulose imparts a crystalline appearance and thus become resistant to attack by chemical reagents (Singhania et al. 2010; Agbor et al. 2011).

9.6.2.1.2 Hemicellulose

Hemicellulose $(C_5H_8O_4)$n is a second major constituent of agricultural extract, having long branched chains and is a heteropolymer containing five carbon sugar (xylanose and arabinose) and six carbon sugar (galactose and mannose) with acetate units (Kumar et al. 2008) (Figure 9.4). The structure of hemicellulose is less complicated and can be simply changed into fermentable sugars like pentose sugars and hexose sugars (Saha and Cotta 2007).

9.6.2.1.3 Lignin

Lignin $(C_6H_{11}O_2)$ is the third major component of agricultural biomass. It is aromatic in nature and contains phenyl propane units (p-coumaryl, sinapyl, and coniferyl alcohol), which are joined through the covalent bonds like C–C and C–O–C bonds (Figure 9.4) (Howard et al. 2003). Lignin is resistant to attack by chemicals since it is strongly bound to cellulose and hemicelluloses units. Since it is an amorphous

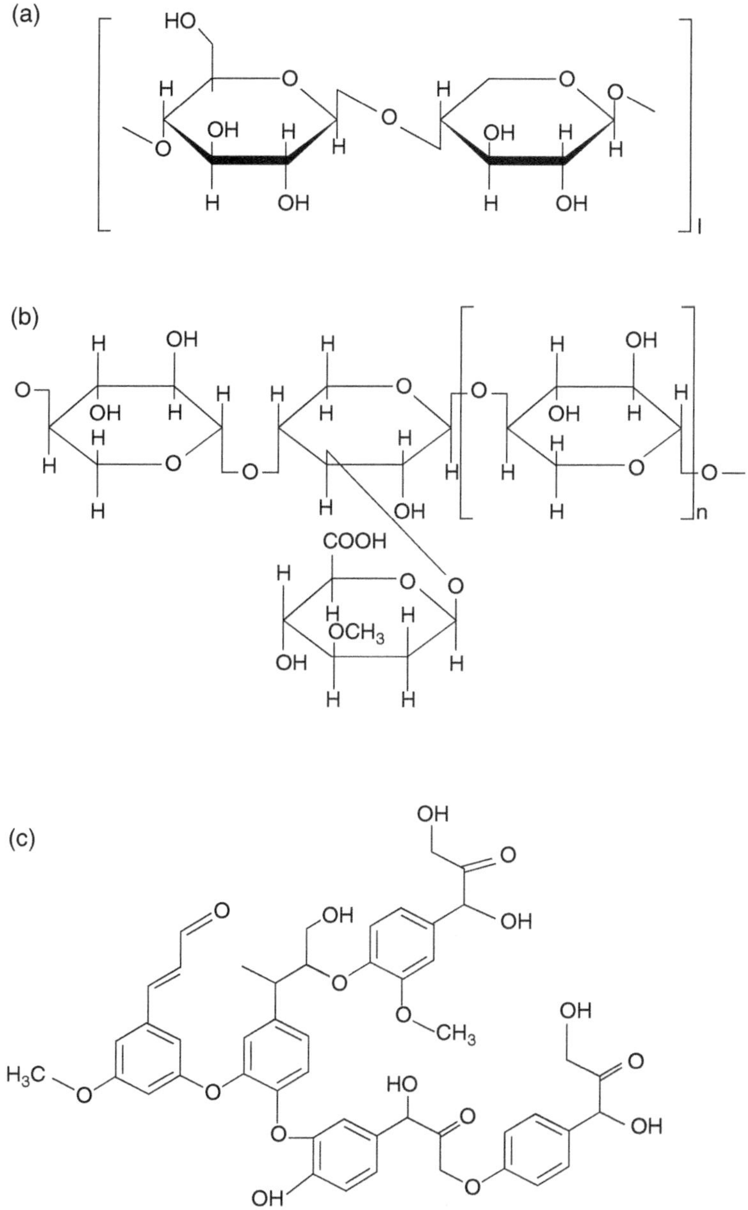

Figure 9.4 Structure of (a) cellulose, (b) hemicellulose, and (c) lignin as components of agricultural waste.

heteropolymer or copolymer, it can resist microbial attack through a hydrophobic interaction with the cell wall. This defensive nature of lignin hinders the alteration of agricultural debris into bioethanol and thus delignification of agricultural biomass accelerates the conversion of the agricultural substrate into ethanol (Vanholme et al. 2010).

9.6.3 Conversion of Agricultural Waste to Ethanol

Agricultural waste can be converted into ethanol with the help of below-mentioned four stages:

- pretreatment,
- hydrolysis of pretreated waste to sugars using enzymes,
- fermentation of sugars in order to synthesize ethanol, and
- recovery of ethanol.

It is noted that when lignin and hemicelluloses are removed from the cellulose, the tendency of hydrolysis increases; crystalline pattern decreases thereby increasing the porosity (Sun and Cheng 2002; Kumar et al. 2009) of cellulose. The enzymatic hydrolysis is catalysed with the help of enzymes like cellulase, followed by fermentation which is again supported with the help of micro-organisms like yeast and bacteria (Taherzadeh and Karimi 2008).

9.6.3.1 Pretreatment

Pretreatment methods include segregation and dissolution of the different components of agriculture materials. The primary components of agricultural biomass are cellulose and lignin. Pretreatment is important because of the following reasons:

i) disintegration of hemicelluloses and breaking of the lignin layer from the biomass,
ii) decomposition of cellulose unit,
iii) improvement of the pore volume and available surface area of the substrate,
iv) in preventing the sugar loss,
v) prevention of growth of inhibitors inhibiting the fermentation process,
vi) reduction in energy demands, and
vii) minimizing the budget.

The above pretreatment process in agriculture biomass facilitates enzymatic hydrolysis and yields greater quantity of fermentable sugars such as glucose, arabinose, mannose, galactose, and xylose, which ultimately has a broader view with respect to the entire analytical methods (Sun and Cheng 2002; Yang and Wyman 2008).

The conditions of an economical and efficient pretreatment methods are

i) simple/hassle free operation,
ii) convenient and reduced energy usage as well as water and chemicals,
iii) minimal corrosion of apparatus,
iv) maximum retrieval of the important derivatives of lignin and hemicelluloses,
v) minimal production of secondary degradation products,
vi) minimizing waste production (Taherzadeh and Karimi 2008).

Four different pretreatments are used for the purpose:

a) physical methods
b) chemical methods
c) physicochemical methods
d) biological methods (fungi and bacteria) (Taherzadeh and Karimi 2008; Kumar et al. 2009; Gnanambal et al. 2014; Mohanty and Abdullahi 2016)

9.6.3.1.1 Physical Methods

There are different categories of mechanical and nonmechanical methods that are used in physical pretreatment of agricultural waste. Mechanical methods include milling, grinding, and chipping of agricultural materials. Nonmechanical methods include irradiation techniques using gamma rays, electron beam, microwaves, etc. These pretreatment methods include hydrolysis of agricultural biomass (Zheng et al. 2009). The core benefit of physical method is that usage of harmful chemicals can be curbed, reduction in the quantity of fermentation inhibitors as well as environmentally friendly. Despite being efficient, these methods have huge time constraint and require man-power, expertise, and high-energy tools (Hendriks and Zeeman 2009).

9.6.3.1.2 Mechanical Methods

The mechanical methods include various techniques like chipping, grinding, shearing, or milling, which has an impact on the particle size as well as surface area so that the enzyme cellulase can attack on the substrate and facilitates ethanol generation from cellulose.

9.6.3.1.2.1 Milling In the process of milling, preliminary treatment of agricultural biomass and the entire process is conducted mechanically. By this method, the particle size of the agricultural material is reduced to a considerable extent, thereby the surface area increases that makes the enzymes more active for the hydrolysis of the cellulose present in the cell wall of the agricultural materials (Harmsen et al. 2010). This mechanical method can be subclassed into some unique techniques like hammer milling, ball milling, two-roll milling, disk milling, and colloid milling in order to produce bioethanol from the obtained biomass. Depending on the method used for milling, the final size of the particle is determined, for e.g. after chipping, the size of particle ranges between 10 and 30 mm, while after milling, the size becomes 0.2–2 mm (Sun and Cheng 2002). However, the process involves high temperature and is very time consuming.

On comparing ball milling (BM) and wet disk milling (WDM), pretreatment method that is applicable on sugarcane bagasse and straw, it was observed that the crystallinity decreases if BM is applied, but at the same time, it results in faster hydrolysis. The fibres that are obtained as a by-product from waste are needed for enzymatic hydrolysis and can be discarded by the method of WDM (Da Silva et al. 2010). When centrifugal roller mill pretreatment was given to corn straw, surface area increased by approximately five times, thereby the crystallinity of the polymer decreased and improved the output of lower molecular weight carbohydrates by a factor of 2 (Bychkov et al. 2014).

9.6.3.1.2.2 Pyrolysis or Thermal Cracking In pyrolysis, the agricultural waste is incinerated at elevated temperatures so that cellulose gets degraded and thus produces gaseous substances like hydrogen, carbon monoxide gas along with some tarry products. However, at lower temperature, the breakdown process gets decelerated and a greater number of volatile products are being formed (Prasad et al. 2007b; Mtui 2009). In spite of all these advantages, this method of pretreatment is very expensive and uneconomical since attainment of higher temperature needs a lot of fuel that needs to be burnt or else electrical heating is necessary, hence the overall cost of the process increases. In recent researches, it is found that pyrolysis method could convert 80% weight of the dry lignin coming from wheat straw into 40–60% bio-oil and 30–40% biochar (Wilda et al. 2012).

9.6.3.1.3 *Nonmechanical Methods*

9.6.3.1.3.1 Irradiation Pretreatment High-energy radiations are utilized in the method of irradiation treatment for enhancing the hydrolysis of agricultural waste by enzymes. The various methods that could be used for high-energy radiations are ultrasound (Rehman et al. 2013), microwave heating (Keshwani et al. 2007; Ma et al. 2009), ultrasound electron beam (Bak et al. 2009; Henniges et al. 2013), and γ-ray (Chunping et al. 2008). Although these methods are expensive when carried out on a large scale (Verma et al. 2011), nevertheless it helps in enhancing surface area, decreases the cellulose crystallinity, thereby changing lignin structure and facilitating the hydrolysis of hemicelluloses.

Out of all the methods available for irradiation, pretreatment used in agricultural waste, the most preferred and common method is irradiation using microwaves. The microwave irradiation offers many benefits that includes the following:

i) Uses less energy
ii) The process is simple and easy to use
iii) The process generates low levels of inhibitors
iv) The process disrupts and degrades the cellulose structure (Kumar and Sharma 2017)

9.6.3.1.4 *Chemical Methods*

Many types of chemical agents are used in the chemical pretreatment of agricultural wastes. The various chemicals used are acids (sulphuric acid, hydrochloric acid, organic acids, etc.), oxidizing agents (ozone and hydrogen peroxide), alkalis (sodium hydroxide, sodium carbonate, calcium hydroxide, ammonia, etc.), organic solvents, sulphur dioxide, and carbon dioxide and other chemicals that help in degrading hemicelluloses and removing the lignin from waste material that have lignocellulosic content (Nwosu-Obieogu et al. 2016).

9.6.3.1.4.1 Acid Pretreatment The agricultural waste is treated using diluted or concentrated acids like nitric acid, sulphuric acid, hydrochloric acid, and per acetic acids. Acid treatment helps in hydrolysing hemicelluloses to monosaccharides which makes the cellulose accessible for further process of treatment. The process is carried out at high temperature keeping the concentration of acid low or at low temperature using higher concentration of acid (Taherzadeh and Karimi 2008). The acid pretreatment process is not without any disadvantage. The corrosion of the equipment and formation of by-products like carboxylic acids, phenolic acids, and furans which impede the process of fermentation is some of the disadvantages of the process (Singh and Trivedi 2013).

9.6.3.1.4.2 Alkaline Pretreatment The process of alkaline pretreatment utilizes the bases like ammonium hydroxide, sodium hydroxide, calcium hydroxide, and potassium hydroxide for breaking the ester bonds between the three main components of agricultural waste such as lignin, hemicellulose, and cellulose. The alkalis also carry out the disruption of the lignin structure as well as decreasing the crystallinity of cellulose and enhancing the accessibility of the enzymes to cellulose by increasing the accessible surface area. The method of pretreatment with alkali has some advantages and disadvantages. The requirement of low temperature and pressure is advantageous, while high concentration of bases and long residence time are some limitations of the process. Calcium hydroxide is a favoured base over sodium hydroxide because it is cheap, safe, and can be removed after hydrolysis (Moiser 2005b).

9.6.3.1.4.3 Organosolvent Pretreatment Aqueous or organic solvents (ethanol, methanol, acetone, triethylene glycol, tetrahydrofurfuryl alcohol, and ethylene glycol) are used to pretreat agricultural waste at temperature of 150–200 °C in the presence or absence of catalysts. The catalysts used are hydrochloric acid, oxalic acid, sulphuric acid, acetylsalicylic acid, and salicylic acid (Taherzadeh and Karimi 2008). The lignin and hemicellulose bonds are broken by the action of the solvents (Joshi et al. 2011). This process has both advantages and disadvantages, the benefits being the easy recovery of organic solvents by distillation process (Harmsen et al. 2010), and risks includes corrosion, formation of toxic inhibitors, and moreover, the process is expensive (Bensah and Mensah 2013).

9.6.3.1.4.4 Ozonolysis Pretreatment Powerful oxidizing agent such as ozone is used in the pretreatment process of ozonolysis by which the lignin is degraded and hence the digestibility of biomass is increased. Cellulose is not degraded by ozone, while hemicellulose is partially degraded (Travaini et al. 2016). Some of the advantages of using this method are that it does not produce any harmful substances that inhibit the reactions and is analysed at room temperature and pressure (Vidal and Molinier 1988). The amount of ozone used is quite high, which makes it unsuitable at large industrial scale (Kumar et al. 2009).

9.6.3.1.4.5 Wet Oxidation Pretreatment The wet oxidation pretreatment involves treating of agricultural waste in presence of oxygen either with water or air at 120 °C for 30 minutes (Varga et al. 2003).When the lignin content of agricultural waste is low than this method is preferred, because the efficiency of this process reduces with increased concentration of lignin. The shortcoming involves the high cost as it involves the use of oxygen at high pressure and usage of chemicals (Biswas et al. 2015). The process can degrade lignin and hemicelluloses at low temperature, the recovery of chemicals are not needed, and efficiency of enzymatic hydrolysis increases after this process that adds to the benefits of wet oxidation pretreatment process.

9.6.3.1.5 Physicochemical Methods

This category is a mixed bag wherein both the physical and chemical pretreatment methods are combined together.

9.6.3.1.5.1 Auto Hydrolysis/Steam Explosion Pretreatment Steam ranging within the temperature range 180–240 °C is used under pressure (1–3.5 MPa) to make the components of agricultural waste accessible for hydrolysis. The pressure is then instantaneously reduced causing the agricultural biomass to explode (Sun and Cheng 2002). The steam pre treatment method is generally used in pre treatment. Previously it was called as 'steam explosion' method (Moiser 2005b). This method requires less energy which is an advantage (Sun and Cheng 2002; Kumar et al. 2009) but the formation of inhibitory substances like phenolic compounds, furaldehyde etc. and incomplete cleavage of lignocellulose matrix are some of the limitations of this method (García-Aparicio et al. 2006).

9.6.3.1.5.2 Liquid Hot Water Pretreatment (LHW) The hot liquid water having high pressure above the optimum level is used to hydrolyse hemicelluloses. This is an eco-friendly method that does not use any chemical or catalyst and moreover gives high recovery of xylose (Hamelinck et al. 2005), but the method has some limitations in terms of production of inhibitory substances that inhibit the fermentation process by limiting the growth of micro-organisms (Taherzadeh and Karimi 2008).

9.6.3.1.5.3 Ammonia Fibre Explosion (AFEX) Ammonia fibre explosion process uses liquid ammonia between temperature range of 60–120 °C and 1.72–2.06 MPa pressure for more than half an hour (Kumar et al. 2009) to explode the lignin–carbohydrate complex for increasing enzymatic hydrolysis of agricultural biomass (Chundawat et al. 2006). Using ammonia fibre explosion, enzymatic digestion of cellulose and hemicellulose increases (Krishnan et al. 2010). The method has many advantages first in terms of cost and energy requirement, and second, inhibitors like furans that are generated and finally greater depolymerization of agricultural waste (Chundawat et al. 2006; Sanchez 2009).

9.6.3.1.6 Biological Pretreatment

Biological method of pretreatment of agricultural waste makes use of biological agents like microbes, fungi, and earthworms. The biological treatment of waste materials principally aims at stabilizing the organic or biodegradable fraction of the agricultural waste so that no harmful gases are emitted while the process of final disposal takes place. It also focuses on minimizing the volume of the waste produced.

This treatment method has been a traditional practice in comparison to municipal solid waste treatment methods. Biological treatment has found a revival in the recent past. This has become possible due to increased efforts to improve recycling of nutrients and organic fraction to soil and in particular to reduce land filling of biodegradable waste to minimize the production of greenhouse gases. Energy extraction from degradation of organic matter through microbial action has also drawn attention these days. Thus, the ultimate aim of biological treatment of agricultural waste is to optimize the resource and energy usage and also reduce the harmful effects on the environment at the lowest possible investment and operational costs. Various biological methods of treatment include aerobic and anaerobic methods of treatment. This pretreatment method can be done by either using lignin-degrading microbes, for example white rot fungi, brown-rot fungi, and soft-rot fungi, or enzymes (Lee 1997). Brown-rot fungi attacks on cellulose and degrades it (Ray et al. 2010). Whereas white-rot fungi and soft rot fungi chiefly decompose lignin, cellulose, and hemicellulose component and increase the rate of hydrolysis of waste through enzymatic action. There are fungal species like *Trichoderma reesei* and *Aspergillus niger* that yield numerous extracellular cellulosic enzymes namely endoglucanases, cellobiohydrolases, and β-glucosidases, which are highly capable of degrading residues rich in cellulose. Basidiomycetes like *Phanerochaete chrysosporium* degrade agriculture matter (Dashtban et al. 2009; Sanchez 2009). Biological pretreatment has also been studied with the help of white rot fungi and *Streptomyces* in pulp to take away lignin. In one of the studies, in soft wood 10.5% and hard wood 23.5% loss of lignin was seen (Saritha et al. 2011). Various other experiments have been conducted using biological method of treatment on agricultural waste with the help of enzymes formed by a variety of microbes (Saravanakumar and Kathiresan 2013).

Biological pretreatments are environmentally compatible processes, as there is no use of chemicals. Energy involvement is comparatively less. There are no corrosion issues involved and also production of inhibitors is quite low (Sanchez 2009). In such pretreatment techniques, brown, white, and soft rot fungi destroy lignin and hemicellulose. But in case of cellulose disruption, they are not found to be very promising (Sanchez 2009). With the help of *Ceriporiopsis subvermispora* pretreatment of wheat, corn, and soybean as well as switch grass and hardwood was studied by Wan and Li (Wan and Li 2011). Enzymes are also employed for hydrolysis of raw materials containing lignocellulose. Efficacy of the process in totality depends on the rate of lignocellulosic feedstock hydrolysis to sugars that are fermentable (Binod et al. 2010). Enzymes like cellulases act on cellulose and accomplish enzymatic process of hydrolysis, where bacteria or yeasts are able to ferment the reducing sugars into ethanol (Sun and Cheng 2002). When hydrolysis of cellulose takes place, primarily three enzymatic clusters are involved. These three groups are endoglucanases possessing the ability of attacking areas where there is low crystallinity in the cellulose fibre giving rise to ends with free chains. Second group of enzymes are exoglucanases that remove cellobiose units by degrading its molecules contained in free chain ends. Third group is β-glucosidases having the capability to hydrolyse cellobiose and give glucose (Balat et al. 2008). Enzymatic hydrolysis has been categorized in twofold phases, one is primary and another is secondary. Primary hydrolysis phase relies on the action of exoglucanases and endoglucanases on the surface of solid substrate, yielding oligosacharides which is the liquid phase. Secondary phase is about the hydrolysis of oligosacharides to glucose and cellobiose (Binod et al. 2010). The cellulose access for hydrolysis through enzymes relies on taking away xylan from lignocellulose with the help of enzyme xylanases. Due to less tightly packed crystalline structure in case of xylan unlike cellulose, it is highly prone to enzymatic hydrolysis. Whole hydrolysis of xylan needs a combined action of endo-1,4-β-xylanase, β-xylosidase, α-arabinofuranosidase, and α-glucuronidase. Enzyme esterases attack the ester linkages present in xylose units of the xylan and acetic acid or between arabinose side chain residues and phenolic acids such as ferulic acid and *p*-coumaric acid (Binod et al. 2010). Many bacterial and fungal species secrete cellulases and hemicellulases (Rabinovich et al. 2002; Sun and Cheng 2002). Some of them are *Trichoderma reesei,* which are found to be highly competent cellulose-hydrolysing organism (Galbe and Zacchi 2002). Lignin could not be broken down by cellulases as it is closely bound to cellulose. The chief feature of lignin decomposition is the action of peroxidases that involves the action between lignin peroxidase and manganese peroxidase (Binod et al. 2010). Laccase is yet another enzyme that participates in the decomposition of lignin (Binod et al. 2010). It can alone hydrolyse the lignin completely and also in combination with peroxidases. The oxidation of phenolic units in lignin gets oxidized with the catalytic action of enzyme laccase. Enzymatic hydrolysis is very peculiar. It takes place in moderate situations involving less energy. Environmental impact is also less than what is observed in case of acid hydrolysis of lignocellulose. Glucose yield is also high and by-product formation is low. This is suitable for further use of hydrolysate in fermentation. Here corrosion problems are also found to be less (Mussatto and Teixeira 2010). The end product formed from this enzymatic hydrolysis stops the action of enzymes and therefore it has to be removed quickly to minimize its adverse influence on the hydrolysis kinetics. Various techniques are studied to lessen the obstruction of glucose by hydrolysis. Some such methods include usage of high enzyme concentrations, addition of β-glucosidases during the process of hydrolysis, removal of sugar by ultrafiltration, or we can also go for simultaneous saccharification and fermentation (Soccol et al. 2011; IvančićŠantek et al. 2016). Substrate concentration has a vital part in the whole procedure particularly on the initial rate and the outcome of cellulose. Low substrate concentration enhances yield and rate of hydrolysis. Though cellulase price has come down tremendously yet it signifies above 20% of bioethanol production costs from lignocellulosic feedstocks (Bhaskar et al. 2011). Even after hydrolysis cellulases remain active, therefore their recycling could bring some usefulness besides being an economically practicable method.

9.6.3.2 Enzymatic Hydrolysis

Agricultural biomass is hydrolysed into fermentable sugars by enzymes, and it is gaining attention in comparison to the method of acid hydrolysis as corrosion can be prevented; moreover, energy requirement is less and inhibitory products formed are few (Sanchez 2009). The enzymes that are used in hydrolysing cellulose are endoglucanase, exoglucanase, and cellobiase. Cellobiase plays a key role in converting cellobiose to D-glucose (Joshi et al. 2011). The exoglucanase enzymes form cellobiose units by attacking the nonreducing end of cellulose. The endoglucanase enzymes hydrolyse the cellulose into glucose, cellobiose, and cellotriose. Since hemicelluloses have complex structure, hence many enzymes are required for their hydrolysis. Enzymes from bacterial and fungal sources are employed in enzymatic degradation. Bacterial sources like *Bacillus, Clostridium, Microbispora, Streptomyces, Erwinia, Thermomonospora, Cellulomonas, Bacteriodes, Acetovibrio,* and *Ruminococcus* have been used to produce enzymes required in hydrolysing the agricultural biomass. Hemicellulase and cellulase have been produced from fungal sources such as *Aspergillus, Schizophyllum, Trichoderma, Humicola, P. chrysosporium, Penicillium,* and *Acremonium*. The scientists are preferring enzymes obtained from fungal sources than bacterial strains because they do not produce adequate amount of the desired product (Sun and Cheng 2002). The production of bioethanol requires high enzyme load that becomes a costly affair. The additives such as surfactants help in increasing the efficiency of hydrolysis and also in lowering the enzyme dosage (Qing et al. 2010; Cui et al. 2011).

9.6.3.3 Fermentation

The hexose and pentose sugars are converted to bioethanol and carbon dioxide under both aerobic and anaerobic conditions during the process of fermentation. The most common micro-organism used in fermenting sugars such as sucrose, glucose, maltose, and fructose is the baker's yeast (*Saccharomyces cerevisiae*) but is unable to ferment pentose sugar, which is the main limitation. Hence, non-Saccharomyces yeasts and genetically modified *S. cerevisiae* are used to ferment pentose sugars such as arabinose and xylose, which are obtained from second-generation feedstocks (Oreb et al. 2012; Sinumvayo et al. 2015). Micro-organisms like *Candida shehatae, Pachysolen tannophilus, Pichia stipitis* possess an efficiency to ferment both pentose ad hexose sugar (Lin and Tanaka 2005).

Thermophilic bacteria such as *Thermoanaerobacter ethanolicus* and *Thermoanaerobacterium saccharolyticum* have been also used in fermentation process and offers many advantages like high bioconversion, enhanced mass transfer and diffusion rate, better substrate solubility, and low contamination risk. Moreover, such micro-organisms are capable enough to ferment both pentose and hexose sugars. The cost of the process is also reduced (Sinumvayo et al. 2015) as these bacteria produce hemicelluloses and cellulase enzymes internally; hence, additionally external enzymes are not needed.

9.6.3.4 Separation and Distillation

Based on differences in volatilities, ethanol is separated from water by the process of distillation. Ethanol is concentrated from the fermentation broth in a distillation column and the solids are taken out from the bottom of the apparatus, whereas ethanol is passed to a rectification column to obtain azeotropic composition of ethanol–water mixture 95% by weight (Hamelinck et al. 2005). Centrifugation of the bottom solids helps in further concentrating. Azeotropic mixture is used when ethanol is used as hydrous fuel. When gasoline is blended with anhydrous ethanol, the distillation column should be filled up with cyclohexane or benzene in the distillation column. The process is expensive, and ethanol is recovered with impurities having similar boiling point are some disadvantages of the process (Onuki et al. 2008).

9.7 Prospects

On the basis of fuel blend, the bioethanol market has been classified into E10, E5, E75, E85, E15, and E70 categories, and in the year 2020, the market was dominated by E10 category with a prediction of rapid expansion in the future. The extraordinary compatibility of E10 with the number of petrol-powered motor vehicles worldwide makes it a highly reliable fuel. The bioethanol market has been categorized into transportation, pharmaceuticals, cosmetics, and alcoholic beverages depending on end use. The transport sector has dominated the market in 2020, and this trend is expected to continue. The burgeoning population and superfluous income have impacted the lifestyle with more vehicles on road are the determinants, which are propelling the growth of transportation sector of the bioethanol market. Governments of various countries have ordered the usage of bioethanol fuel blends. The bioethanol market value is likely to touch 77.53 billion by 2026 (Global Bioethanol Market Size By Feedstock Type, By End Use, By Geographic Scope and Forecast 2020). The major share of bioethanol market is being held by North America in 2020, and the situation is going to be same in next five years. Bioethanol blending helps in increasing fuel efficiency of the vehicles and lowering the greenhouse gas emission which acts as a driving force toward the growing demand of bioethanol.

9.7.1 Bioethanol as Savior of Environment

When bioethanol obtained from agricultural biomass is combusted it produces carbon dioxide gas and that producer community can assimilate, hence no net gain of carbon dioxide occurs (Marszalek and Kaminski 2008). It also lowers greenhouse gas emissions. If humans are exposed to bioethanol either by inhalation or skin contact, it does not cause any harm. Atmospheric degradation of bioethanol is fast too. Diesel blended with ethanol helps is decreasing octane number, at the same time increases heating value and aromatic fractions as well as changes the temperature of distillation. All these factors are instrumental in facilitating the complete combustion of ethanol and lowering greenhouse gas emissions (Chan et al. 2007). Ethanol has low photochemical reactivity in the atmosphere and also reduces the ozone formation (Wyman 1994).

9.8 Challenges

The production of bioethanol from agricultural wastes comes with high advantage as it is ecofriendly. It sorts out the problem of management of agricultural waste to large extent. The major advantage is seen in the absence of chemicals from the whole picture. Energy input is quite low. It requires high specificity to lignin, moderate functional conditions, and is environment compatible (Sanchez 2009).The major setback as assessed is the long-time duration as the rate of hydrolysis by micro-organisms is slower (Sun and Cheng 2002). These limitations make it commercially less viable and hinder its gross productivity (Eggeman and Elander 2005).

Challenges involved in bioethanol production from the agricultural waste are multifarious. Though a better fuel in comparison to conventional fuels, yet bioethanol production is not free from shortcomings. There are some issues that we face in connection with the production of bioethanol. Some of them are:

As per latest estimates, when the cost of biofuel production is compared with second-generation biofuels, these are found to be costlier than traditional fuels (Carriquiry et al. 2011).

Feedstock production, feedstock logistics, development of energy efficient technologies, coproducts development, establishment of biofuel and biochemical standards, biofuel distribution, societal acceptance, and environmental impact minimization are among the variety of challenges that come in the way of converting lignocellulosic biomass to biofuels (Hoekman 2009; Luo et al. 2010; Menon and Rao 2012) and require attention and redressal mechanism.

High cost of the hydrolysis and low production in terms of final yield are other concerns. Several efforts are employed to improve this whole process of hydrolysis of lignocelluloses' materials by pretreatment for removal of lignin and hemicelluloses.

9.9 Conclusion

The extraction of ethanol from agricultural wastes is a promising technique. The techniques discussed are very much compatible with the environmental principle, and hence it is better to produce ethanol from agricultural wastes. The need of the hour is to develop ecofriendly pretreatment processes, extremely efficient system of enzyme to convert the pretreated waste to fermentable sugars, and finally efficacious micro-organisms to transform various sugars to ethanol. Bioethanol ensures lesser harmful impact on the environment compared to traditional fuels like coal and petroleum. This holds true predominantly in terms of atmospheric and greenhouse gas emissions. These challenging areas require interdisciplinary approach wherein expertise would be required in the fields of microbiology, agronomy, chemical engineering, genetic engineering, economics, etc. Using bioethanol can help mankind to emerge from the current fuel crisis and at the same time it could help in sustainable development. A lot of development has been done in the biomass pretreatment processes, cellulase production, and cofermentation of sugars and also in the separation and purification of bioethanol, but still the production cost for bioethanol from other sources except sugar cane is not yet competitive to the fossil fuels. Thus, we can say that still a big challenge remains in terms of the cost of production. Hence, we need to focus on the concept of biorefinery that makes use of the renewable feedstocks in an accomplished manner and encourages value-added coproducts like bio-based materials from the lignin that would curtail bioethanol production cost.

References

Agbor, V.B., Cicek, N., Sparling, R. et al. (2011). Biomass pretreatment: fundamentals toward application. *Biotechnology Advances* 29 (6): 675–685. 10.1016/j.biotechadv.2011.05.005 PMID:2162445.

Ameri, M., Ghobadian, B., and Baratian, I. (2008). Technical comparison of a CHP using various blends of gasohol in an IC engine. *Renewable Energy* 33 (7): 1469–1474. 10.1016/j.renene.2007.09.015.

Ayadi, M., Sarma, S.J., Pachapur, V.L. et al. (2016). History and global policy of biofuels. In: *Green Fuels Technology* (ed. C.R. Soccol, S.K. Brar, C. Faulds, et al.). Switzerland: Springer International Publishing 10.1007/978-3-319-30205-8_1.

Bak, J.S., Ko, J.K., Han, Y.H. et al. (2009). Improved enzymatic hydrolysis yield of rice straw using electron beam irradiation pretreatment. *Bioresource Technology* 100 (3): 1285–1290. 10.1016/j.biortech.2008.09.010 PMID:18930388.

Balat, M. (2009). Gasification of biomass to produce gaseous products. *Energy Sources* 31 (6): 516–526. 10.1080/15567030802466847.

Balat, M., Balat, H., and Cahide, O.Z. (2008). Progress in bioethanol processing. *Progress in Energy & Combustion* 34 (5): 551–573. 10.1016/j.pecs.2007.11.001.

Bensah, E.C. and Mensah, M. (2013). Chemical pretreatment methods for the production of cellulosic ethanol: technologies and innovations. *International Journal of Chemical Engineering* 2013: 1–21. 10.1155/2013/719607.

Bhaskar, T., Bhavya, B., Singh, R. et al. (2011). Thermochemical conversion of biomass to biofuels. In: *Biofuels – Alternative Feedstocks and Conversion Processes* (ed. A. Pandey, C. Larroche, S.C. Ricke, et al.), 51–77. Oxford, UK: Academic Press 10.1016/B978-0-12-385099-7.00003-6.

Binod, P., Sindhu, R., Singhania, R.R. et al. (2010). Bioethanol production from rice straw: an overview. *Bioresource Technology* 101 (13): 4767–4774. 10.1016/j.biortech.2009.10.079.

Global Bioethanol Market Size By Feedstock Type, By End Use, By Geographic Scope And Forecast (2020). Report ID: 33537, 94.

Biswas, R., Uellendahl, H., and Ahring, B.K. (2015). Wet explosion: a universal and efficient pretreatment process for lignocellulosic biorefineries. *Bioenergy Research* 8 (3): 1101–1116. 10.100712155-015-9590-526. Agricultural Waste Management for Bioethanol Production.

Bušić, A., Marđetko, N., Kundas, S. et al. (2018). Bioethanol production from renewable raw materials and its separation and purification: a review. *Food Technology and Biotechnology* 56 (3): 289.

Bychkov, A.L., Buchtoyarov, V.A., and Lomovsky, O.L. (2014). Mechanical pretreatment of corn straw in a centrifugal roller mill. *Cellulose Chemistry and Technology* 48 (5-6): 545–551.

Cardona, C.A. and Sanchez, O.J. (2007). Fuel ethanol production: process design trends and integration opportunities. *Bioresource Technology* 98 (12): 2415–2457. 10.1016/j.biortech.2007.01.002 PMID:17336061.

Carriquiry, M.A., Du, X., and Timilsina, G.R. (2011). Second generation biofuels: economics and policies. *Energy Policy* 39 (7): 4222–4234.

Chandel, A.K., Chan, E.S., Rudravaram, R. et al. (2007). Economics and environmental impact of bioethanol production technologies: an appraisal. *Biotechnology and Molecular Biology Reviews* 2 (1): 14–32.

Chundawat, S.P., Venkatesh, B., and Dale, B.E. (2006). Effect of particle size based separation of milled corn stover on AFEX pretreatment and enzymatic digestibility. *Biotechnology and Bioengineering* 96 (2): 219–231. 10.1002/bit.21132 PMID:16903002.

Chunping, Y., Zhiqiang, S., Guoce, Y.U., and Jianlong, W. (2008). Effect and after effect of gama radiation pretreatment on enzymatic hydrolysis of wheat straw. *Bioresource Technology* 99 (14): 6240–6245. 10.1016/j.biortech.2007.12.008 PMID:18248982.

Cui, L., Liu, Z., Hui, L.F., and Si, L.F. (2011). Effect of cellobiase and surfactant supplementation on enzymatic hydrolysis of pretreated wheat straw. *BioResources* 6 (4): 3850–3858.

Da Silva, A.S., Inoue, H., Endo, T. et al. (2010). Milling pretreatment of sugarcane bagasse and straw for enzymatic hydrolysis and ethanol fermentation. *Bioresource Technology* 101 (19): 7402–7409. 10.1016/j.biortech.2010.05.008.

Dashtban, M., Schraft, H., and Qin, W. (2009). Fungal bioconversion of lignocellulosic residues; opportunities & perspectives. *International Journal of Biological Sciences 5* (6): 578–595. 10.7150/ijbs.5.578 PMID:19774110.

Dufey, A. (2006). *Biofuels Production, Trade and Sustainable Development: Emerging Issues*. London: International Institute for Environment and Development.

Edeh, I., 2020. Bioethanol production: an overview. *Bioethanol*. Bioethanol Technologies, Freddie Inambao, IntechOpen, 10.5772/intechopen.94895. https://www.intechopen.com/chapters/74319 (accessed 8 November 2021).

Eggeman, T. and Elander, R.T. (2005). Process and economic analysis of pretreatment technologies. *Bioresource Technology 96* (18): 2019–2025. 10.1016/j.biortech.2005.01.017 PMID:16112490.

Faustine, A.S. and Rustini, A.D. (2021). Bioethanol production from various agricultural waste substrate using saccharomyces cerevisiae. *IOSR Journal Of Pharmacy And Biological Sciences* 1 (1): 7–13 www.Iosrjournals.Org. (accessed 8 November 2021).

Forster, P., Ramaswamy, V., Artaxo, P. et al. (2007). Changes in atmospheric constituents and in radiative forcing. In: *Climate Change 2007: The Physical Science Basis* (ed. S. Solomon, D. Qin, M. Manning, et al.), 129–234. Cambridge, UK: Cambridge University Press.

Foyle, T., Jennings, L., and Mulcahy, P. (2007). Compositional analysis of lignocellulosic materials: evaluation of methods used for sugar analysis of waste paper and straw. *Bioresource Technology* 98 (16): 3026–3036. 10.1016/j.biortech.2006.10.013 PMID:17142038.

Galbe, M. and Zacchi, G. (2002). A review of the production of ethanol from softwood. *Applied Microbiology and Biotechnology* 59 (6): 618–628. 10.1007/s00253-002-1058-9.

García-Aparicio, M.P., Ballesteros, I., González, A. et al. (2006). Effect of inhibitors released during steam-explosion pretreatment of barley straw on enzymatic hydrolysis. *Applied*

Biochemistry and Biotechnology 129 (1-3): 278–288. 10.1385/ABAB:129:1:278 PMID:16915647.

Gnanambal, V.S., Jabastin, J., and Swaminathan, K. (2014). Biological pretreatment of six lignocellulosic wastes for bioethanol production. *International Journal of Science and Research* 3: 1860–1864.

Goh, C.S., Tan, K.T., Lee, K.T., and Bhatia, S. (2010). Bioethanol from lignocellulose: status, perspectives and challenges in Malaysia. *Bioresource Technology* 101 (13): 4834–4841. 10.1016/j.biortech.2009.08.080 PMID:19762229.

Gray, K.A., Zhao, L., and Emptage, M. (2006). Bioethanol. *Current Opinion in Chemical Biology* 10 (2): 141–146. 10.1016/j.cbpa.2006.02.035 PMID:16522374.

Hamelinck, C.N., Hooijdonk, G.V., and Faaij, A.P. (2005). Ethanol from lignocellulosic biomass: techno-economic performance in short-, middle- and long-term. *Biomass and Bioenergy* 28 (4): 384–410. 10.1016/j.biombioe.2004.09.002.

Harmsen, P., Huijgen, W., Bermudez, L., and Bakker, R. (2010). Literature review of physical and chemical pretreatment processes for lignocellulosic biomass (Tech. Rep. 1184). Biosynergy, Wageningen UR Food & Biobased Research.

Hendriks, A. and Zeeman, G. (2009). Pretreatments to enhance the digestibility of lignocellulosic biomass. *Bioresource Technology* 100 (1): 10–18. 10.1016/j.biortech.2008.05.027 PMID:18599291.

Henniges, U., Hasani, M., Potthast, A. et al. (2013). Electron beam irradiation of cellulosic materials-opportunities and limitations. *Materials (Basel)* 6 (5): 1584–1598. 10.3390/ma6051584 PMID:2880923028. Agricultural Waste Management for Bioethanol Production.

Hoekman, S.K. (2009). Biofuels in the U.S. – challenges and opportunities. *Renewable Energy* 34 (1): 14–22.

Howard, R., Abotsi, E., Jansen, V.R., and Howard, S. (2003). Lignocellulose biotechnology: issues of bioconversion and enzyme production. *African Journal of Biotechnology* 2 (12): 602–619. 10.5897/AJB2003.000-1115.

IvančićŠantek, M., Miškulin, E., Beluhan, S., and Šantek, B. (2016). New trends in the ethanol production as a biofuel. *Kemija u Industriji* 65 (1–2): 25–38. https://doi.org/10.15255/KUI.2014.032.

Joshi, B., Bhatt, M.R., Sharma, D. et al. (2011). Lignocellulosic ethanol production: current practices and recent developments. *Biotechnology and Molecular Biology Reviews* 6 (8): 172–182.

Keshwani, D.R., Cheng, J.J., Burns, J.C. et al. (2007). Microwave pretreatment of switchgrass to enhance enzymatic hydrolysis. Paper presented at the 2007 ASAE Annual Meeting.

Koltermann, A., Kraus, M., Rarbach, M. et al. (2014). Cellulosic ethanol from agricultural residues – an advanced biofuel and biobased chemical platform. *JSM Biotechnology Bioengineering* 2 (1): 1024.

Krishnan, C., Sousa, L.D., Jin, M. et al. (2010). Alkali-based AFEX pretreatment for the conversion of sugarcane bagasse and cane leaf residues to ethanol. *Biotechnology and Bioengineering* 107 (3): 441–450. 10.1002/bit.22824 PMID:20521302.

Kumar, A.K. and Sharma, S. (2017). Recent updates on different methods of pretreatment of lignocellulosic feedstocks: a review. *Bioresources and Bioprocessing* 4 (1): 7. 10.118640643-017-0137-9 PMID:28163994.

Kumar, R., Singh, S., and Singh, O. (2008). Bioconversion of lignocellulosic biomass: biochemical and molecular perspectives. *Journal of Industrial Microbiology & Biotechnology* 35 (5): 377–391. 10.100710295-008-0327-8 PMID:18338189.

Kumar, P.D., Barrett, D.M., Delwiche, M.J., and Stroeve, P. (2009). Methods for pretreatment of lignocellulosic biomass for efficient hydrolysis and biofuel production. *Industrial & Engineering Chemistry Research* 48 (8): 3713–3729. 10.1021/ie801542g.

Kwon, E.E., Jean, Y.J., and Yi, H. (2012). New candidate for biofuel feedstock beyond terrestrial biomass for thermo-chemical process (pyrolysis/gasification) enhanced by carbon dioxide (CO_2). *Bioresource Technology* 123: 673–677. 10.1016/j.biortech.2012.07.035. PMID:2293959729. Agricultural Waste Management for Bioethanol Production.

Lee, J. (1997). Biological conversion of lignocellulosic biomass to ethanol. *Journal of Biotechnology 56* (1): 1–24. 10.1016/S0168-1656(97)00073-4 PMID:9246788.

Lin, Y. and Tanaka, S. (2005). Ethanol fermentation from biomass resources: current state and prospects. *Applied Microbiology and Biotechnology* 69 (6): 627–642. 10.100700253-005-0229-x PMID:16331454.

Lin, Y. and Tanaka, S. (2006). Ethanol fermentation from biomass reseorces: current state and prospects. *Applied Microbiology and Biotechnology* 69: 627–642.

Linde, M., Galbe, M., and Zacchi, G. (2008). Bioethanol production from non-starch carbohydrate residues in process streams from a dry-mill ethanol plant. *Bioresource Technology* 99 (14): 6505–6511. 10.1016/j.biortech.2007.11.032 PMID:18281213.

Luo, L., van der Voet, E., and Huppes, G. (2010). Biorefining of lignocellulosic feedstock – technical, economic and environmental considerations. *Bioresource Technology* 101 (13): 5023–5032.

Ma, H., Liu, W., Chen, X. et al. (2009). Enhanced enzymatic saccharification of rice straw by microwave pretreatment. *Bioresource Technology* 100 (3): 1279–1284. 10.1016/j.biortech.2008.08.045 PMID:18930389.

Marszalek, J. and Kaminski, W. (2008). Environmental impact of bioethanol production. In *Proceedings of ECOpole*. Wolczanska

Menon, V. and Rao, M. (2012). Trends in bioconversion of lignocellulose: biofuels, platform chemicals & biorefinery concept. *Progress in Energy and Combustion Science* 38 (4): 522–550.

Ministry of New and Renewable Energy, Government of India (2009). Annual report.

Mohanty, B. and Abdullahi, I.I. (2016). Bioethanol production from lignocellulosic waste-a review. *Biosciences, Biotechnology Research Asia* 13 (2): 1153–1161. 10.13005/bbra/2146.

Moiser, N. (2005b). Features of promising technologies for pretreatment of lignocellulosic biomass. *Bioresource Technology* 96 (6): 673–686. 10.1016/j.biortech.2004.06.025 PMID:1558877030. Agricultural Waste Management for Bioethanol Production.

Mtui, G.Y.S. (2009). Recent advances in pretreatment of lignocellulosic wastes and production of value added products. *African Journal of Biotechnology* 8 (8): 7–13.

Mussatto, S.I. and Teixeira, J.A. (2010). Lignocellulose as raw material in fermentation processes. In: *Current Research, Technology and Education Topics in Applied Microbiology and Microbial Biotechnology*, vol. 2 (ed. A. Méndes-Vilas), 897–907. Badajoz, Spain: Formatex Research Center.

Nwosu-Obieogu, K., Chiemenem, L.I., and Adekunle, K.F. (2016). Utilization of agricultural waste for bioethanol production- a review. *International Journal of Current Research & Review* 8 (19): 1–5.

Onuki, S., Koziel, J.A., Leeuwen, V.J. et al. (2008). Ethanol production, purification, and analysis techniques: a review. In Proceedings of the ASABE Annual International Meeting, Rhode Island.

Oreb, M., Dietz, H., Farwick, A., and Boles, E. (2012). Novel strategies to improve co-fermentation of pentoses with D-glucose by recombinant yeast strains in lignocellulosic hydrolysates. *Bioengineered* 3 (6): 347–351. 10.4161/bioe.21444 PMID:22892590.

Prasad, S., Singh, A., Jain, N., and Joshi, H.C. (2007a). Ethanol production from sweet sorghum syrup for utilization as automotive fuel in India. *Energy & Fuels* 21 (4): 2415–2420. 10.1021/ef060328z.

Prasad, S., Singh, A., and Joshi, H.C. (2007b). Ethanol as an alternative fuel from agricultural, industrial and urban residues. *Conservation & Recycling* 50 (1): 10.1016/j.resconrec.2006.05.007.

Qing, Q., Yang, B., and Wyman, C.E. (2010). Impact of surfactants on pretreatment of corn stover. *Bioresource Technology* 101 (15): 5941–5951. 10.1016/j.biortech.2010.03.003 PMID:20304637.

Rabinovich, M.L., Melnik, M.S., and Bolobova, A.V. (2002). Microbial cellulases (review). *Applied Biochemistry and Microbiology* 38 (4): 305–322. 10.1023/A:1016264219885.

Ratnavathi, C.V., Suresh, K., Kumar, B.S.V. et al. (2010). Study on genotypic variation for ethanol production from sweet sorghum juice. *Biomass and Bioenergy* 34 (7): 947–952. 10.1016/j.biombioe.2010.02.002.

Ray, M.J., Leak, D.J., Spanu, P.D., and Murphy, R.J. (2010). Brown rot fungal early stage decay mechanism as a biological pretreatment for softwood biomass in biofuel production. *Biomass and Bioenergy 34* (8): 1257–1262. 10.1016/j.biombioe.2010.03.015.

Rehman, M.S.U., Kim, I., Chisti, Y., and Han, J.I. (2013). Use of ultrasound in the production of bioethanol from lignocellulosic biomass. *Energy Education Science and Technology, Part A. Energy Science and Research* 30 (2): 1391–1410.

Renewable Fuels Association-ethanolrfa (2012). http://www.org/statistics/annual-ethanol-production (accessed 1 November 2021).

Roehr, M. (2000). *Biotechnology of Ethanol, Classic and Future Application*. Wiley 10.1002/3527602348.

Saha, B.C. and Cotta, M.A. (2007). Enzymatic saccharification and fermentation of alkaline peroxide pretreated rice hulls to ethanol. *Enzyme and Microbial Technology* 41 (4): 528–532. 10.1016/j.enzmictec.2007.04.006.

Sanchez, C. (2009). Lignocellulosic residues: biodegradation and bioconversion by fungi. *Biotechnology Advances* 27 (2): 185–194. 10.1016/j.biotechadv.2008.11.001 PMID:19100826.

Saravanakumar, K. and Kathiresan, K. (2013). Bioconversion of lignocellulosic waste to bioethanol by *Trichoderma* and yeast fermentation. *Biotech 4* (5): 493–499. PMID:28324381.

Saritha, M., Arora, A., and Lata (2011). Biological pretreatment of lignocellulosic substrates for enhanced delignification and enzymatic digestibility. *Indian Journal of Microbiology 52* (2): 122–130. 10.100712088-011-0199-x PMID:23729871.

Singh, D.P. and Trivedi, R.K. (2013). Acid and Alkaline pretreatment of lignocellulosic biomass to produce ethanol as biofuel. *International Journal of ChemTech Research* 5 (2): 727–734.

Singhania, R.R., Sukumaran, R.K., Patel, A.K. et al. (2010). Advancement and comparative profiles in the production technologies using solid-state and submerged fermentation for microbial cellulases. *Enzyme and Microbial Technology* 46 (7): 541–549. 10.1016/j.enzmictec.2010.03.010.

Sinumvayo, J.P., Ishimwe, N., Komera, I., and Niyomukiza, S. (2015). Ethanol biofuel production from lignocellulosic biomass by Engineered *Saccharomyces cerevisiae*. *Journal of Academia & Industrial Research* 3 (10): 463–471.

Soccol, C.R., Faraco, V., Karp, S. et al. (2011). Lignocellulosic bioethanol: current status and future perspectives. In: *Biofuels – Alternative Feedstocks and Conversion Processes* (ed.

A. Pandey, C. Larroche, S.C. Ricke, et al.), 101–122. Oxford, UK: Academic Press 10.1016/B978-0-12-385099-7.00005-X.

Solomon, B.D., Barnes, J.R., and Halvorsen, K.E. (2007). Grain and cellulosic ethanol: history, economics, and energy policy. *Biomass and Bioenergy* 31 (6): 416–425.

Sun, Y. and Cheng, J. (2002). Hydrolysis of lignocellulosic materials for ethanol production: a review. *Bioresource Technology* 83 (1): 10.1016/S0960-8524(01)00212-7 PMID:1205882632Agricultural Waste Management for Bioethanol Production.

Taherzadeh, M.J. and Karimi, K. (2008). Pretreatment of lignocellulosic wastes to improve ethanol and biogas production: a review. *International Journal of Molecular Sciences* 9 (9): 1621–1651. 10.3390/ijms9091621 PMID:19325822.

Tan, L., Sun, Z.Y., Okamoto, S. et al. (2015). Production of ethanol from raw juice and thick juice of sugar beet by continuous ethanol fermentation with flocculating yeast strain KF-7. *Biomass and Bioenergy* 81: 265–272. 10.1016/j.biombioe.2015.07.019.

The ICCT (2021). India, don't fall for ethanol: Roadmap leads National Policy on Biofuels off track | International Council on Clean Transportation (theicct.org). https://theicct.org/india-dont-fall-for-ethanol-roadmap-leads-national-policy-on-biofuels-off-track/

Torres-Jimenez, E., Jerman, M.S., Gregorc, A. et al. (2011). Physical and chemical properties of ethanol-diesel fuel blends. *Fuel* 90 (2): 795–802. 10.1016/j.fuel.2010.09.045.

Travaini, R., Martin-Juarez, J., Lorenzo-Hernando, A., and Bolado-Rodriguez, S. (2016). Ozonolysis: an advantageous pretreatment for lignocellulosic biomass revisited. *Bioresource Technology* 199: 2–12. 10.1016/j.biortech.2015.08.143 PMID:26409859.

Turner, D., Xu, H., Cracknell, R.F. et al. (2011). Combustion performance of bioethanol of various blend ratios in a gasoline direct injection engine. *Fuel* 90 (5): 1999–2006. 10.1016/j.fuel.2010.12.025.

United States Department of Agriculture (2009). *Colombia: A New Ethanol Producer on the Rise*. Economic Research Service: USDA.

Vanholme, R., Demedts, B., Morreel, K. et al. (2010). Lignin biosynthesis and structure. *Plant Physiology* 153 (3): 895–905. 10.1104/pp.110.155119PMID:20472751.

Varga, E., Schmidt, A.S., Reczey, K., and Thomsen, A.B. (2003). Pretreatment of corn stover using wet oxidation to enhance enzymatic digestibility. *Applied Biochemistry and Biotechnology* 104 (1): 37–50. 10.1385/ABAB:104:1:37 PMID:12495204.

Verma, A., Kumar, S., and Jain, P.K. (2011). Key pretreatment technologies on cellulosic ethanol production. *Journal of Scientific Research* 55: 57–63.

Vidal, P.F. and Molinier, J. (1988). Ozonolysis of lignin improvement of in vitro digestibility of poplar sawdust. *Biomass* 16 (1): 1–17. 10.1016/0144-4565(88)90012-1.

Wan, C. and Li, Y. (2011). Effectiveness of microbial pretreatment by *Ceriporiopsis subvermispora* on different biomass feedstocks. *Bioresource Technology* 102 (16): 7507–7512. 10.1016/j.biortech.2011.05.026.

Wang, M., Saricks, C., and Santini, D. (1999). *Effect of Fuel Ethanol Use on Fuel-Cycle Energy and Greenhouse Gas Emission*. Argonne, IL: Argonne National Laboratory.

Wilda, P.J.D., Huijgena, W.J.J., and Heeres, H.J. (2012). Pyrolysis of wheat straw-derived organosolv lignin. *Journal of Analytical and Applied Pyrolysis* 93: 95–103. 10.1016/j.jaap.2011.10.00233 Agricultural Waste Management for Bioethanol Production.

World Bioenergy Association (2017). WBA Global Bioenergy Statistics. www.worldbioenergy.org/global-bioenergy-statistics (accessed 8 November 2021).

Wyman, C.E. (1994). Ethanol from lignocellulosic biomass: technology, economics, and opportunities. *Bioresource Technology* 50 (1): 3–15. 10.1016/0960-8524(94)90214-3.

Yang, B. and Wyman, C.E. (2008). Characterization of the degree of polymerization of xylooligomers produced by flowthrough hydrolysis of pure xylan and corn stover with water. *Bioresource Technology* 99 (13): 5756–5762.

Yüksel, F. and Yüksel, B. (2004). The use of ethanol–gasoline blends as a fuel in an SI engine. *Renewable Energy* 29 (7): 1181–1191. https://doi.org/10.1016/j.renene.2003.11.0127.

Zheng, Y.I., Pan, Z., and Zhang, R. (2009). Overview of biomass pretreatment for cellulosic ethanol production. *International Journal of Agricultural and Biological Engineering* 2 (3): 51–68.

10

Bioethanol Production from Lignocellulose Agricultural Waste Biomass

Subham Roy[1], Ratan Chowdhury[1], Hangma Boro[1], Biman Kumar Sarma[2], and Sanjib Kalita[3]

[1] *Department of Botany, Rangapara College, Rangapara, Assam, India*
[2] *Department of Zoology, Rangapara College, Rangapara, Assam, India*
[3] *Department of Botany, Gauhati University, Guwahati, Assam, India*

CHAPTER OUTLINE

Agriculture Waste Management and Bioresource: The Circular Economy Perspective, First Edition.
Edited by Suruchi Singh, Pardeep Singh, Anu Sharma, and Moharana Choudhury.

10.1 Introduction

All over the world, the world's economy is entirely reliant on nonrenewable energy sources that have become the only source for the generation of fuels, electricity, and food. Some of such examples are petroleum, diesel, natural gases, etc. But now, due to depleting fossil fuel resources, renewable energy now capturing a good share of worldwide headlines, and it compelled humanity to rely on bioenergy or renewable energy sources such as biofuel or bioethanol. Biofuel is composed of gaseous or liquid fuel developed primarily from lignocellulose and used in the shipping sector. Biofuels may be classified under first-generation and second-generation biofuels (Kandasamy et al. 2021). First-generation biofuels are made from carbohydrate, lipid, oils, or agroindustrial waste, while second-generation biofuels are derived from cellulosic plant biomass like wood, stalk, stem, and lignocellulosic biomass. Biomass includes primarily of straws, fallen leaves, wood, and also can be household biowaste. This substance can be reasonably transformed into bioethanol. In the environment, biomass is annually produced naturally, estimated to be around 10–50 × 109 tons.

Ethanol as a biofuel has lots of advantages over conventional fuel: low toxicity, biodegradability, and fewer air pollutants than petroleum fuel. Bioethanol can also be used as a pure fuel for the operation of different motor vehicles. Also, they can be incorporated with petrol or diesel as a fuel extender to run smoothly (Estevez et al. 2019). Most of the biofuel is produced from sugar and starch products. However, the continuous use of this feedstock may lead to food competition. Lignocellulose mass represents the abundant source for the development of bioethanol. The complex structure comprises cellulose, hemicelluloses, and lignin (Dey et al. 2018). Lignocellulose biomass evolves chemical mechanisms that can resist assaults on its structural sugar from microbial and chemical degradation (Zhao et al. 2012). From the past 80 years, many technologies have been developed regarding the conversion process of biomass into ethanol. The technology involved in the bioconversion process consists of three crucial steps: first, the pretreatment process that helps reduce the particle size; then second, the hydrolysis; and finally, the fermentation is done. The pretreated lignocellulosic materials are subjected to enzymatic hydrolysis, which results in forming of hydrolysates. These are then fermented to ethanol using fermenting microbes (Kuhad et al. 2011a). Pretreatment is done to remove the lignin content and the fraction of hemicellulose, due to which it leads to reduction in cellulose crystallinity and increase in the permeability of the materials. One such efficient pretreatment strategy is the thermochemical hydrolysis of the lignocellulosic biomass. It helps in biological transformation of the biomass to form sugars and finally to bioethanol. The microbial cell metabolism during fermentation is very much affected by the chemical harshness and higher temperature. The generation of toxic by-products from the products such as furfural, hydroxymethylfurfural (HMF), and phenolics occurs (Zhang et al. 2006; Jönsson et al. 2013). Therefore, to make the hydrolysates amenable to fermenting microbes, detoxification of acid hydrolysate is required (Kuhad et al. 2011a, Gupta et al. 2011a). Although the process of enzymatic hydrolysis has exhibited many superior products as no degradation constituents of glucose are formed for the subsequent fermentation, but comparatively, the process is very much slower and also not much economic (Du et al. 2018). The competition to lower the effective enzyme cost, maximizing the enzyme production, and using genetically engineered systems to make enzyme usage have been investigated in more than the last three decades (Wiltschia et al. 2020; Kuhad et al. 2011a); although it is not successful, there is still a desire to search for more relevant results.

The efficient fermentation of sugars present in the lignocellulose biomass, such as structural sugars like hexose and pentose sugar, has become a necessity for the low-cost production of bioethanol. The efficient utilization of microbes can fulfil this to ferment these sugars. However, compared to hexose sugar, the pentose sugar fermentation done by microbes is significantly less in numbers. The most common microbes for the fermentation of pentose sugars are *Candida shehatae*, *Pichia stipites*, and *Pachysol entannophilus* (Kuhad et al. 2011a; Selim et al. 2018). However, no promising results were found after implementing those microbes to ferment sugars. To make use of all the sugars present in the hydrolysates, efforts have been made to improve the fermentation approaches and carry out genetic manipulation. Overall, using lignocellulose biomass to generate biofuel holds excellent potential because an alternative to gasoline-based fuel can be a viable product. However, there is a great scope of improvement in every step of processing, including selecting feedstock, pretreatment, hydrolysis, and fermentation.

Considering the potentiality and practicality of bioethanol and all the given facts of cellulosic ethanol production, the present investigation aimed to produce bioethanol from the agricultural waste product (rice straw).

10.1.1 Bioethanol

There are many such advantages of using bioethanol as a biofuel. The price of it is significantly lower compared to gasoline or diesel. The thermal energy is also standard, only approximately 45% less than the traditional fuel or diesel. The matter of preignition is also less in ethanol due to its higher octane number of about 99 than petrol (Kasibhatta 2019). Also, the creation of bioethanol is considered from the sun's energy that is with the help of photosynthesis, the feedstock is produced and then converted into using renewable energy.

For biofuel production, lignocellulosic biomass is considered the amplest and low-cost material. Bioethanol is an eco-friendly fuel used as a substitute for fossil fuel and is also considered a sustainable and renewable transportation fuel. Harmful gases are less omitted by ethanol as it contains 355 oxygen, resulting in incomplete combustion of fuel, and thus, it minimizes the ejection of greenhouse gases.

In ethanol, octane number is high in number, due to which there is a decrease in the vapour pressure, and the fuel thus produced has clean burning properties (Dhillon et al. 2007).

Most bioethanol is produced from the starchy crop; however, during biofuel production, it needs to be considered that it does not create competition among food sources; thus, it needs to be practical in terms of economy and be sustainable in production. One of the important components of lignocellulose, lignin, and hemicelluloses, that is, polysaccharides are not readily broken. The effective fermentation process needs to be hydrolysed to hexose and pentose sugars to be split easily to produce sugars. There are several obstructions during the conversion process; one major drawback is the hydrolysis process. Hydrolysis of lignocellulose by enzymes is superior to that of acid hydrolysis because acid hydrolysis has several environmental impacts, and the by-products released by acid hydrolysis are minimized during enzymatic hydrolysis; however, cost-effective measure of this process still needs improvement. Although the enzymatic cycle in ethanol yield is considered cost effective, the drawback in evaluating the process is that the product of ethanol fermentation is minimized to 20% of the conversion efficiency of glucan. (Kim and Lee 2007). Bioethanol can be obtained from various sources. One of the most used resources to produce ethyl alcohol (CH_3CH_2OH) are woody crops and delicate starchy crops. Also, forest residue and agricultural by-products can help in this regard to a great extent. It is a liquid produced by the distillation of fermented sugar. Bioethanol is considered by many countries as much practicable for the replacement of fossil fuel. The source in the transport sector due to the distress felt on climate crisis also the reduction in nonrenewable fuels that leads to energy security and less dependent on the import and export load of petroleum product (Dufey 2006).

We know that bioethanol is mainly extracted from crops with starch like rice, wheat bran, and beetroot, whereas biodiesel is obtained from crops that produce oil like palm trees and Jatropha plants, rapeseed oil, and the sunflower. They are blended with petroleum-based diesel and then it is grown commercially (Dufey 2006).

10.1.2 Properties of Bioethanol

Ethanol is noncarcinogenic fuel that boils at 78.5 °C and liquefies at −117.3 °C, i.e. it is a volatile and flammable colourless liquid. Its vapourizers are denser than air and tend to settle in lower areas, due to which it is miscible in both nonpolar solvents and water. Still, at a high water concentration, the ethanol will separate from the gasoline and is a clear monohydric primary alcohol. Ethanol is superior to traditional gasoline in all respect except one that the pure bioethanol and bioethanol blends are heavier than gasoline.

10.1.3 Environmental Impact on Bioethanol

Bioethanol is harmless to the environment. The main advantage of bioethanol is that it is a renewable energy source. CO_2 in the atmosphere is fixed by photosynthesis and metabolized to building blocks of plants. The energy of sunlight is used to prepare carbohydrates stored in crops and whole plants. Bioethanol is eco-friendly as almost no net CO_2 is produced and if produced is almost negligible as it is again assimilated by plants. Almost no net CO_2 is produced by using bioethanol. After numerous studies of the life cycle of changes from fossil fuel to biofuel, it was found that the CO_2 emissions reduced very drastically by a factor of ½ or ⅕th also the degradability of ethanol in groundwater and soil mixtures is very fine with both aerobic and anaerobic conditions depending on how much noteworthy the use of renewable fuels is at all phases in the process (Kimming et al. 2015). (100 mg/l in 3–25 days).

10.1.4 Composition of Biomass

Since rice is one of the staple foods globally, its straw is one of the most bountiful renewable resources. After wheat, it has held in third place as an important grain crop in the whole world. It is also the most attractive lignocellulosic material in the world for bioethanol production. Rice straw annually produced globally is about 650–975 million tons, of which the world's annual rice production as per FAO statistics in 2007 was about 650 million tons. Usually, for every kilogram of rice harvested straw produces is about 1–1.5 kg (Maiorella 1985). Due to higher mineral content also slow degradation due to lower bulk density, there is a problem in the disposal of rice straw. Although field burning of rice straw is prevalent in some places, the consequences of field burning are that it increases air pollution due to the release of CO_2 and other greenhouse gases, which health wise the public is affected. Day by day, the environment is affected due to an increase in the burning of fossil fuels due to which the development process remains concerned by the threat to the environment. Thus, researchers have a growing interest in developing alternative sources of agricultural residues for the generation of energy. Lignocellulosic biomass contains a very good amount of cellulose and hemicelluloses content that can be easily converted by hydrolysis into sugars that can be converted to biofuel (Figure 10.1).

Lignocellulosic materials are an enormous substrate for producing various biotechnology-based products (Kuhad and Singh 1993; Kuhad et al. 1997; Kuhad et al. 2007). In each cellulosic material is present a polymer of glucan, β-1 linked polymer, and 4-glucosidic bonds, this helps the cellulose to from a hard bond, thus the degree of polymerization of wood cellulose is nearly at an average of 9000–10 000 that can go up to

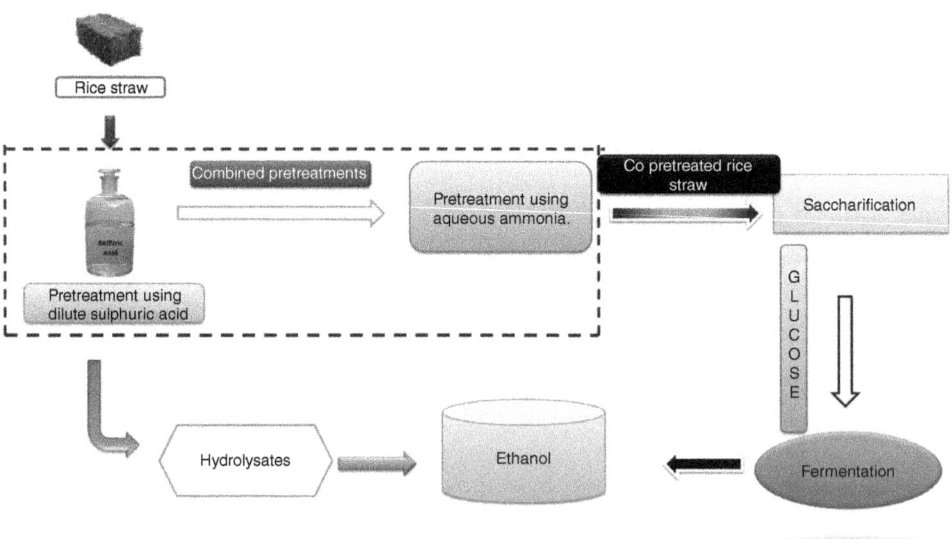

Figure 10.1 Bioconversion of lignocellulosic biomass (rice straw) to fermented sugar pretreatment of biomass.

15 000 with the highest polymerization degree. Compared to that hemicelluloses that have both linear and branched heteropolymers, achieving an average DP of 100–200, they are composed of polysaccharides mainly of sugars like the D-glucopyranose, D-Xylopyranose, D-mannopyranose, glucopyranosiduronic acid, L-arabinofuranose, D-galactopyranose, and D-galactopyranosyl-uronic acid and other simple sugars like the D-glucose, D-glucuronic acid, D-Xylose, D-mannose, and L-arabinose (Kuhad et al. 1997). Cellulose consists of one repeating sugar unit as a backbone and has a molecular weight within the range of about approximately 10 000–150 000 Da (Goring and Timell 1962). The hemicelluloses also contain acetyl- and methyl-substituted groups (Rowell et al. 2005). Hemicelluloses are composed of heterogenous pentose polymers, for, e.g.: galactose, hexoses, xylose, arabinose, mannose, glucose, and sugar acids are all pentose polymers. There is a higher amount of xylan found in the hardwood and a very high content of mannan in the softwood of the agricultural residues of hemicelluloses (Kuhad et al. 1997). The homopolymeric backbone chains of Xylan consist of 1,4-linked β D xylopyranose units and arabinose, glucuronic acid or its 4-O-methyl ether, acetic, ferulic, and p-coumaric acids. About 10–25% of plant materials consist of lignin content. Lignin, an important component of plant biomass, is considered to be an encrusting substance composed of an amorphous and highly complex polymer of phenylpropanoid units (Kuhad et al. 1997). Three different alcohols ρ-hydroxyphenyl alcohol, guaiacyl alcohol, and syringyl alcohol are the precursor for lignin synthesis, and they link with each other to form a massive network of phenolic compounds. The chemical structure of lignin depends upon various environmental factors such as climatic conditions, the plant biomass's age, and the amount of sunlight. p-coumaryl alcohol, coniferyl alcohol, and sinapyl alcohol are the precursors of lignin biosynthesis (Kuhad and Singh 1993; Kuhad et al. 1997). Rice straw seems to be one of the most available and feasible biomass sources for ethanol production. Still, high ash and silica content in rice straw show the inferiority in using them as feedstock. The cell wall of plants consists of a high amount of cellulose, so it is considered the most abundant chemical found on the earth's surface. The considerable strength of cellulose makes it reasonably resistant to biological attack during enzymatic pretreatment, which is a significant limitation to cell wall hydrolysis. It is due to the linear β-1,4 linked D-glucopyranose residues. Hydrogen bonds bonded these residues to form a linear chain and enable the neighbouring parallel and antiparallel chain to become condensed. Thus, the structure formed is more vigorous, and a thin cellulose chain is included (Perez et al. 2002). The type of protection in the plant body serves as an effective barrier to pests and disease, increasing the stability and rigidity of cellulose and hemicelluloses structure. This condition is due to the enhancement of the strength of the hydrogen bond between polysaccharides. However, in bioprocessing to the internal polysaccharide, this protection is not at all preferred (Perez et al. 2002).

Pretreatment is very much necessary for bioconversion to remove hemicelluloses, increase the porosity of rice straw materials, and reduce the cellulose crystallinity to convert into ethanol (Saha 2003; Mosier et al. 2005). Several approaches to get polymeric components of lignocellulosic materials are being applied. Among them includes physical, chemical, biological, thermal, or combination of these to get maximum yield and purity of the ethanol. While sugar and lignin degradation, the resulting hydrolysate that is formed contains pentose and hexose sugars, a reaction by-product, and in subsequent fermentation steps may have an inhibitory effect on the micro-organism residing there. So the fermentation yield will also be decreased. Thus, an appropriate pretreatment can also prevent the formation of inhibitors to hydrolysis and fermentation (Sun and Cheng 2002).

10.1.4.1 Physical Pretreatment

This type of pretreatment mechanical comminution, steam explosion, and microwave radiation, etc., are generally included. For better enzyme access, a larger substrate surface area is required, which is achieved by cutting and grinding the substrate in the mechanical pretreatment method – physical treatment results in reduced size and a lesser degree of polymerization and crystallite of the substrate.

10.1.4.2 Auto-Hydrolysis (Steam Explosion)

Steam explosion is considered one of the most common methods used for the physical pretreatment of lignocellulosic biomass. In the physical pretreatment method, the temperature is increased and subjected to lignocellulosic material to break down hemicellulose and alter lignin and thus make hydrolysis more efficient (Lee et al. 2009; Boluda-Aguilar et al. 2010). In a steam explosion pretreatment, the lignocellulosic substrate is subjected to a high temperature ranging from 160 to 260 °C and soaked steam for ten seconds to a few minutes trailed by a blazing cycle to deliver the moisture violently. This treatment brings about an unstable disturbance of the lignocellulose material, hence 'opening up' the substrate to increase digestibility (Mosier et al. 2005). Steam explosion gives warm extension, which opens up the design of target biomass. It is considered an appropriate pretreatment for horticultural build-ups. The variables influencing the aftereffects of the steam blast are home time, temperature, molecule size, and just as dampness content (Duff and Murray 1996).

10.1.5 Acid Hydrolysis

Among all pretreatments, acidic hydrolysis is widely used since it is exceptionally dynamic, cheap, and simple to perform. It solubilizes hemicelluloses into monomeric sugars (xylose, arabinose, galactose, glucose, and mannose) and soluble oligomers, thus improving cellulose conversion. Acid pretreatment hydrolyses plant cell walls, especially their hemicelluloses component. Sulphuric acid, nitric acid, and hydrochloric acid are generally used for acid pretreatment. The solubilized hemicellulose can be changed over to xylose in acidic media. Xylose can be overdegraded in solid acidic climate.

Concentrated mineral acids, for example, H_2SO_4 and HCl, are incredible specialists for cellulose hydrolysis. They have been utilized to pretreat the lignocellulosic materials. However, these acids are harmful, destructive, and perilous. And one of the essential things is that the reactors used must be impervious to erosion. Besides, to make the cycle financially plausible, the recuperation of undiluted corrosive is sufficiently tricky (Sivers and Zacchi 1995; Torget et al. 2000). At the same time, diluted acid is used for the hydrolysis of cellulose and is effective for the pretreatment of lignocellulosic materials. The weakened sulphuric acid can accomplish very high response rates for the pretreatment of the material. It also further develops the hydrolysis of the cellulosic biomass (Esteghlalian et al. 1997; Gupta et al. 2011a) fundamentally.

10.1.6 Alkali Hydrolysis

Alkali pretreatment is considerably used for changing the composition of cell walls. During such a procedure, various types of reactions like solution and saponification take place. As a result, the lignin portion is removed, and with the help of the saccharification enzyme, the biomass swells and enhancement of the access to its inner space takes place. Alkali pretreatment is much more time consuming and depends on the operation temperature selected (Mosier et al. 2005). Among the different bases, potassium hydroxide, ammonia, and sodium hydroxide are used for alkali pretreatment. Whereas for commercial purposes, the potentiality of sodium hydroxide is the highest due to its high volatility so that it can be recycled and recovered. Due to which its cost in chemical and waste treatment is reduced (Mosier et al. 2005).

One of the strategies used basically for the pretreatment of the plant biomass is alkali hydrolysis. However, the pretreatment technique used in hydrolysis of plant biomass is solely dependent upon the lignin content of the materials (Kumar et al. 2009; Gupta et al. 2011a). The mechanism of alkaline hydrolysis is believed to be saponification of intermolecular ester bonds, cross-linking xylan hemicelluloses, and other components, for example, lignin and other hemicellulose. Dilute NaOH treatment of lignocellulosic materials caused swelling of lignocellulosic materials, leading to an increase in internal surface area, a decrease in the degree of polymerization, and crystallinity, separation of structural linkages between lignin and carbohydrates, and disruption of the lignin structure, which are commonly reported (Sun and Cheng 2002; Carrillo et al. 2005).

10.1.7 Biological Pretreatment

To increase the hydrolysis of cellulosic materials, the pretreatment has turned into a need and eventually, and biological pretreatment has the ability the expedite ethanol creation. There are various advantages in the biological pretreatment of plant biomass over alkaline hydrolysis, (I) the reaction conditions in biological pretreatment are gentle and mild, (II) the utilization of poisonous and destructive synthetic substances like chemicals are kept away from this process, (III) yield is increased after the completion of the process, (IV) energy requirement in the whole treatment process is significantly less due to which it results in reduced after effects, and (V) it also results in mere resistance from the pressure and corrosion of the biological reactor (Lee 1997; Sánchez 2009).

10.1.7.1 Fermentation

The fermentation of ethanol is a natural process where microbes interact with sugars present in the plant's cellulosic biomass to produce ethanol and release gases like CO_2. There is a perplexing interaction in the fermentation of plant biomass hydrolysates compared to starch and molasses. Hemicellulose and cellulose generate separately two notable effluxes of sugars such as hexose-rich sugars and pentose-rich sugar syrup. An organism incorporated for ethanol production from lignocellulosic biomass should be able to give higher productivity of ethanol yield and be able to ferment at lower pH and tolerant against inhibitors produced during the process when ethanol concentrations are higher able to withstand it.

A variety of micro-organisms ranging from bacteria, fungi, and yeasts are known to ferment hexose sugars. The yeast and various bacterial species identified so far during the last three decades have demonstrated the utilization of pentose sugars to produce alcohols and other fermentation products. *Pichia stipitis*, *Candida shehatae*, and *Pachysolen tannophilus* are among those that ferment pentose sugar, whereas *Saccharomyces cerevisiae* and *Zymomonas mobilis* are used for hexosefermentation (Abbi et al. 1996; Hahn-Hagerdal et al. 2007; Kuhad et al. 2011).

Although several yeast species can produce ethanol by fermenting monosaccharides and oligosaccharides, some facultative bacterium *Z. mobilis* are also used for alcohol production. However, yeast ferment ethanol using glucose that acts as exclusive material. It has a remarkable ability to tolerate insensitivity to temperature and control the fermentation rates and substrate concentration (Lindén and Hahn-Hägerdal 1989).

Additional inhibitors produced during pretreatment of biomass like furfural (2-furaldehyde), 5-hydroxymethyl-2-furaldehyde (5-hydroxymethylfurfural HMF) diminish ethanol production by breaking down DNA and RNA combination, decreases activities of enzymes, cell development hindering, restrain protein, and harm cell division (Modig et al. 2002; Liu and Blaschek 2009). *Z. mobilis* possess advantages over *S. cerevisiae* concerning bioethanol productivity and high bioethanol tolerance.

10.1.7.2 Ethanol Recovery: Distillation and Dehydration

The mixtures of ethanol and water can be isolated under favourable conditions based on the difference in both volatilities. Ethanol vapourizes at 78 °C; thus, ethanol is more volatile than water. And during heating in an ideal distillation column separation, the ratio of ethanol-to water in the vapour phase will become higher than that in the liquid phase, thus water will remain in the bottom and the overhead product will mainly be ethanol. An azeotropic combination of ethanol (95.6%) water (4.4%) will be an endless supply of refining activity, which is controlled by the distinction in the limits between water and ethanol (Wang 2010). They involve a very profound coordinated recuperation for the measurement of genuine ethanol content (Green and Perry 1997).

In this way, in the current examination, an endeavour has been accomplished for catalyst creation from ecological yeast strain and its application in the bioconversion of rice straw build-ups into glucose for product of second-generation bioethanol.

10.1.8 Sample Collection Site

The collection of the samples were done by Dr H. K. Baruah, Memorial Department of Botany. The experiment was conducted in the laboratory of Microbiology, Botany Department of Gauhati University, Guwahati, Assam. The straw samples were collected from a nearby paddy field.

10.1.9 Sample Preparation

The agrowaste sample, such as rice straw as substrate, was collected paddy straw and cut into small pieces. They were then allowed to dry in the oven at 65 °C for 48 hours. Pieces of raw materials were used as a carbon source.

10.1.10 Isolation of Yeasts

Yeast was isolated from soil samples by serial dilution methods using different media such as – yeast extract peptone dextrose agar (YEPDA), Rose Bengal agar base media, Peptone yeast extract agar, Malt yeast extract agar (MYA), and Sobouraud dextrose agar (SDA). Each of the media was supplemented with chloramphenicol (200 mg/l) to control the bacterial growth. Incubation of prepared plate's was done at 27 °C for 3–4 days in a BOD incubator. Although different media are used for the isolation of yeast, YEPDA media (peptone 20 g, yeast extract 10 g, agar 20 g, and dextrose 20 g in 1 l of water) was able to show maximum growth of yeast. The plates were incubated for 2–3 days – the colony which is singly formed was selected and observed under the microscope.

10.1.10.1 Culture Maintenance

The maintenance of cultured yeast strains was achieved by subculturing on YEPDA slants, incubated at 30 °C for 40–48 hours and for long time preservation, the culture was stored in a refrigerator at 4 °C for future work.

10.1.11 Pretreatment of Substrate

The substrate that is physically treated were again exposed to other pretreatments by acid, alkali, and steam explosion technique.

10.1.12 Steam Explosion

In a 250 ml conical flask, 20 g of the substrate was taken. Then powdered substrates were exposed to a rapid steam explosion in an autoclave for 15–30 minutes at 121 °C and 15 psi by completely release the steam exhaust regulator with the focus of getting the concentrated output of fermentable sugars spending minimum duration for pretreatment.

10.1.13 Alkali Pretreatment

Autoclaving 20 g of each substrate in 100 ml of 0.5 M NaOH solution was done for 10 minutes at 115 °C and kept overnight. The alkali-treated substrates were filtered through a nylon cloth and were neutralized. Additional water molecules were removed by pressing the substrate in polyester cloth, after which the pretreated substrate was admitted to dry in the oven.

10.1.14 Acid Pretreatment

Twenty of the pretreated substrate were taken in 100 ml of 1% sulphuric acid, autoclaved at 115 °C for 15 minutes, and left at room temperature for 24 hours. The acid-treated substrate was also filtered through nylon cloth and neutralized. Excess water was removed, and solid residues were dried overnight.

10.1.14.1 Fermentation

Totally, 400 ml of yeast extract broth was prepared in 2 conical flasks by adding yeast extract (10 g/l), peptone (20 g/l) to distilled water, plugged with cotton, and sterilized at 15 lb. pressure for 15 minutes. In each flask, a pretreatment straw was added as a source of carbohydrate. The media was then inoculated with a yeast strain and allowed for fermentation for five days in a shaker incubator at 150 rpm at 30 °C. After fermentation, it was centrifuged for ten minutes at 10 000 rpm distilled for further process.

10.1.15 Distillation of Fermented Product

Based on volatilities, i.e. boiling point of the discrete constituents of a mixture separation was done. It is often used only if one product is required. After centrifuge for 10 minutes at 10 000 rpm, the supernatant fluid is distilled using a distillation bath at 78 °C. As ethanol has a lower boiling point than water, it will be converted into a vapour state first then condensed and separated from the water.

10.1.16 Estimation of Ethanol

Ethanol determination was done by using the spectrophotometric technique (Caputi et al. 1968). For estimation of ethanol, 2 g potassium dichromate was mixed in 20 ml of 5 M sulphuric acid. Then 2 ml of distillate was mixed with 6 ml of distilled water, and 2 ml of potassium dichromate solution was added. For 1 minute, the mixture was shaken gently and allowed to stand for 15 minutes. Then the mixture was heated to boil for 20 minutes. The absorbance was read on a spectrophotometer at 600 nm. Using a spectrophotometer at 600 nm, the alcohol content in fermenting sugar medium was estimated as per the methods given by Natu et al. (1986). The spectrophotometer at 600 nm measured the green-coloured substance in alcohol formed after the reaction of a strong acid like concentrated sulphuric acid (H_2SO_4) and potassium dichromate ($K_2Cr_2O_7$).

10.1.17 Screening of Yeast for Extracellular Enzyme Production Yeast

The yeast isolates were tested for the production of hydrolytic enzymes such as cellulase and amylase by using the following protocol.

10.1.18 Cellulase Production

For cellulase activity, yeasts were cultivated on potato dextrose agar (PDA) plates complemented by 1% carboxymethylcellulose. The culture plates were incubated at 30 °C for three days and finally drowned with 0.1% congored for 30 minutes tailed by 1 M NaCl usage for extra 10 minutes. The clear zones were observed from the background, and by deducting the colony diameter from the total diameter of the clear zone, solubilizing index (SI) was calculated.

10.1.19 Amylase Production

For amylase activities, yeast are cultivated in PDA medium (39 g/l) supplemented with 2 g/l starch at their optimal temperature of growth. The plates where the culture was done were then incubated at 30 °C for seven days and lastly drowned by 1% IKI solution.

10.1.20 Estimation of the Overall Carbohydrate Content of Substrates

The carbohydrate content of pretreated and untreated raw materials of the culture broth was measured using the phenol sulphuric acid method (Thimmaiah 1999). In the boiling tube, 100 mg of the sample was taken, and 5 ml of 2.5 N HCl sample was hydrolysed for three hours in a water bath and then cooled to room temperature. The solid sodium carbonate was used to neutralize it until the bubbliness stopped. The volume is made up to 20 ml and centrifuged at 10 000 for five minutes. One and two of the sample were pipette out in two different test tubes, and capacity is made up to 1 ml in each test tube with water. 0.2, 0.4, 0.6, 0.8, and 1 ml of the working standards were pipette out into a series of test tubes. A blank is set with a test tube contains 1 ml of water. Then 1 ml of phenol is taken in each test tube, and 5 ml of 96% sulphuric acid is added to each tube, and after the next 10 minutes of shaking, the test tubes were placed at 25 °C for 20 minutes in water bath. The absorbance was read at 490 nm. The total present carbohydrate amount in the sample solution was estimated using the standard graph (Fig. 10.3–10.4).

Concentration (μl)	Glucose(μl)	Distilled water(μl)	Phenol (μl)	Sulphuric acid (μl)	Water (μl)
10	10	90	100	500	300
25	25	75	100	500	300
50	50	50	100	500	300
75	75	25	100	500	300
100	100	0	100	500	300

Glucose concentration	Glucose (μl)	Water (μl)	DNS (μl)	Water (μl)
0.1	0.18	99.82	100	800
0.2	0.36	99.64	100	800
0.3	0.54	99.46	100	800
0.4	0.72	99.28	100	800
0.5	0.9	99.1	100	800
0.6	1.08	98.92	100	800
0.7	1.26	98.74	100	800
0.8	1.44	98.56	100	800
0.9	1.62	98.38	100	800
1	1.8	98.2	100	800

10.1.21 Determination of Reducing Sugar

3,5-Dinitro salicylic acid (DNS) reagent estimates reducing sugar-like glucose (Miller 1959). A sample 0.05 ml was collected from the filtrate, where 0.6 ml of DNS and 0.35 ml citrate buffer (pH 6.5) was added, and then to stop the reaction immediately, the sample was boiled for 5–10 minutes. The spectrophotometer absorbance was recorded at 540 nm for reducing sugar using (Bennett and Tornabene 1974).

10.1.22 Lignin Measurement

For two hours, the lignin was hydrolysed using 1.25% sulphuric acid in treated and untreated samples. The deposits were sifted and wash away with purified water for 10 minutes to neutralize sulphuric acid and afterwards oven-dried at 105 °C for 10 hours for consistent weight. The following equation communicated the measure of lignin.

$$\textbf{Lignin} = \textbf{Lignin weight} / \textbf{biomass} \times \textbf{100}$$

A total of 3 yeast strains were selected for the production of bioethanol by fermenting rice straw based on their morphology and enzyme production activity.

10.1.22.1 Morphological Characterization

The morphology of the vegetative cells of yeast was observed grown on solid media.

10.1.22.2 Growth on Solid Medium

On YEPDA medium yeast strains makes butyrous and smooth white raised colonies. The growing phase of the potent yeast strains was seen under (40X) magnifying lens and affirmed them to be yeast.

10.1.23 Microscopic Observation

The cell morphology of yeast strain saw under compound magnifying instrument showed an ovoidal to stretch having single, matches, or maturing cells (budding cell). The strain multiplies vegetatively by budding.

10.1.24 Cellulase Production

Table 10.1 Cellulase production test for different yeast strains.

Yeast strain	Cellulose production
B7	+ve
B20	+ve
ARI	+ ve

10.1.25 Amylase Production

Three isolates B7, B20, and ARI showed positive results for cellulase (Table 10.1), as well as amylase production in terms of clear zone formation in PDA media supplemented with potassium iodide and iodine (Table 10.2).

Table 10.2 Amylase production test for different strains of yeast.

Yeast strain	Amylase production
B7	+ ve
B20	+ve
ARI	+ ve

10.1.25.1 The Yield of Bioethanol Production

The results of ethanol yield from the pretreated and untreated substrate by different strains were illustrated in the figure. The highest ethanol was produced by yeast strain B7 0.336 at 595 nm and 339 at 600 nm (Fig. 10.2).

The current studies revealed that B7 produced the maximum yield of ethanol. The lowest ethanol was created from the fermentation of rice straw by the ARI strain. During the acid pretreatment, more reducing sugars are released as the substrate becomes less complex due to the higher accessibility to the enzyme. Then they are converted to bioethanol by yeast strain (Table 10.3).

The lignin content of pretreated rice straw measured

$$\text{Lignin content} = 14.5 / 20 \times 100$$
$$= 72.5\%$$

It was found that as the initial concentration of substrate increases, the lignin content was also increased. The substrate with high lignin content produces less ethanol due to the resistance of lignin in the substrate to the pretreatment. The lignin degradation is essential for better performance of the production of ethanol. After pretreatment, the decrease in lignin content indicates that acid pretreatment reduced lignin

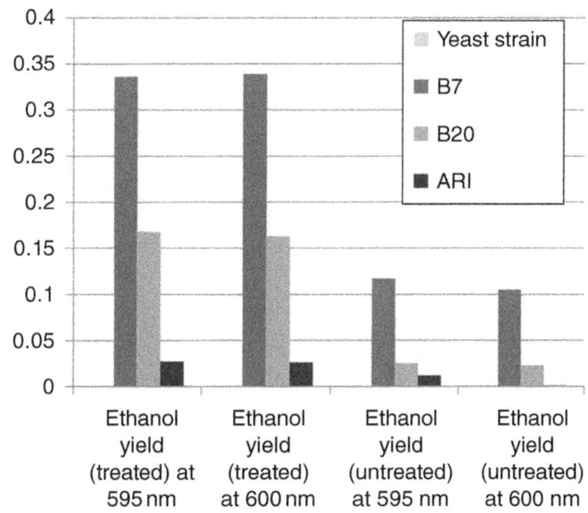

Figure 10.2 Estimation of ethanol from pretreated and untreated rice straw at 595 and 600 nm.

Table 10.3 Optical density of the pretreated and untreated substrate.

Yeast strain	Ethanol yield (treated) at 595 nm	Ethanol yield (treated) at 600 nm	Ethanol yield (untreated) at 595 nm	Ethanol yield (untreated) at 600 nm
B_7	0.336	0.339	0.117	0.105
B_{20}	0.168	0.163	0.025	0.023
ARI	0.027	0.026	0.012	0.001

content and converted the noncarbohydrate polymers into lower molecular weight compounds (Lee et al. 2015).

10.1.26 The Estimation of Reducing Sugar

Dinitrosalicyclic acid (DNS) method is used to determine the amount of reducing sugars in pretreated raw material in the culture broth, taking glucose as a standard (Table 10.4).

Therefore, the yield of reducing sugar after pretreatment of rice straw is 0.604.

The pretreatment has the superiority over other methods to convert hemicelluloses, which is ultimately solubilized to fermentable sugars.

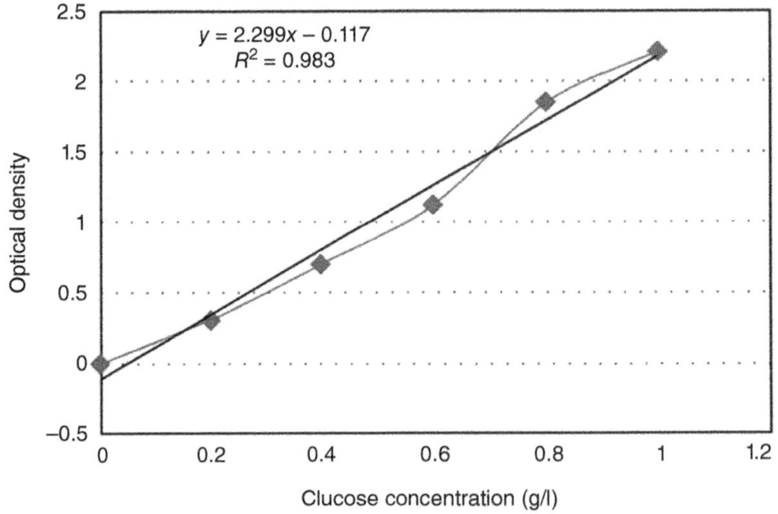

Figure 10.3 Glucose standard curve at 535 nm.

Table 10.4 The reducing sugar yield from rice straw after pretreatment was presented as follows.

Sample	535 nm	540 nm	545 nm	575 nm	580 nm
Rice straw	1.924	1.669	1.466	0.686	0.616

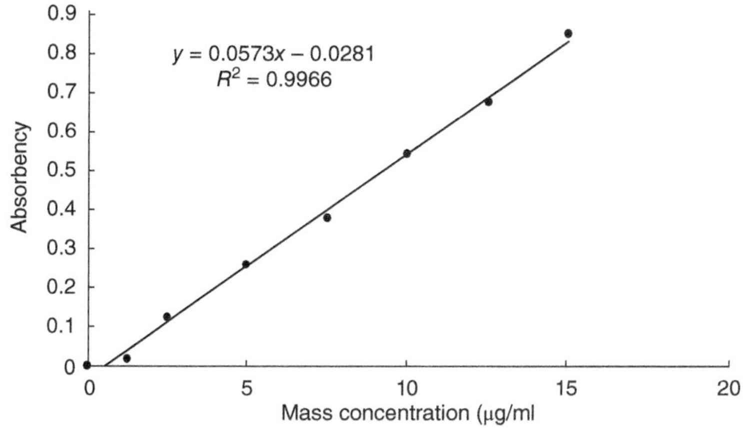

Figure 10.4 The standard curve of glucose at 490 nm.

Table 10.5 Optical density of pretreated rice straw for carbohydrate estimation.

Sample	475 nm	480 nm	485 nm	488 nm	490 nm
Rice straw	**0.**461	0.477	0.429	0.390	**0.357**

10.1.27 Carbohydrate Estimation

The carbohydrate content of pretreated crude materials in the stock culture broth was estimated by phenol sulphuric acid strategy taking standard as glucose (Table 10.5).

Therefore, the estimation of carbohydrate from pretreated rice straw is 7.86 μg/μl.

10.2 Discussion

A huge amount of ethanol was produced in the world. About 1000 g of paddy straw comprises 390 g of cellulose, and it is tentatively enough to make 200 g of ethanol. Bioethanol is a renewable fuel that can be mixed with gasoline or used as a sole fuel. The present study's main objective was to determine ethanol from paddy straw by effective fermentation by yeast.

Bioethanol technology is a modern and eco-friendly technology based on the fermentation of organic materials in anaerobic conditions at suitable temperatures and pH by yeast. The critical step in lignocellulosic bioethanol production is to convert lignocellulosic biomass into fermentable sugar.

In the present investigation, the B7 strain was among the three identified strains that showed higher reducing sugar and cellulosic productivity from paddy straw than the other two strains – clearly indicating B7 strain as high potential to degrade rice straw than the other. Therefore, this strain was considered efficient for producing extracellular and reducing sugar and was selected for further studies.

Pretreatment of rice straw is an essential step to open up the structure of biomass. Acid pretreatment removes hemicelluloses from the biomass, while the alkali causes delignification of biomass to provide bioethanol. These studies show that during pretreatment of rice straw, there was a positive effect of steam explosion, 1% alkali and 1% acid, on B7 that helps inefficient conversion of polysaccharides into sugars. Sugar content was increased as a result of sterilization done by the method of autoclaving.

The critical factor regarding the productive feasibility of the process is that the availability of cellulose that hinders the process of hydrolysis of cellulosic materials. B7 strain shows a positive test for cellulose production and efficiently perform acid enzymatic conversion, alkali and steam explosion pretreated cellulose from rice straw into sugar followed by fermentation into ethanol. Cellulose has many industrial applications, including manufacturing of medicines, food, and help break the waste plant supplies to clean up the environment.

10.3 Conclusion

One essential chemical intermediate produced in the steam cracking of natural gases and petroleum liquid is ethylene. Catalytic dehydrogenation of bioethanol is another way to produce ethanol, decreasing greenhouse gas emissions and limiting reliance on nonrenewable fuels. Pretreatment of rice straw can effectively increase cellulose production and improve the fermentation for bioethanol production. This study proves that the substantial dumping problem of agricultural wastes can be overcome by processing wastes to fermentation and saccharification and producing bioethanol through the process as a by-product. Thus, the present investigation has unlocked new opportunities for exploiting commercial ethanol production for industrial use.

References

Abbi, M., Kuhad, R.C., and Singh, A. (1996). Bioconversion of pentose sugars to ethanol by free and immobilized cells of Candida shehatae (NCL-3501): Fermentation behaviour. *Process Biochemistry* 31 (6): 555–560. https://doi.org/10.1016/s0032-9592(95)00104-2.

Bennett, L.G. and Tornabene, T.G. (1974). Characterization of the antigenic subunits of the envelope protein of *Yersinia pestis*. *Journal of Bacteriology* 117 (1): 48–55.

Boluda-Aguilar, M., García-Vidal, L., González-Castañeda, F.d.P. et al. (2010). Mandarin peel wastes pretreatment with steam explosion for bioethanol production. *Bioresource Technology 101*: 3506–3513.

Caputi, A., Ueda, M., and Brown, T. (1968). Spectrophotometric determination of ethanol in wine. *The American Journal of Enology and Viticulture* 19: 160–165.

Carrillo, F., Lis, M.J. et al. (2005). Effect of alkali pretreatment on cellulase hydrolysis of wheat straw: kinetic study. *Process Biochemistry* 40 (10): 3360–3364.

Dey, P., Pal, P., Kevin, J.D., and Das, D.B. (2018). Lignocellulosic ioethanol production: prospects of emerging membrane technologies to improve the process – a critical review. *Reviews in Chemical Engineering* 36 (3): 333–367.

Dhillon, G.S., Bansal, S., and Oberoi, H.S. (2007). Cauliflower waste incorporation into cane molasses improves ethanol production using *Saccharomyces cerevisiae* MTCC 178. *Indian Journal of Microbiology* 47: 353–357.

Du, R., Su, R., Qi, W., and He, Z. (2018). Enhanced enzymatic hydrolysis of corncob by ultrasound-assisted soaking in aqueous ammonia pretreatment. *3 Biotech* 8 (3): 166.

Dufey, A. (2006). *Biofuels Production. Trade and Sustainable Development: Emerging Issues.* London: International Institute for Environment and Development.

Duff, S.J. and Murray, W.D. (Jan. 1996). *Rioresource Technology* 55 (I): l–33.

Esteghlalian, A., Hashimoto, A.G., Fenske, J.J. et al. (1997). Modeling and optimization of the dilutesulfuric acid pretreatment of corn stover, poplar and switchgrass. *Bioresource Technology* 59: 129–136.

Estevez, R., Aguado-Deblas, L., Bautista, F.M. et al. (2019). Biodiesel at the crossroads: a critical review. *Catalysts* 9: 1033.

Goring, D.A.I. and Timell, T.E. (1962). Molecular weight of native cellulose. *Tappi* 45 (6): 454–460.

Green, D.W. and Perry, R.H. (1997). Separation processes. In: *Perry's Chemical Engineers' Handbook* (ed. Don W. Green Deane E. Ackers Professor of Chemical and Petroleum Engineering, University of Kansas), 3–89. McGraw Hill.

Gupta, R., Mehta, G., Khasa, Y.P., and Kuhad, R.C. (2011a). Fungal delignification of lignocellulosic biomass improves the saccharification of cellulosics. *Biodegradation* 22: 797–804.

Gupta, R., Khasa, Y.P., and Kuhad, R.C. (2011b). Evaluation of pretreatment methods in improving the enzymatic saccharification of cellulosic materials. *Carbohydrate Polymers* 84: 1103–1109.

Hahn-Hagerdal, B., Karhumaa, K., Jeppsson, M. et al. (2007). Metabolic engineering for pentose utilization in *Saccharomyces cerevisiae*. *Biofuels* 108: 147–177.

Jönsson, L.J., Alriksson, B., and Nilvebrant, N.-O. (2013). Bioconversion of lignocellulose: inhibitors and detoxification. *Biotechnology for Biofuels* 6 (1): 16. https://doi.org/10.1186/1754-6834-6-16.

Kandasamy S.K., Kumar, N., Manickam, K. et al. (2021). Application in biofuels and bioenergy production systems. 511–518.

Kasibhatta, S. (2019). Alcohol Fuels as an Alternative Fuels - Bringing New Heights in Sustainability. In (Ed.), Alcohol Fuels - Current Technologies and Future Prospect. IntechOpen. https://doi.org/10.5772/intechopen.86626

Kim, T.H. and Lee, Y.Y. (2007). Pretreatment of corn stover by soaking in aqueous ammonia at moderate temperatures. In: *Applied Biochemistry and Biotechnology*, ABAB Symposium (ed. J.R. Mielenz, K.T. Klasson, W.S. Adney and J.D. McMillan). Humana Press.

Kimming, M., Sundberg, C., Nordberg, A. et al. (2015). Replacing fossil energy for organic milk production – potential biomass sources and greenhouse gas emission reductions. *Journal of Cleaner Production* 106: 400–407.

Kuhad, R.C. and Singh, A. (1993). Lignocellulose biotechnology: current and future prospects. *Current Reviews in Biotechnology* 13: 151–172.

Kuhad, R.C., Singh, A., and Ericksson, K.E. (1997). Microorganisms and enzymes involved in the degradation of plant fiber cell walls. *Advances in Biochemical Engineering/Biotechnology* 57: 45–125.

Kuhad, R.C., Singh, A., and Eriksson, K.E.L. (1997). Microorganisms and enzymes involved in the degradation of plant fiber cell wall. *Advances in Biochemical Engineering Biotechnology* 57: 47–125.

Kuhad, R.C., Kuhar, S., Kapoor, M. et al. (2007). Lignocellulolytic microorganisms, their enzymes and possible biotechnologies based on lignocellulolytic microorganisms and their enzymes. In: *Lignocellulose Biotechnology: Future Prospects* (ed. R.C. Kuhad and A. Singh), 3–32. New Delhi: IK International.

Kuhad, R.C., Gupta, R., Khasa, Y.P. et al. (2011a). Bioethanol production from pentose sugars: current status and future prospects. *Renewable & Sustainable Energy Reviews* 15: 4950–4962.

Kumar, P., Barrett, D.M., Delwiche, M.J. et al. (2009). Methods for pretreatment of lignocellulosic biomass for efficient hydrolysis and biofuel production. *Industrial and Engineering Chemistry Research* 48: 3713–3729.

Lee, J. (1997). Biological conversion of lignocellulosic biomass to ethanol. *Journal of Biotechnology* 56: 1–24.

Lee, J.M., Shi, J., Venditti, R.A. et al. (2009). Autohydrolysis pretreatment of coastal bermuda grass for increased enzyme hydrolysis. *Bioresource Technolology* 100: 6434–6441.

Lee, J.W., Kim, J.Y., Jang, H.M. et al. (2015). Sequential dilute acid and alkali pretreatment of corn stover: sugar recovery efficiency and structural characterization. *Bioresource Technology* 182: 296–301.

Lindén, T. and Hahn-Hägerdal, B. (1989). Fermentation of lignocellulose hydrolysates with yeasts and xylose isomerase. *Enzyme and Microbial Technology* 11: 583–589.

Liu, Z.L. and Blaschek, H.P. (2009). Lignocellulosic biomass conversion to ethanol by saccharomyces. In: *Biomass to Biofuels* (ed. A. Vertes, N. Qureshi, H. Yukawa and H. Blaschek), 17–36. West Sussex: Wiley.

Maiorella, B.L. (1985). Ethanol. In: *Comprehensive Biotechnology* (ed. M. Moo-Young), 861–914. Oxford: Pergamon Press.

Miller, G.L. (1959). Use of dinitrosalicylic acid reagent for determination of reducing sugar. *Analytical Chemistry* 31 (3): 426–428.

Modig, T., Liden, G., and Taherzadeh, M.J. (2002). Inhibition effects of furfural on alcohol dehydrogenase, aldehyde dehydrogenase and pyruvate dehydrogenase. *Biochemical Journal* 363 (3): 769.

Mosier, N., Wyman, C., Dale, B. et al. (2005). Features of promising technologies for pretreatment of lignocellulosic biomass. *Bioresource Technology* 96 (6): 673–686.

Natu, R.B., Sawant, A.D., and Jadhav, S.J. (1986). Spectrophotometric assay of ethanol in fermentedmolasses and sugarcane juice. *Bharatiya Sugar* 11: 41–43.

Perez, J., Dorado, J.M., Rubia, T.D. et al. (2002). Biodegradation and biological treatment of cellulose, hemicellulose and lignin: an overview. *International Microbiology* 5: 53–63.

Rowell, R.M., Pettersen, R., Han, J.S. et al. (2005). Chapter 3: Cell wall chemistry. In: *Handbook of Wood Chemistry and Wood Composites* (ed. R.M. Rowell), 35–74. Boca Raton: CRC Press.

Saha, B.C. (2003). Hemicellulose bioconversion. *Journal of Industrial Microbiology and Biotechnology* 30: 279–291.

Sánchez, C. (2009). Lignocellulosic residues: biodegradation and bioconversion by fungi. *Biotechnology Advances* 27: 185–194.

Selim, K.A., El-Ghwas, D.E., Easa, S.M., and Hassan, M.I.A. (2018). Bioethanol a microbial biofuel metabolite; new insights of yeasts metabolic engineering. *Fermentation* 4: 16.

Sivers, M.V. and Zacchi, G. (1995). A techno-economical comparison of three processes for the production of ethanol from wood. *Bioresource Technology* 51: 43–52.

Sun, Y. and Cheng, J. (2002). Hydrolysis of lignocellulosic materials for ethanol production: a review. *Bioresource Technology* 83 (1): 1–11.

Thimmaiah, S.K. (1999). *Standard Method of Biochemical Analysis*. New Delhi: Kalyani Publishers.

Torget, R.W., Kim, J.S., and Lee, Y.Y. (2000). Fundamental aspects of dilute acid hydrolysis/fractionation kinetics of hardwood carbohydrates. Cellulose hydrolysis. *Industrial & Engineering Chemistry Research* 39 (8): 2817.

Wang, Y.G. (2010). Separation of ethanol/water azeotrope using compound starch-based adsorbents. *Bioresource Technology 101*: 6170–6176.

Wiltschia, B., Cernava, T., Denniga, A. et al. (2020). Enzymes revolutionize the bioproduction of value-added compounds: from enzyme discovery to special applications. *Biotechnology Advances* 40: 107520.

Zhang, H., Hong, H.Z., Xiao, Y.Z. et al. (2006). Efficient production of laccases by *Trametes* sp. AH28-2 in cocultivation with a *Trichoderma* strain. *Applied Microbiology and Biotechnology* 73: 89–94.

Zhao, X., Zhang, L., and Liu, D. (2012). Biomass recalcitrance. Part I: the chemical compositions and physical structures affecting the enzymatic hydrolysis of lignocellulose. *Biofuels, Bioproducts and Biorefining* 6 (4): 465–482.

11

Hydrothermal Liquefaction of Waste Agricultural Biomass for Biofuel and Biochar

Vishal Rajput[1], *Vinod Kumar*[2,3], *M.S. Vlaskin*[4], *Manisha Nanda*[5], *and Monu Verma*[6]

[1] *Himalayan School of Bio-sciences, Swami Rama Himalayan University, Dehradun, Uttarakhand, India*
[2] *Algal Research and Bioenergy Lab, Department of Chemistry, Uttaranchal University, Dehradun, Uttarakhand, India*
[3] *Department of Environmental Monitoring and Forecasting, Ecological Faculty, RUDN University, Moscow, Russia*
[4] *Laboratory for Energy Accumulating Materials, Joint Institute for High Temperatures of the Russian Academy of Sciences, Moscow, Russia*
[5] *Department of Biotechnology, Dolphin (P.G.) Institute of Biomedical and Natural Sciences, Dehradun, India*
[6] *Department of Environmental Engineering, The University of Seoul, Seoul, Republic of Korea*

CHAPTER OUTLINE

Agriculture Waste Management and Bioresource: The Circular Economy Perspective, First Edition.
Edited by Suruchi Singh, Pardeep Singh, Anu Sharma, and Moharana Choudhury.

11.1 Introduction

In the current context, biomass refers to a variety of organic substances such as green plants, living organisms, micro-organisms, organic matter derived from excretory systems, and metabolic activities as well as various residues from agriculture and forest, human population, plants in aquatic environments, and organic waste generated by industries (Gardner et al. 2015, pp. 4418–4422; Magdeldin et al. 2017, pp. 679–695). Biomass has been counted in top five places on the basis of consumption followed by traditional sources of energy such as coal, gas, and petroleum (Zheng et al. 2015, pp. 230–238). The contribution of biomass may be measured around 14% of the total basic utilization of energy (Haarlemmer et al. 2016, pp. 180–188).

Biomass energy has advantages over fossil fuel in various ways including: (Deniel et al. 2016, pp. 4895–4904; Raikova et al. 2016, pp. 219–227) (i) biomass related to lignocellulosic, which is a renewable as it is obtained from plants with the capacity of solar energy storage; (ii) it is believed to be an environmentally friendly energy source, producing low pollution and causing low generation of toxic gases; (iii) biomass may be obtained from the ocean as well as dry land. That is why the development of biomass energy is comparatively easier than traditional renewable energy forms, such as sunlight, wind, and tidal energy (Van Doren et al. 2017, pp. 67–74).

Researchers firmly believe that biomass is a potential alternative energy source, which may replace fossil fuel, supported by large quantities of forestry and agricultural biomass (Durak and Aysu 2016, pp. 123–135). Hydrothermal liquefaction (HTL) has been obtained from the technique of thermochemistry, which may have been thoroughly investigated to achieve good production of bio-oil, from the waste generated from agricultural and forest (Langholtz et al. 2016, pp. 1–411; Tekin 2015, pp. 4382–4392). HTL has shown various prospects when compared with existing techniques. (i) HTL does not need prior thermal drying, which causes lowering of wet material price (Tungal and Shende 2014, pp. 401–412). (ii) Warm and pressure treated water is applied in the form of reactant (Yang et al. 2016, pp. 191–198). Normally, in the traditional process of HTL, lignocellulosic materials have been depolymerized in biogas, biochar, matter (water-soluble), and bio-oil (Madsen et al. 2016, pp. 137–145). Bio-crude oil obtained may be utilized as diesel, gasoline, lubricants, and fuels in turbines, boilers, and engines (Duan et al. 2016, pp. 73–81). During the past few years, HTL has dragged the attention of researchers in the form of promising technology for the conversion of waste obtained from forests and agricultural areas. Due to such reasons, the present chapter targets to accumulate the recent advancements in the area of HTL.

11.2 Composition of Waste Obtained from Agriculture

Forest and agricultural wastes are composed of plant-based materials, specifically known as lingocellulosic biomass. The cell wall of plants contains carbohydrate polymers and some other polymers with non-carbohydrate background (Ekpo et al. 2016, pp. 951–960). Forestry and agricultural biomass is a specific lingocellulosic biomass, majorly composed of various organic polymers, for instance cellulose, lignin, and hemicelluloses (Zhang et al. 2016, pp. 117–122; Jindal and Jha 2016, pp. 157–171). Further, this biomass is also composed of nitrogen compounds, inorganic components, and pectin.

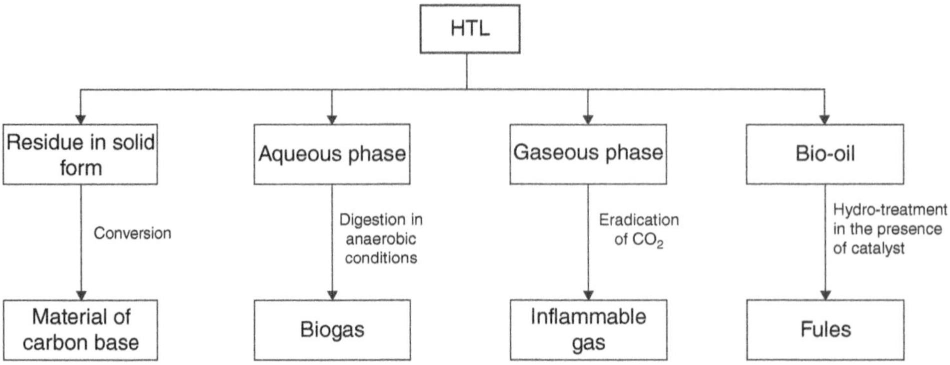

Figure 11.1 Hydrothermal liquefaction process and products.

11.3 Hydrothermal Liquefaction

Ever-increasing demand for energy and improper supply provoke the demand to search for new options, especially those that may fulfil the ever-increasing demand of civilization. One of the major concerns of the human population is the energy source.The present scenario searches for new options of commercial sources derived from biomass. Biomass is made up of several components such as waste from wood, crops, plants (aquatic), and agriculture; also, accumulated waste from animal excreta and municipal trash may be regarded as a valuable source of fuel, for instance, several studies have found that biomass may be used to produce first-generation biofuels by pyrolysis of lignocellulose. Nowadays, HTL is regarded as a novel strategy for generating energy from agriculture and forest biomass that can be converted into bio-crude oil and utilized in various engines as transport fuel 'See Figure 11.1'.

According to current research, HTL is a thermochemical conversion process that results in the form of liquid fuels as a result of processing in a controlled temperature and pressure environment. This process results in the form breakage of solid biopolymeric structure. Typically, hydrothermal processing takes place at temperatures ranging from 523 to 647 K and pressures ranging from 4 to 23 MPa. So far, the primary emphasis has been on the generation of biofuel from solid biomass; however, biomass in liquid state has outperformed solid biomass (Zhu et al. 2017, pp. 9–16).

11.4 Recent Development in HTL Technology

The process of production of biofuel through the biomass by HTL dragged attention at the beginning of the century due to the shortage of energy as well as pollution (Ching et al. 2017, pp. 1794–1800). During the past years, HTL is mainly utilized in biomass (wet), especially biomass of algal origin (Iris and Tsang 2017, pp. 716–732). The rich content of lipids in algae causes the yield of bio-oil loaded by increased energy levels. Later, catalysts have been started to apply to improve HTL conversion yield (Grigoras et al. 2017, pp. 116–123).

11.5 Process for Degradation of Cellulose

Cellulose has been considered as a major ingredient of lingocellulosic mass. Initially, cellulose is divided into monosaccharides and oligosaccharides, and obtained monosaccharide undergoes changes and forms derivatives, by the elevation of temperature in HTL. Due to some reasons, the primary produce undergoes multiple condensation and fragment formation which causes the precipitation process at cellulose surface, blocks water access, and causes dehydration, which finally results in the form of char-like substance in the reactor of HTL reactor (Rahimi et al. 2014, pp. 249–252). 'See Figure 11.2'.

Cellulose conversion and yield of selected products may be elevated by the help of temperature increment. However, the production of hydroxymethylfurfural (HMF) and levulinic acid may be observed at the temperature of 175 °C. Almost, selected products may be obtained at absolute selectivity at temperatures ranging from 125 to 155 °C. On the other hand, at 185 °C, reduced selectivity of needed products was recorded. The reason behind this reduction may have been the process of condensation products of HMF. In the absences of catalyst, glucose decomposed and converted into the derivatives of furan, where the temperature of HTL elevates upto 250 °C. At this point, various components such as carbonic acid and acetic acid appeared with the elevation of temperature (Caporgno et al. 2016, pp. 404–410).

11.5.1 Degradation of Hemicellulose

Hemicellulose contains various types of sugars such as xylose, lactose, glucose, and mannose. The hemicellulose reported in herbaceous plants majorly contains xylan, while hemicellulose in woody plants majorly contains mannose, chitosan, and glucose. Hemicellulose shows weaker crystallinity when compared with cellulose, as a result of branched chains

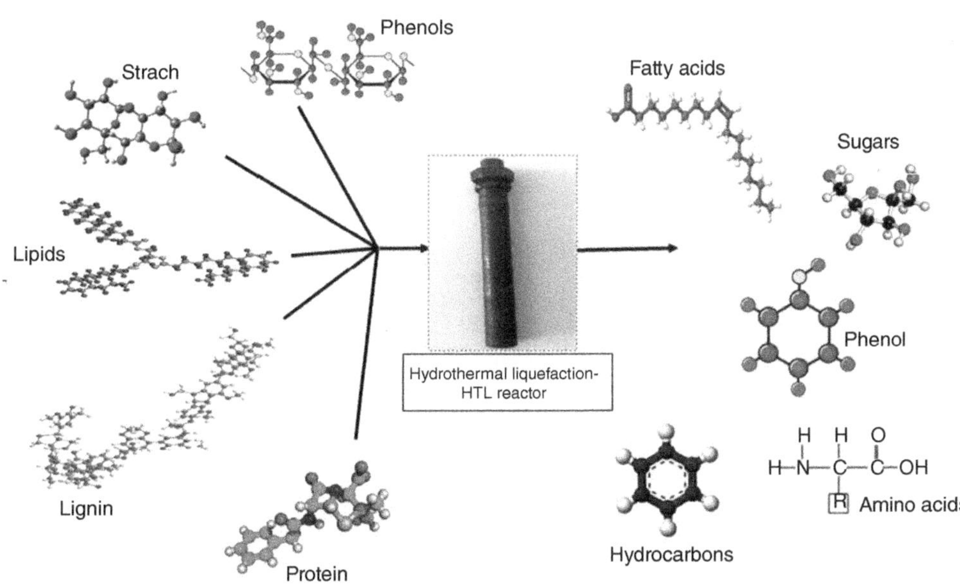

Figure 11.2 Degradation of macromolecules into useful products.

and lower regularity of structure. At 120 °C or higher temperature, hemicellulose hydrolyses easily, which may be catalysed through alkali or acid. In the process of hydrolysis of hemicellulose, the process of sugars decomposition takes place, which seems very similar to the process of cellulose hydrolysis. The process of dehydration is performed on d-xylose and straight xylan conversion to furfural through the catalytic base of betaine acid. On the other hand, 172 °C temperature causes maximum yields of 76 and 80% of polysaccharide and pentose, respectively (Shuai et al. 2016, pp. 329–333).

11.5.2 Degradation of Lignin

Lignin is considered as the second majorly reported natural polymer and found in large quantities in wood tissue and plant cells (Zhu et al. 2017, pp. 9–16). Generally, hardwood contains nearly 19–26% of dry weight lignin; on the other hand, softwood maintains 26–37% (McVeigh et al. 2016, pp. 4142–4150; Kang et al. 2013, pp. 546–558). Nowadays, HTL is considered as a promising approach for changing lignin into the form with low molecular weight. Scientifically, lignin is recorded as an aromatic heteropolymer containing *p*-hydroxy-phenyl-propanoid subunits stacked collectively through C–O–C and C–C bonds. Here, the conversion of the products having low molecular weight into the process of HTL is based on the split of C–O–C or C–O bond (Rahimi et al. 2014, pp. 249–252). During the cycle of liquefaction, the cycle of various phenols shows that fragments' reaction may not be playing major responsibility in the decomposition of lignin. Further, rings of aromatic compounds may convert into alkaline form, majorly to obtain large compounds in condensed form; on the other hand, the bounds of ether tend to break than C–C bonds (Barbier et al. 2012, pp. 479–491). The addition of formaldehyde (FA) during the treatment of biomass obtains lignin (soluble form) fragments that may change into the monomers of syringyl and guaiacyl. Due to the depolymerization of cellulose, hemicelluloses, and lignin one by one, the yield of monomer was reported approximately 77 and 92 mol% (Barbier et al. 2012, pp. 479–491; Cao et al. 2016, pp. 15260–15270).

11.6 Products of HTL

11.6.1 Bio-Crude Oil

Bio-crude oil is a viscous, black-textured, viscous liquid product obtained through HTL of various biomass components which exhibit energy content of approximately 72–96% compared to fossil fuel (Ekpo et al. 2016, pp. 951–960). Feedstock and production process majorly affect the chemical and physical attributes of bio-crude oil. Various constituents of bio-crude oil have a broad molecular weight range that may be characterized through gas chromatography. Indeed, the majority of heavy compounds may not be loaded in the separation column and remain unseparated (Barbier et al. 2012, pp. 479–491; Cao et al. 2016, pp. 15260–15270).

11.6.2 Gaseous Phase

Major gas yield of HTL is CO_2 followed by various other gasses produced such as H_2, CO, and CH_4. Quantity of CO_2 is higher when compared with other gases. The reduced amount of CO recorded previous results cited may exhibit the removal of O_2 process during HTL majorly due to decarboxylation, in the place of decarbonylation (Barbier et al. 2012, pp. 479–491; Cao et al. 2016, pp. 15260–15270).

11.6.3 Aqueous Phase

Huge amount of aqueous phase is yielded in the form of by-product during the HTL process. Normally, the yield in aqueous part ranges between 22 and 51% of the biomass in raw form, but this process largely depends on different factors such as pressure, catalysts, temperature, residence time, and category of biomass (Tommaso et al. 2015, pp. 139–146). The analysis of aqueous phase derived from different lingocellulosic feedstocks exhibited the majority of components such as organic acids and various forms of alcohols containing acetone and cyclopenta (Panisko et al. 2015, pp. 162–171).

11.6.4 Biochar

As per the findings, based on proportion and composition of the target products, the process of hydrothermal biomass may be distributed in hydrothermal gasification as well as hydrothermal carbonation (Panisko et al. 2015, pp. 162–171). Ash and char appear as the by-products of liquefaction. Residues in solid form, obtained from liquefaction of forestry and agricultural wastes, may possess high percentages of carbon, nitrogen, and hydrogen. Here, it may be stated that obtained ash may work as a potential soil supplement. Hydrochar contains low structure of aromatic compounds and temperature recalcitrance, and also lower area of surface and with lower porosity. These factors may affect the exploitation of hydrochar in the field of environmental applications (Zhu et al. 2014, pp. 627–636). Indeed, the energy capacity of hydrochar is lower than biogas and bio-oils; however, it is found easier to store due to its non-toxic and stable nature. Hydrochar may also be utilized in the process of carbon sequestration, and the carbonaceous materials may be utilized in the form of low-cost adsorbent, magnetic carbon composite, and carbon catalyst (Jaiswal et al. 2021, pp. 1–14; Zhu et al. 2016, pp. 338–347).

11.7 Factors Affecting the Productivity of Bio-Crude Oil

Bio-crude oil has been believed to be a most vital and major product obtained in the HTL of forestry and agriculture. Recent investigations have clearly mentioned that the bio-crude oil yield may be majorly altered by various parameters of operations. Here, in the present section, the effects of various parameters during the production of bio-oil through the HTL of forestry and agricultural waste material are discussed.

11.7.1 Type of Biomass

This parameter may impact the processing of the HTL of biomass (Chen et al. 2017, pp. 127–133; Wikberg et al. 2017, pp. 70–78). It has been recorded in various biomass components that a major variation in the liquefaction difficulty and thermal decomposition intervals were evident (Table 11.1). Similar patterns are reported in similar processing conditions; moreover, variations were recorded in product yield as well as in the composition

Table 11.1 Yield of HTL bio-oil with different biomass.

Biomass	Temp. (°C)	Pressure (MPa)	Product (%)	References
De-oiled *Jatropha curcas* cake	250	4.5	41	Alhassan et al. (2016, pp. 375–381)
Laminaria saccharina	350	16.5	79.1	Bach et al. (2014, pp. 271–276)
Nannochloropsis oceanica	300	32–34	60	Caporgno et al. (2016, pp. 404–410)
Oil mill waste water	200–300	190	58	Hadhoum et al. (2016, pp. 9–17)
Nannochloropsis sp.	400	400	46	Hietala et al. (2016, pp. 102–111)
Arthrospira platensis, tetraselmis sp.	200–350	180	18	Lavanya et al. (2016, pp. 228–235)
Water hyacinth	250–300	1300 psi	23	Singh et al. (2015a, pp. 70–75)
Rice straw	280–320	6–9	17.1–78.2	Tian et al. (2015, pp. 336–343)
Algal blooms of Dianchi lake	260–340	3	18.5	Xu and Etcheverry (2008, pp. 335–345)
Jack pine wood	473–623	2–10	64	Zhang et al. (2014, pp. 1961–1964)
Duckweed	350	5	26	Jain et al. (2016, pp. 789–805)
Chlorella pyrenoidosa	280	1	57.5	Cao et al. (2016, pp. 15260–15270)

of HTL. Lignocellulosic biomass has exhibited the noticeable potential which may have been utilized in terms of economical and renewable feedstock.

In some investigations, a liquefaction process on various feedstocks was commenced (white pine, white spruce bark, and birch bark) and outputs in terms of bio-oil yields were recorded as 37, 56, and 66%, respectively. Such outcomes may have appeared due to the following reasons: lignin is considered as more stable and the molecular chain is tough to break; however, it may be cooked with ease under the process of liquefaction. Whereas, structures of hemicelluloses and cellulose are comparatively simpler combined with weaker bonds of hydrogen, which exhibits lower stability (thermal) and higher rate of degradability during HTL (Alhassan et al. 2016, pp. 375–381). In general, it may be stated that, during the high hemicellulose and cellulose of biomass, the yield of bio-oil through the process of HTL is higher. Few previous efforts commenced on various plants for bio-oil production through HTL exhibited different effects on the hemicellulose, cellulose, and lignin contents, during the production of biochar and bio-crude oils (Zhu et al. 2015, pp. 183–192).

11.7.2 Effect of Temperature

Temperature of the reaction has a vital impact on the HTL process of biomass. This holds a vital role on the quality as well as the quantity of bio-oil. As reported by previous workers, lowering of temperature causes gradual fracture in chemical bonds of every constituent found in the biomass and results in the form of splicing into smaller molecules. When concentration of free radical elevates, high possibility of the polycondensation in small molecules appears, causing increased bio-oil yield. As reported earlier, when the temperature of reaction remains closed to the critical point of water, further large molecules of bio-oil get polymerized or decomposed into gas products; at last, it leads to elevation residues of solid form and yield of gas but reductive produce of bio-oil (Meryemoglu et al. 2014, pp. 278–283).

In one of the attempts, HTL performed on barley straw in multiple reaction temperatures, i.e. 280–300 °C reported the highest yield upto 36.3%. On the other hand, when temperature elevated up to 410 °C, the yield reduced up to 16.4%, whereas the yield of solid residue increased up to 7.3%. Indeed, it may be stated that temperature at lower level plays a vital role during the degradation of biomass into the form of bio-oil; on the other hand, elevated temperature cause reduction in the yield of bio-oil as a result of polymerization at molecular level (Brand et al. 2014, pp. 420–427).

Appropriate changes in temperature of reaction may benefit bio-oil yield, quality, and loss of energy. This may be obtained through the effectiveness of liquefaction. To evaluate the effect of temperature during the reaction (220, 240, 260, 280, and 300 °C) on the yield of bio-oil, at last, the maximum yield of the bio-oil was recorded (46.5%) at 280 °C, water and mass feedback ratio was 20 : 1, with controlled pressure of 2 MPa. The final product exhibited a high burning value of 32.0 MJ/kg, recorded at a higher rate when compared with raw material (21.3 MJ/kg). Further, carbon content in bio-oil elevates with the increment of temperature. On the basis of this, it may be stated that exact establishment of temperature for fixed liquefaction may not be performed. In fact, optimum temperature is dependent on selected biomass, precise catalyst, and various other parameters (Tommaso et al. 2015, pp. 139–146).

11.7.3 Heating Rate

Heating rate is considered as a major factor that may affect the HTL biomass. Indeed, supporting literature based on scenarios is still not sufficient. The effect of cooling as well as heating rate on the process of liquefaction of lingocellulosic biomass, under the temperature of reaction (260–36 °C), with the time of residence (1–45 minutes). At the end of the process, the heating rate exhibited a slight effect on the liquefaction. Heating rate made a minor effect on the dispersion of pyrolysis products. Present scenario may exhibit the stability of dispersion and dissolution of molecules at higher water pressure during the process of HTL. Higher rate is capable of initiating the process of decomposition and also enhancing the cleavage of bond as well as breakdown of molecules, ultimately resulting in the form of high number of gaseous products with higher energy (Zhu et al. 2016, pp. 338–347).

11.7.4 Retention Time

Residence time may be defined as the process tenure of HTL reaction at selected temperature, without adding the time used in cooling and heating. Residence tenure may affect the yield of bio-oil as shorter duration causes incomplete degradation; on the other hand, longer time results in the intermediate stage of polymerization reaction, ultimately lowers the yield of bio-crude oil for instance. Further, as a result of breaking and polymerization, later elevation in residue time negatively impacts the yield of bio-oil. The production of residue (solid) was altered with the reaction time, and the decrease in biochar was reported from 16.2% at 40 minutes to 14.5% at 70 minutes in the HTL of *Litseacubeba* (Wang et al. 2013a, pp. 509–515). Moreover, the yield of gas was elevated with the increase of residence time (28.4% for 130 minutes). Such findings exhibited that the yield of bio-crude oil approached the highest level multiple time slots and various temperatures of reaction. Here, it may be stated that the impact reaction temperature and residence time were closely related to the reactions (Shuai et al. 2016, pp. 329–333).

11.7.5 Effect of Pressure

Pressure of reaction is a vital parameter in HTL of forestry and agricultural waste. High pressure may provoke the presence of water molecules during one phase at the supercritical as well as subcritical stage. Such a scenario holds stage transition in the molecules of water and further minimizes the needed reaction energy, which results in the form of reduction in the utilization of energy. On the other hand, higher pressure may elevate the density of water, which ultimately increases the penetration capacity of water and also elevates the process of decomposition as well as biomass conversion rate (Meryemoglu et al. 2014, pp. 278–283; Brand et al. 2014, pp. 420–427). Researchers reported that the yield of bio-oil was unaltered during the pressure 8–13 MPa. Further, pressure and temperature were closely interrelated in the reaction of liquefaction. However, with the increases of temperature, the pressure increases simultaneously, further provoking the process of polymerization (of liquefied yield) and elevating the yield of coke (Chen et al. 2017, pp. 127–133).

11.7.6 Effect of Catalysts

During the process of HTL for biomass, it is believed that catalysts may stop parallel reactions, resulting in the reduction of temperature and pressure and increment of the yield quality (Sharma et al. 2021, pp. 810–822. During the HTL process, catalysts (homogenous) as well as catalysts (heterogeneous) are widely applied. However, most commonly used catalysts are phosphoric acid, acetic acid, formic acid, and sulphuric acid. Although strong acids exhibited noticeable effects, robust corrosive effects hamper industry applications. On the other hand, catalysts (alkali) are being applied in several studies, including Na_2CO_3, K_2CO_3, and KOH.

During the analysis of (Grigoras et al. 2017, pp. 116–123) the response of catalyst (alkali) on the HTL obtained from sawdust of furniture waste in water conditions at subcritical stage, in a reactor at 275 °C for 18 minutes. At the completion of reaction, total yield of bio-crude oil was 36.9 wt% and minimum yield was reported as 8.2 wt% under the presence of K_2CO_3 (1 M). Rate of reaction order for alkalis depended on the conversion of biomass, and yield in liquid form was reported as $K_2CO_3 > KOH > Na_2CO_3 > NaOH$. However, the selection of catalyst is completely based on the feedstocks. The method of selection is vital as it may exhibit negative or positive impact on the needed decomposition of reactions. Here, future analysis is required to identify the right type of catalysts so that desired yield may be obtained (Tekin et al. 2014, pp. 673–687).

11.7.7 Ratio of Liquid to Solid

H_2O (water) plays a critical role during the procedure of HTL. Water has a major responsibility in the biomass dispersion, protection of free radicals, availability of hydrogen source, and increment of the quality of yield. While (Wang et al. 2013a, pp. 509–515) reported that the elevation of liquid-to-solid ratio improves the process of biomass dissolution, therefore lowering the formation of gas and solid residue. It has been observed by previous workers that low solid-to-liquid ratio results in the form of solubility reduction of molecular products and elevates the yield of solid residue (Sangon et al. 2006, pp. 201–207; Yan et al. 2016, pp. 229–235). Here, it may be stated that optimum ratio of liquid to solid may elevate the bio-crude oil yield.

11.8 HTL: Future Prospects and Practical Limitations

Production of bio-crude oil through the HTL of biomass may be applied to exhaust the huge biomass for effective energy production. HTL has shown enormous benefits when compared with liquefaction techniques (Kumar et al. 2019, pp. 1–9; Singh et al. 2015b, pp. 157–165). On the other hand, still various issues are pending to be resolved. In the current scenario, investigations in the field of biomass of HTL for obtaining bio-crude oil still confined in the premises of laboratories due to the following limitations:

a) Need of highly advanced instrument is a must to maintain temperature and pressure for the completion of reaction.
b) The implementation of reducing gases results in the cost elevation of production.
c) Tar, coke, and solid maintain low solubility; therefore, their compounds form deposition quickly, leading to the blockage of instruments.
d) Still the selection of operating parameters highly depends on the experience of the operator; therefore it may be stated that the selection of required reaction conditions is still tough.
e) Extraction process faces various hurdles such as low quality and complexity of composition. To overcome these issues, refining of bio-crude oil plants and denitrification process need further investigation.

Most important key in the present situation is to reduce the operating cost as well as optimize the reaction environment so that quality of yield may be improved. On the basis of this review attempt, further research is needed to achieve the following goals:

a) A major milestone is still needed in this field. Better design of reactors with high temperature and high pressure is still needed. Upcoming research should equip the reactor in such a way that it may sustain corrosion and fouling problems.
b) To lower the processing cost, reducing gas may be replaced by more efficient hydrogen, which may cause major cost reduction of the reaction.
c) Next challenge is to look for multifunctional catalysts, by which the performance of reaction may be enhanced as well as the reduction of cost to maintain needed pressure and temperature may be achieved.
d) The role of homogeneous catalysts in the removal of nitrogen has wide potential for further research.
e) More efficient techniques in the field of bio-oil refining are needed to lower the reaction expenditure.
f) On the basis of pilot plots running worldwide, optimum models and mechanisms may be tested to find suitable operating environments to enhance the quality and production of bio-oil; moreover, the HTL is implemented inlarge-scale industries.

References

Alhassan, Y., Kumar, N., and Bugaje, I.M. (2016). Hydrothermal liquefaction of de-oiled *Jatropha curcas* cake using Deep Eutectic Solvents (DESs) as catalysts and co-solvents. *Bioresource Technology* 199: 375–381.

Bach, Q.V., Sillero, M.V., Tran, K.Q. et al. (2014). Fast hydrothermal liquefaction of a Norwegian macro-alga: screening tests. *Algal Research* 6: 271–276.

Barbier, J., Charon, N., Dupassieux, N. et al. (2012). Hydrothermal conversion of lignin compounds. A detailed study of fragmentation and condensation reaction pathways. *Biomass and Bioenergy* 46: 479–491.

Brand, S., Hardi, F., Kim, J. et al. (2014). Effect of heating rate on biomass liquefaction: differences between subcritical water and supercritical ethanol. *Energy* 68: 420–427.

Cao, L., Luo, G., Zhang, S. et al. (2016). Bio-oil production from eight selected green landscaping wastes through hydrothermal liquefaction. *RSC Advances* 6 (18): 15260–15270.

Caporgno, M.P., Pruvost, J., Legrand, J. et al. (2016). Hydrothermal liquefaction of Nannochloropsis oceanica in different solvents. *Bioresource Technology* 214: 404–410.

Chen, B.Y., Zhao, B.C., Li, M.F. et al. (2017). Fractionation of rapeseed straw by hydrothermal/dilute acid pretreatment combined with alkali post-treatment for improving its enzymatic hydrolysis. *Bioresource Technology* 225: 127–133.

Ching, T.W., Haritos, V., and Tanksale, A. (2017). Microwave assisted conversion of microcrystalline cellulose into value added chemicals using dilute acid catalyst. *Carbohydrate Polymers* 157: 1794–1800.

Deniel, M., Haarlemmer, G., Roubaud, A. et al. (2016). Bio-oil production from food processing residues: omproving the bio-oil yield and quality by aqueous phase recycle in hydrothermal liquefaction of blackcurrant (*Ribes nigrum* L.) pomace. *Energy & Fuels* 30 (6): 4895–4904.

Duan, P., Zhang, C., Wang, F. et al. (2016). Activated carbons for the hydrothermal upgrading of crude duckweed bio-oil. *Catalysis Today* 274: 73–81.

Durak, H. and Aysu, T. (2016). Structural analysis of bio-oils from subcritical and supercritical hydrothermal liquefaction of *Datura stramonium* L. *The Journal of Supercritical Fluids* 108: 123–135.

Ekpo, U., Ross, A.B., Camargo-Valero, M.A. et al. (2016). A comparison of product yields and inorganic content in process streams following thermal hydrolysis and hydrothermal processing of microalgae, manure and digestate. *Bioresource Technology* 200: 951–960.

Gardner, D.W., Huo, J., Hoff, T.C. et al. (2015). Insights into the hydrothermal stability of ZSM-5 under relevant biomass conversion reaction conditions. *ACS Catalysis* 5 (7): 4418–4422.

Grigoras, I.F., Stroe, R.E., Sintamarean, I.M. et al. (2017). Effect of biomass pretreatment on the product distribution and composition resulting from the hydrothermal liquefaction of short rotation coppice willow. *Bioresource Technology* 231: 116–123.

Haarlemmer, G., Guizani, C., Anouti, S. et al. (2016). Analysis and comparison of bio-oils obtained by hydrothermal liquefaction and fast pyrolysis of beech wood. *Fuel* 174: 180–188.

Hadhoum, L., Balistrou, M., Burnens, G. et al. (2016). Hydrothermal liquefaction of oil mill wastewater for bio-oil production in subcritical conditions. *Bioresource Technology* 218: 9–17.

Hietala, D.C., Faeth, J.L., and Savage, P.E. (2016). A quantitative kinetic model for the fast and isothermal hydrothermal liquefaction of *Nannochloropsis* sp. *Bioresource Technology* 214: 102–111.

Iris, K.M. and Tsang, D.C. (2017). Conversion of biomass to hydroxymethylfurfural: a review of catalytic systems and underlying mechanisms. *Bioresource Technology* 238: 716–732.

Jain, A., Balasubramanian, R., and Srinivasan, M.P. (2016). Hydrothermal conversion of biomass waste to activated carbon with high porosity: a review. *Chemical Engineering Journal* 283: 789–805.

Jaiswal, K.K., Kumar, V., Vlaskin, M.S. et al. (2021). Hydropyrolysis of freshwater macroalgal bloom for bio-oil and biochar production: kinetics and isotherm for removal of multiple heavy metals. *Environmental Technology and Innovation 22*: 101440.

Jindal, M.K. and Jha, M.K. (2016). Catalytic hydrothermal liquefaction of waste furniture sawdust to bio-oil. *Indian Chemical Engineer* 58 (2): 157–171.

Kang, S., Li, X., Fan, J. et al. (2013). Hydrothermal conversion of lignin: a review. *Renewable and Sustainable Energy Reviews* 27: 546–558.

Kumar, V., Kumar, S., Chauhan, P.K. et al. (2019). Low-temperature catalyst based hydrothermal liquefaction of harmful macroalgal blooms, and aqueous phase nutrient recycling by microalgae. *Scientific Reports 9* (1): 1–9.

Langholtz, M.H., Stokes, B.J., and Eaton, L.M. (2016). 2016 Billion-ton report: advancing domestic resources for a thriving bioeconomy, Volume 1: economic availability of feedstock. *Oak Ridge National Laboratory, Oak Ridge, Tennessee, managed by UT-Battelle, LLC for the US Department of Energy* 2016: 1–411.

Lavanya, M., Meenakshisundaram, A., Renganathan, S. et al. (2016). Hydrothermal liquefaction of freshwater and marine algal biomass: a novel approach to produce distillate fuel fractions through blending and co-processing of biocrude with petrocrude. *Bioresource Technology* 203: 228–235.

Madsen, R.B., Lappa, E., Christensen, P.S. et al. (2016). Chemometric analysis of composition of bio-crude and aqueous phase from hydrothermal liquefaction of thermally and chemically pretreated *Miscanthus* x *giganteus*. *Biomass and Bioenergy* 95: 137–145.

Magdeldin, M., Kohl, T., and Järvinen, M. (2017). Techno-economic assessment of the by-products contribution from non-catalytic hydrothermal liquefaction of lignocellulose residues. *Energy* 137: 679–695.

McVeigh, A., Bouxin, F.P., Jarvis, M.C. et al. (2016). Catalytic depolymerisation of isolated lignin to fine chemicals: part 2–process optimisation. *Catalysis Science & Technology* 6 (12): 4142–4150.

Meryemoglu, B., Hasanoglu, A., Irmak, S. et al. (2014). Biofuel production by liquefaction of kenaf (*Hibiscus cannabinus* L.) biomass. *Bioresource Technology* 151: 278–283.

Panisko, E., Wietsma, T., Lemmon, T. et al. (2015). Characterization of the aqueous fractions from hydrotreatment and hydrothermal liquefaction of lignocellulosic feedstocks. *Biomass and Bioenergy* 74: 162–171.

Rahimi, A., Ulbrich, A., Coon, J.J. et al. (2014). Formic-acid-induced depolymerization of oxidized lignin to aromatics. *Nature* 515 (7526): 249–252.

Raikova, S., Smith-Baedorf, H., Bransgrove, R. et al. (2016). Assessing hydrothermal liquefaction for the production of bio-oil and enhanced metal recovery from microalgae cultivated on acid mine drainage. *Fuel Processing Technology* 142: 219–227.

Sangon, S., Ratanavaraha, S., Ngamprasertsith, S. et al. (2006). Coal liquefaction using supercritical toluene–tetralin mixture in a semi-continuous reactor. *Fuel Processing Technology* 87 (3): 201–207.

Sharma, N., Jaiswal, K.K., Kumar, V. et al. (2021). Effect of catalyst and temperature on the quality and productivity of HTL bio-oil from microalgae: a review. *Renewable Energy* 174: 810–822.

Shuai, L., Amiri, M.T., Questell-Santiago, Y.M. et al. (2016). Formaldehyde stabilization facilitates lignin monomer production during biomass depolymerization. *Science* 354 (6310): 329–333.

Singh, R., Balagurumurthy, B., Prakash, A. et al. (2015a). Catalytic hydrothermal liquefaction of water hyacinth. *Bioresource Technology* 178: 157–165.

Singh, R., Chaudhary, K., Biswas, B. et al. (2015b). Hydrothermal liquefaction of rice straw: effect of reaction environment. *The Journal of Supercritical Fluids* 104: 70–75.

Tekin, K. (2015). Hydrothermal conversion of russian olive seeds into crude bio-oil using a CaO catalyst derived from waste mussel shells. *Energy & Fuels* 29 (7): 4382–4392.

Tekin, K., Karagöz, S., and Bektaş, S. (2014). A review of hydrothermal biomass processing. *Renewable and Sustainable Energy Reviews* 40: 673–687.

Tian, C., Liu, Z., Zhang, Y. et al. (2015). Hydrothermal liquefaction of harvested high-ash low-lipid algal biomass from Dianchi Lake: effects of operational parameters and relations of products. *Bioresource Technology* 184: 336–343.

Tommaso, G., Chen, W.T., Li, P. et al. (2015). Chemical characterization and anaerobic biodegradability of hydrothermal liquefaction aqueous products from mixed-culture wastewater algae. *Bioresource Technology* 178: 139–146.

Tungal, R. and Shende, R.V. (2014). Hydrothermal liquefaction of pinewood (*Pinus ponderosa*) for H_2, biocrude and bio-oil generation. *Applied Energy* 134: 401–412.

Van Doren, L.G., Posmanik, R., Bicalho, F.A. et al. (2017). Prospects for energy recovery during hydrothermal and biological processing of waste biomass. *Bioresource Technology* 225: 67–74.

Wang, F., Chang, Z., Duan, P. et al. (2013a). Hydrothermal liquefaction of *Litsea cubeba* seed to produce bio-oils. *Bioresource Technology* 149: 509–515.

Wikberg, H., Grönqvist, S., Niemi, P. et al. (2017). Hydrothermal treatment followed by enzymatic hydrolysis and hydrothermal carbonization as means to valorise agro-and forest-based biomass residues. *Bioresource Technology* 235: 70–78.

Xu, C. and Etcheverry, T. (2008). Hydro-liquefaction of woody biomass in sub-and super-critical ethanol with iron-based catalysts. *Fuel* 87 (3): 335–345.

Yan, W.H., Duan, P.G., Wang, F. et al. (2016). Composition of the bio-oil from the hydrothermal liquefaction of duckweed and the influence of the extraction solvents. *Fuel* 185: 229–235.

Yang, L., Nazari, L., Yuan, Z. et al. (2016). Hydrothermal liquefaction of spent coffee grounds in water medium for bio-oil production. *Biomass and Bioenergy* 86: 191–198.

Zhang, J., Zhang, Y., and Luo, Z. (2014). Hydrothermal liquefaction of chlorella pyrenoidosa in ethanol-water for bio-crude production. *Energy Procedia* 61: 1961–1964.

Zhang, D., Liu, P., Lu, X. et al. (2016). Upgrading of low rank coal by hydrothermal treatment: coal tar yield during pyrolysis. *Fuel Processing Technology* 141: 117–122.

Zheng, J.L., Zhu, M.Q., and Wu, H.T. (2015). Alkaline hydrothermal liquefaction of swine carcasses to bio-oil. *Waste Management* 43: 230–238.

Zhu, X., Liu, Y., Zhou, C. et al. (2014). A novel porous carbon derived from hydrothermal carbon for efficient adsorption of tetracycline. *Carbon* 77: 627–636.

Zhu, Z., Rosendahl, L., Toor, S.S. et al. (2015). Hydrothermal liquefaction of barley straw to bio-crude oil: effects of reaction temperature and aqueous phase recirculation. *Applied Energy* 137: 183–192.

Zhu, X., Qian, F., Liu, Y. et al. (2016). Controllable synthesis of magnetic carbon composites with high porosity and strong acid resistance from hydrochar for efficient removal of organic pollutants: an overlooked influence. *Carbon* 99: 338–347.

Zhu, Z., Si, B., Lu, J. et al. (2017). Elemental migration and characterization of products during hydrothermal liquefaction of cornstalk. *Bioresource Technology* 243: 9–16.

12

Biogas Production through Anaerobic Digestion of Agricultural Wastes: State of Benefits and its Future Trend

Anwesha Gohain[1], *Navamallika Gogoi*[2], *and Rani Jha*[2]

[1] *Department of Botany, Arunachal University of Studies, Namsai, Arunachal Pradesh, India*
[2] *Department of Chemistry, Arunachal University of Studies, Namsai, Arunachal Pradesh, India*

CHAPTER OUTLINE

Agriculture Waste Management and Bioresource: The Circular Economy Perspective, First Edition.
Edited by Suruchi Singh, Pardeep Singh, Anu Sharma, and Moharana Choudhury.

12.1 Introduction

Unmanaged millions of wastes generated every year from various agricultural, industrial, and municipal sources are alarming to the environment. Methane (CH_4) and carbon dioxide (CO_2) are released into the atmosphere as the result of uncontrolled anaerobic reactions from all the dumping sites that responsible for rise in the earth atmospheric temperature (Chynoweth et al. 2001). However, controlled anaerobic digestion (AD) utilizing the microbial community can yield at least 50% of methane and nutrients-rich fertilizers.

Discomposing organic waste is becoming a difficult task in recent times. The US and the European Union members are forced to decrease biodegradable organic decay through recent legislation, which ultimately enters landfills. The increased charge of sewage and the reluctance of municipal authority towards the waste treatment plants are creating havoc to dispose of solid waste with a very high amount of ecological organic decay. In the agricultural field, after harvesting, the residual parts of crops are returned back into the soil where natural degradation takes place through the microbial community.

As a result, eutrophication takes place in water bodies and lakes due to the release of nutrients. Furthermore, this uncontrolled anaerobic reaction increases the greenhouse gases on the Earth. Researchers expected that a substantial reduction of greenhouse gases could be controlled by changing the waste treatment technology within a short time. Methane can replace fossil fuels through controlled digestion of waste anaerobically, thus lowering CO_2 emission. Compared to other fuels, less number of atmospheric pollutants were observed from methane, i.e. using better techniques for managing waste will direct environmental benefits (Kashyap et al. 2003).

12.2 Anaerobic versus Aerobic Digestion

Studies revealed some advantages of the anaerobic process over aerobic process, for example, use of minimum energy, production of low sludge, minimum utilization of space, and cost-effective. Conversely, the energy requirement is necessary for AD, which ultimately provides aeration. CH_4 formed by the anaerobic method is generally a combustible gas. The boilers might be utilizing as an eco-friendly fuel to generate hot water for refrigerators as gas fired absorption chillers. Consequently, anaerobic-digestion-derived methane is both energy and cost-efficient in biogas production (Chynoweth et al. 2001).

Although AD has disadvantages, however, improved and controlled AD can solve those problems such as generation of odour from H_2S gas and other sulphur compounds. In spite of having this disadvantages, this method reduces organic pollution to 85–90% and guarantees high sewage quality.

12.3 Biochemical and Microbiological Aspect of Anaerobic Digestion

According to Lema and Omil 2001, for correct designing and application of anaerobic systems in sewage treatment, a better understanding of technical features and process, the biochemistry and microbiology of AD is essential.

There are several independent but complex sequential and parallel biological reactions in the AD process. The transformation of inorganic matter, a combination of CH_4 and CO_2, takes place from one group of micro-organisms to another (Noykova et al. 2002).The AD process involves the following four stages:

12.3.1 Hydrolysis

The extracellular enzymes such as protease, cellulose, amylase, and lipase, produced by micro-organisms, help depolymerize the organic matter. As suggested by Kaseng et al. 1992, the hydrolases produced in this process might be cell-bound and they are either belonging to endohydrolases or exohydrolases.

The key concept here is that the microbial community, with the help of these hydrolases, converts the organic particles and polymers to transportable molecules. Hydrolysis plays a crucial role in degrading solid waste; thus, considering the capacity of hydrolytic enzymes would help in the working principle of solid waste through AD.

In hydrolysis, the complexes like cellulose, which are insoluble, are converted into the soluble substances (e.g. amino acids, fatty acids, and proteins), i.e. from polymeric complexes to monomers. In this process, the key role is played by the fermentative bacteria. To enhance the reaction rate, chemicals can be added, ultimately increasing methane yield by decreasing the digestion time.

12.3.2 Acetogenesis

This process is the second step where simple organic acids, carbon dioxide, and hydrogen are formed from products obtained from the hydrolysis step. Acetogenic bacteria perform this conversion.

Mosey and Fernandes 1989 implied that in anaerobic transfer of complex organic matter, acetogenesis plays a key role as it is the fastest reaction in the liquid phase digestion stage. The main pathway in the steady state of anaerobic reaction includes the production of reduced fermented products via acetate, carbon dioxide, and hydrogen, and these can be utilized by the methanogens (Schink 1997).

12.3.3 Methanogenesis

The final product, methane, is produced by the application of methanogenic bacteria, also known as methane formers. It can be produced in two different ways, either by breaking down two molecules of acetic acid to produce carbon dioxide and methane or by reducing carbon dioxide with hydrogen.

The methanogens initiate the methanogenesis process by utilizing carbon dioxide, hydrogen, and acetic acid to form methane and carbon dioxide. Hwang et al. 2001 explained that these methanogens could not utilize other substrates such as methanol, methylamines, and alcohols. To produce form methane, the methanogens use carbon dioxide as an electron acceptor to reduce it via methenyl, formyl, and methyl in association with some definite coenzymes for the final methane production.

On the other hand, aceticlastic pathway (Solera et al. 2002) is well known as the methane yield in this process is 70%. For example, Methanosarcina and Methanosaeta are two well-known methanogenic genera that grow by this aceticlastic reaction (Zinder 1984). Although few known species have the capability of methane production, almost all the known species of methanogens can produce methane by utilizing carbon dioxide and hydrogen.

This condition is an accurate balanced anaerobic decomposition as the products formed in the preceding metabolic phase are transformed completely into the following products. This mechanism's significant point is the total conversion of biodegradable organic matter into biogas without any addition of harmful intermediate products. However, the accumulation of volatile fatty acids might happen when the hydrogen partial pressure increases. This accumulation causes a decline in the pH, influencing the pH-sensitive methanogens and eventually hindering the process. This condition leads to the failure of the whole AD process. According to Archer et al. 1986, hydrogen is considered an influencing parameter in the anaerobic waste digestion method, but it is not easy to detect it.

In this context, the chapter will discuss the methanogenic bacteria, that is, a part of archaea domain. It is not a common group, diverse phylogenetically from the main prokaryotic micro-organisms. This group of prokaryotes shares some general characters with eukaryotes, such as the occurrence of introns in the genome of methanogens, organization of RNA polymerase subunit is similar in both methanogens and prokarytoes, sequences in rRNA and tRNA are homologous in both even though they hold a prokaryotic cell organization and cell structure (Garcia et al. 2000). A vast array of methnogenetic bacteria are found in nature, such as methylotrophs which influence the conversion of methyl compounds to methanol and methylamines. Similarly, acetotrophic methanogens are also found whose function is to convert acetate to methane and carbon dioxide. Another category, the hydrogenotrophic methanogens, utilizes H and CO_2 to generate methane.

The general performance of the AD system depends on concentration and proportion of individual VFAs produced in acetogenesis as some of the acids such as acetic acids and butyric acids act as precursors in methane formation. Moreover, some important parameters influence the AD process. This stage is the second step where simple organic acids, carbon dioxide, and hydrogen are formed from products obtained from the hydrolysis step. Acetogenic bacteria perform this conversion.

12.4 Parameters Involved in the Process

Retention duration, pH and buffering systems, processed architecture, and the solubility of gases, available source of nutrient, and the extent of hazardous substances are among the environmental elements that affect AD, according to studies (Bjornsson et al. 2000).

12.4.1 Temperature

Temperature influences the AD and thus based on temperature, the microbial community is divided into three main categories such as psychrophilic (0–20 °C), mesophilic (20–42 °C), and thermophilic (42–75 °C). Despite of these three categories, variation of optimum temperature can be observed depending on the composition of feedstock and the type of digester used. But it is advisable to keep a stable temperature to sustain the rate of gas production. However, thermophilic digestion is considered as the most efficient and best alternative as the growth rate of bacteria is higher in this process and has better utility of retention time, loading rate as well as nominal production of gas, though the process is sensitive and needs a higher amount of heat to deal with variable environmental conditions. This situation makes thermophilic AD a little more problematic than the other two processes, i.e. mesophlic and psychrophilic AD.

Thus, studies also implied that thermophlic AD effectively treats high-temperature effluents from various industries such as food, agriculture, and canning factories, compared to mesophlic digestion (Dinsdale et al. 1997; Lepisto and Rintala 1997).

12.4.2 Retention Time

The overall time taken by the digestion process for complete digestion of organic matter can be called retention time. But, based on the some parameters such as composition of wastes and temperature, retention time varies. For example, in the mesophilic digestion process, the retention time taken for complete conversion is approximately 15–30 days, whereas in the thermophilic process it is12–14 days.

12.4.3 pH and Alkalinity

The optimal pH for methane and acid-producing bacteria varies in methanogenesis and acidogenesis stages, respectively. pH is one of the most highly influencing factors in the AD process because the microbial groups have a definite pH range for favourable development (Elefsiniotis and Oldham 1994). For methane-producing micro-organisms, the optimal pH is 6.8–7.2, whereas it is 6 for acid-producing microbes (Moosbrugger et al. 1993). Mosey and Fernandes 1989 explained a sharp fall in the growth of microbes below pH 6.6. On the other hand, the microbes that belong to acetoclastic methanogens are more sensitive to low pH than hydrogenotrophic group of methanogens. According to Demirel and Yenigun (2002), variation in pH from 6.0 to 8.0 can influence the growth of dominant populations in acidic phage. Utilization of carbon and other energy sources, protein synthesis, substrate dissimilation efficiency, and other metabolic activities is entirely dependent on pH by an array of microbes (Elefsiniotis and Oldham 1994).

Another important parameter for the AD process is alkalinity. With the help of alkalinity, the chemical buffering capacity of the aqueous solution can be measured. It is important to neutralize any possible volatile fatty acids in the reactor, and for stabilization the pH is maintained at around 6.7–7.4 (Callander and Barford 1983). For example, bicarbonate, hydrogen sulphide, dihydrogen phosphate, and ammonia play a significant role in providing an accurate buffering environment at 7.

12.4.4 Carbon-to-Nitrogen Ratio

In anaerobic digesters, carbon-to-nitrogen ratio ranges from 20 to 30. A higher C:N ratio indicates that the methanogen utilizes the N content and gas production rate. On the contrary, a lower ratio of C:N indicates ammonia accumulation having a pH of 8.5, which is an adverse condition for methanogenic microbes. Moreover, a favourable C:N ratio can be attained by the addition of a low and high ratio of C:N.

For high methane production, the favourable C:N:P ratio is shared to be 100 : 3 : 1, and essential nutrients are found to be nitrogen and phosphorus (Rajeshwari et al. 2000). Nyns 1986 reported that a high C/N ratio increases the threat of nutrient scarcity and low buffering capacity will lead to a sensitive AD process. On the other hand, breakdown of nitrogenous compounds will lead to a discharge of ammonia which increases the nitrogen content and ultimately failure of ammonia inhibition.

12.4.5 Volatile Fatty Acids

Volatile fatty acids (VFAs) and nutrients are also some of the factors that influence the AD process. Proper monitoring of the AD process can be carried out on the concentration of VFAs (Choi et al. 2011; Llamas et al. 2020). It has been suggested that the accumulation of VFA is due to unstable conditions for the AD process (Bjornsson et al. 1997). Besides, low pH favours build-up of VFA, resulting in toxicity and ultimate failure of the AD process (Ahring et al. 1995). This situation shows that the toxicity of VFA is pH-dependent. Moreover, the significant indicators of the metabolic state for the most sensitive microbial groups are the concentrations of acetic, propionic, and butyric acids.

12.4.6 Nutrients

The nutrients, including micro, macro, and trace elements, have a great effect on the growth of micro-organism, and also, lack of these nutrients might harm their growth (Lettinga 1995). These nutrients, especially nitrogen, phosphorus, and trace elements like sulphur, potassium, calcium, and magnesium, are necessary for an adequate amount in the degradation process of wastes during the anaerobic reaction digesters (Rajeshwari et al. 2000).

Moreover, various studies revealed that the accumulation of heavy metals has some inhibitory and toxic effects on acidogenic stage of the AD process. Copper and zinc were reported to be most toxic, whereas the least toxic heavy metal was lead to acidigens (Lin 1993). In an experiment, Alagöz et al. (2015) reported that copper has more toxic effects as compared to zinc. The batch cultures showed inhibitory effects in the range of 1–10 mg/dm^3 and 5–40 mg/dm^3 for copper and zinc, respectively. On the other hand, it has also been reported that heavy metals reduce the breakdown of VFA to methane (Lin 1992).

12.5 Anaerobic Digestion Process

12.5.1 Pre-Treatment

The steps in pre-treatment involve identification and removal of non-biodegradable materials which remain unaffected by the digestion process. For efficient digestion, uniform particle size feedstock should be provided which is followed by removal of unwanted materials which may affect the digestate. Moreover, protection of downstream plant is necessary to overcome the physical damage.

Several pre-treatment processes depend on the wastes. For example, the recyclable and non-digestible wastes are separated from the organic wastes in the pre-treatment process for municipal solid wastes. Two methods can do this separation. First is source separation and second is mechanical separation.

In source separation, unwanted wastes are removed at the source, and if there are mixtures of solid wastes, then mechanical separation is implemented. However, source separation is the highly recommended method for pre-treatment.

12.5.2 Thermal Sterilization

Thermal sterilization, also known as pasteurization of digestate (co-substrates), is based on the specific feedstock. The temperature for thermophilc digesters is considered to be not sufficient for adequate sterilization. Moreover, according to the guidelines of the European Commission, for biological treatment of biowaste, the wastes including agriculture waste, solid waste, municipal waste, etc., should be operated at 55 °C for 22 days. Pre-treatment of substrates should be at 70 °C for 60 minutes or post-treatment of co-substrates at 70 °C for 60 minutes followed by digestate composting.

12.6 By-Products of Anaerobic Digestion

AD process is a cost-efficient way to get rid of biodegradable wastes easily, and the by-products of this process are biogas and digestate. The range of key variables of the AD can be changed depending on the aim as well as the use of AD.

12.6.1 Biogas

The composition of biogas produced during the anaerobic process is as follows: methane and carbon dioxide is present in large amounts, whereas hydrogen sulphide and ammonia are present in fewer amounts. Hydrogen, nitrogen, carbon monoxide, saturated carbohydrates, and oxygen are considered as trace gases found in the biogas. Moreover, there are some similarities between the AD-produced biogas and the landfill gas which contains considerable amounts of saturated compounds and oxygen.

Thus, a variation in the composition of biogas is observed based on the types of wastes. Biogas has various applications, such as it can be used in boilers for heating purposes. In general, high-quality gas is not required. Still, the gas should not contain hydrogen sulphide as hydrogen sulphide will form corrosive sulphuric acid, which is not preferable for the boilers. Water vapour should be condensed if it is raw gas as it can cause trouble in gas nozzles. At this point, while removing the water vapour, many hydrogen sulphides will also be removed.

Although biogas has applications in different fields, there is a strict demand for the quality of biogas. Henceforth, to obtain the quality of the biogas, we need to meet the certain criteria such as a high energy value fuel so that the vehicles can function for a longer period of time, quality of the gas should be consistent for safe driving, better operation of engine, the removal of corrosive gases like hydrogen sulphide, ammonia, and water and the gas should not contain any harmful particles.

The biogas that meets the earlier-mentioned criteria will be considered the most environment-friendly fuel, which has a less harsh effect on the ecosystem.

The biogas with 95% of methane content can be used as in the fuel cells. For the production of DC electricity, fuel cells are used by utilizing oxygen and fuel in an electrochemical reaction.

Moreover, there are several technologies to achieve the criteria to deal with gas from agriculture waste, landfill, and solid waste to become an ideal gas for proper functioning.

12.6.2 Digestate

To treat organic waste, AD is a process to extort maximum usable materials from the wastes. The digestate can play a significant role in increasing the yield as the maximum amounts of nutrients are present in digestate. The criteria based on which digestate can be assessed are biological, chemical, and physical.

To manage the quality of digestate on chemical aspect, contamination of heavy metals and other organic and inorganic pollutants and micronutrients such as nitrogen, phosphorus, and potassium are considered. Agricultural wastes are persistent with organic contaminants, viz. residues of antibiotics and pesticides, while industrial wastes, sludge, and household wastes contain contaminants like aliphatic, aromatic, and other halogenated compounds. Based on the types of contaminants and origin, these wastes can be hazardous to the environment and every living organism. That is why, accessing the quality of this kind of biogas is significant.

As the chapter highlighted the high concentration of nutrients in digestate as compared to the untreated organic wastes, it improves the plant uptake of NPK. The pH value of digestate is more, and accessibility to inorganic nitrogen is 25% more in digestate than the liquid nitrogen. The use of digestate is beneficial to the humous of the soil. Therefore, it can be used as a fertilizer in the field of agriculture because it enhances the nutrient cycle and improves the soil texture and quality.

12.7 Types of Digesters

Following are the different categories of anaerobic digesters:

12.7.1 Single Stage

There are two categories under single-stage digesters: single-stage low solid (SSLS) and single-stage high soild (SSHS). SSLS has been used for decades for any waste for its operational utility and simplicity. Moreover, the types of equipment used in this method are much cheaper. Even then, preventive measure has to be taken in the pre-treatment process. In the process of heavy digestion metals, contaminants sink at the bottom and form three layers of contaminants that affect the reactor's propellers and influence the mixing. Accumulation of heavy metals might cause damage to the pumps, which is why cleaning the feedstock was done previously. In SSHS, the main hurdle is the handling and transport, but the pre-treatment in this process is easy compared to SSLS as the tolerance capacity for all kinds of impurities is high in this process.

12.7.2 Multi-Stage

To improve the AD process, application of separate reactors for different stages is implemented, to increase the flexibility and optimization of each reaction.

The main benefit of the multi-stage process is the constancy in the biological degradation process to degrade to highly degradable agriculture waste including vegetables and fruits.

However, the distinction between the reactors has along with or without biomass retention design. We have discussed the importance of the retention time AD process, so biomass retention has been selected as a significant variable in the multi-stage process. Yet again, multi-stage process has been categorized into multi-stage low solid (MSLS) process and multi-stage high solid (MSHS) process. Similar problems have been observed in multi-stage as those of SSLS reactor, such as foaming and short-circuiting, based on different densities formation of contaminant layers, etc. On the other hand, the techniques of MSLS are quite complex.

12.7.3 Batch Stage

In this stage, fresh wastes are used to fill the digesters and allowed to perform all the degradation steps in dry mode. Three different types of batch processes that are available are single-stage system, sequential system, and upflow anaerobic sludge blanket (UASB) reactor.

12.7.4 Single-Stage System

During the batch process, accumulation of leachate is observed at the bottom. In single-stage system, recycling of this leachate is done at the top of the same reactor which is almost equal to partial mixing.

12.7.5 Sequential System

Two or more reactors are involved in a sequential system. In this stage, recirculation of the leachate obtained from first reactor (containing the highest of organic acids) is done in the second reactor where methanogenesis occurs. In this methanogenic reactor, accumulated leachate (containing little or no acids) is mixed with a buffering agent and proceeds to the first reactor for recirculation. This situation confirms the mixing of two reactors, and this eliminates the addition of fresh waste to the reactor again repeatedly.

12.7.6 Upflow Anaerobic Sludge Blanket (UASB) Reactor

UASB can be considered as a hybrid process sharing similarities with multi-stage processes. According to Elias et al. (1999), for retaining high-density anaerobic sludge, the USAB does not require any carriers or support materials. It is a high-powered methane bioreactor with a layer of micro-organisms settled at the reactor.

The USAB reactors are quite simple. In USAB, an equal distribution of incoming wastewater is observed near the reactor's base. Moreover, in this method, separation of gas is effective; to prevent excessive trapping of biomass, a good cross section is provided in the reactor (Fox and Pohland 1994). There is a three-phase separator, i.e. biogas, liquid, and biomass, in this method. This condition ensures the separation of biogas as well as the microbial mass. This situation ultimately returns to the lower zone of the reactor, which is the highly active zone (Lettinga 1995). Rajeswari et al. (2000) suggested that recirculation of effluent in USAB is cheap as it neither requires a high energy supply nor more energy consumption.

Because of its simple, cost-effective design, AD is useful even in laboratory studies and can determine biogas yields.

On the other hand, temperature ranges, i.e. either mesophilic or thermophilic, and solid content utility are defined.

12.8 Mesophilic and Thermophilic Process

Traditionally operated mesophilic AD process required a high temperature, and it was necessary to set up and regulate this within the digester. Maintaining the uniformity of temperature is found to be complicated; thus, it is observed that the process is self-sustainable and requires less heat to sustain the process of digestion. On the other hand, compared to mesophilic stage, thermophlic stage works at a faster rate, approximately 12–14 days in the case of thermophilc process and 15–30 days in the case of mesophlic process.

Although these processes yield a large amount of methane, it is pretty much costly and needs proper monitoring and high energy input.

12.9 Suitable Wastes for Anaerobic Digestion

AD as compared to aerobic digestion is suitable for digesting agriculture waste, sewage sludge, municipal solid waste, and organic industrial and commercial sludge.

12.9.1 Agricultural Waste Degradation Process

Agricultural waste has become an ecological problem as it raises concern in public health and environmental awareness. If properly treated, these biodegradable wastes might produce a large quantity of energy in the form of biogas production (Zhang et al. 2012). Nowadays, most countries like England and India adopted technologies to utilize biodegradable wastes like agricultural waste in potential energy sources. Implementation of digestion plant in Bombay is a good example of this.

The installation of AD digesters is the most standard technology in the field of agriculture nowadays. Furthermore, manure waste from poultry, cattle, etc., and garden waste, farmyard manure, etc., are some of the agriculture wastes suitable for the AD process.

Manure or slurry from the animals or animal to waste is important waste material; on the other hand, beddings including straw, wood shavings, etc., are the components of farmyard manure. Yet again, the production of biogas varies depending on the types of feedstock used. Figure 12.1 depicts the overall reaction of AD process which involves hydrolysis, fermentation, acetogenesis, and methanogenesis.The by-product produced at the end can be reused in various forms such as fuel, thermal heat, and electricity. (Figure 12.2). In addition to that, this treatment would not influence the nitrogen present

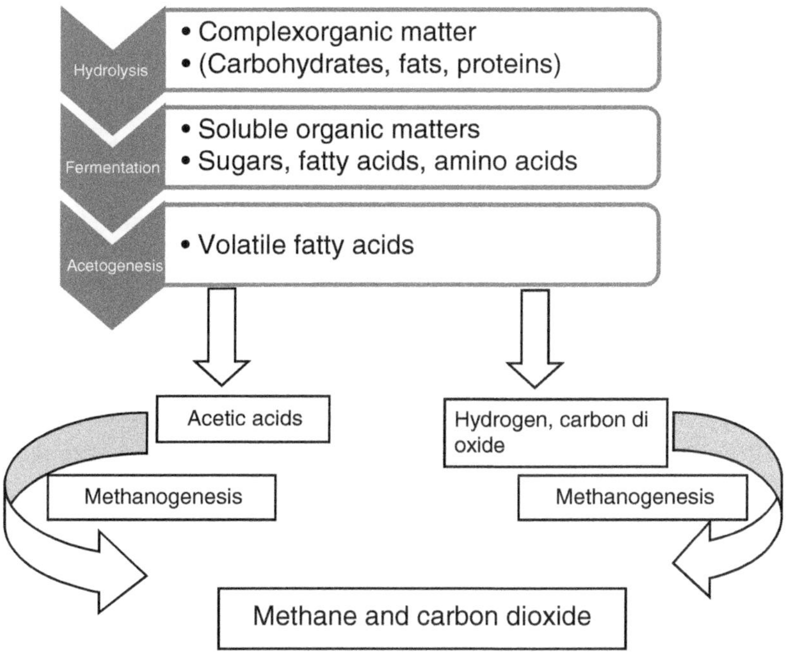

Figure 12.1 Chemical reactions that are involved in anaerobic digestion process.

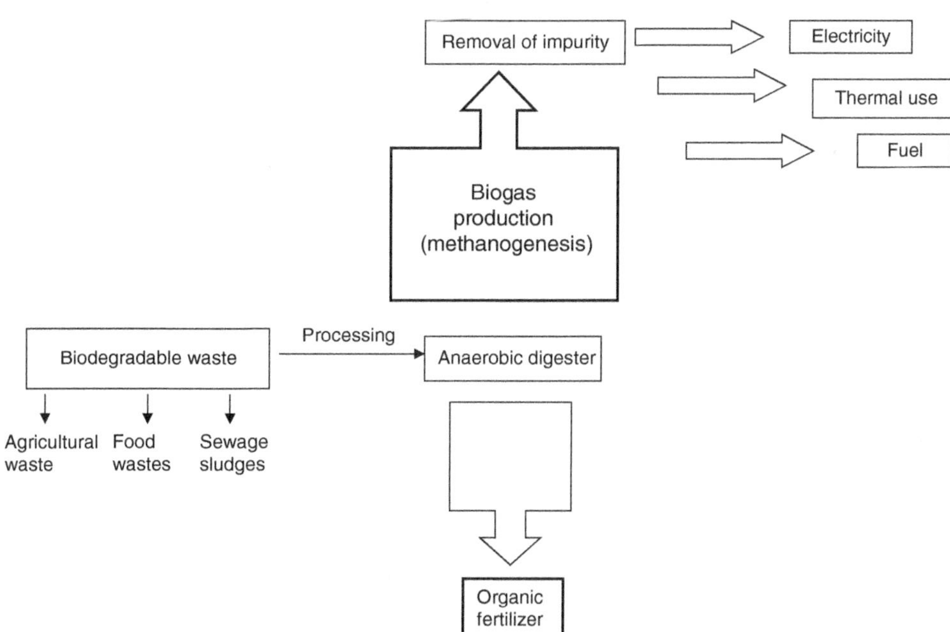

Figure 12.2 Diagrammatic representation of anaerobic digestion process, by-product produced at the end can be used.

in livestocks and others; instead, it is incorporated with energy crops with increasing amount of organic residues. Henceforth, we are supplemented with biogas, and rest of the organic matters can be reused.

12.9.2 Other Wastes

The demand for AD process in degradation of other wastes like organic industrial and commercial sludge, sewage sludge, and municipal solid waste is increasing day by day. More than 30 different types of industries have been identified and are amenable for AD. For example, pharmaceutical companies, food companies, and fermentation industries prefer AD for the digestion of wastes which are produced after processing of organic chemicals and food products. Because of its low operational sludge, this process becomes a standard method for digestion of wastes.

In case of municipal solid wastes, the AD process is considered as the perfect technical method. But source separation should be implemented properly in municipal wastes, i.e. removal of wastes, plastics as well as other unwanted materials should be done properly. If this procedure follows in an appropriate way, then a high-quality product will be achieved.

12.10 Perspective of Anaerobic Digestion

The statistics of a reusable form of energy through the AD process are not available. Still, developing countries like India, and Eastern and Southern Africa signify a constant dependence on this energy. In India and China, biogas is significant in the cooking, lighting, and operation of small engines (Reddy 2004). However, these biogas plants are not technically sound, and the duration is very short. On the other hand, large-scale AD process technology in countries like Eastern and Southern Africa is still in the initial leg, although there are many perspectives in this field.

The technology that involves methane production with waste treatment can improve the technology in managing biodegradable organic wastes including agricultural wastes and reduce the possible errors and defects (Ayalon et al. 2001; Lema and Omil 2001).

Initiation of the long-term AD process and other renewable energy resources is important to promote AD's implementation and proper use. Through appropriate research and scientific work, the experts should provide consistent and applicable information about the method and its utility shortly.

12.11 Policies and Application of Anaerobic Digestion Technology

12.11.1 Outlook

Gijzen (2001) explained the developments in the AD process field and suggested it is the most feasible way to stabilize different types of organic wastes. The most significant application of the AD process is septic tanks to remove wastes such as agriculture wastes, municipal solid wastes, hazardous wastes including aromatic and saturated compounds, and industrial wastes. In around 1995, a country like Europe drew the world's attention in the field of the AD process, and less than 5% of total biological waste treatment capacity was represented through AD. Still, eventually, this rate has increased over the continent (de Baere 2000). Sonesson et al. (2000) explained the facts of gaining interest in anaerobic treatment, and in 2000, about 50% of solid wastes were decomposed, 45% were dumped into landfills, and 5% were mainly decomposed. Production of 1.4 TWh of biogas was reported in Sweden and targeted the production of biogas up to 25 TWh, of which it has been estimated that the agriculture sector would provide 14 TWh (Nordberg and Edstrom 2003). Countries like Germany, Netherlands, and Denmark are the leading countries in this field. It was reported that there are 18 functional centralized anaerobic plants in Denmark.

The main perspective of this idea is the stable, reliable, significant dependence on traditional biomass energy instead of modern energy technology. Developing countries like China and India use biogas on small scales, viz. cooking, lighting, and operating small engines. Other countries like Bangladesh, Thailand, and Ivory Coast rely on Chinese fixed dome digesters and the Indian floating cover digester based on the AD process. But this method's productivity is poor as these plants have been built for schools and small-scale farmers. Thus, the advance of more efficient high-level AD process is much needed.

12.11.2 Policies

The European Union is trying to regulate the carbon dioxide emission by taking several initiatives such as the European Climate Change Programme, Kyoto protocol, and EU emission trading scheme. The AD technology plays a role in this regard to boost up these practices and enhance sustainability.

By implementing various policies and regulations, the UK government is also trying to develop renewable energy technologies.

12.11.3 Economic Benefits of Anaerobic Digestion

AD has found a home worldwide, from small residential digesters in rural emerging countries to massive industrial systems. AD is more able to deal with economic productivity, public health, environmental quality, agricultural efficiency, and environmental sustainability than conventional methods. AD generates renewable energy (biogas and its various applications) and other by-products with several economic advantages, viz. the different micro-/macronutrients used in the digestate of AD process can incorporate algae cultivation which leads to lower cost of CO_2 mitigation(Leu et al. 2021). Moreover, in large scale of AD process, availability and choice of feedstock contribute significantly to the economic growth of the region. Digestate management and biogas utilization could enhance the economic growth of AD plants.

12.11.4 Agricultural Benefits of Anaerobic Digestion

Biogas produced through AD process has a high heating value, and the digestate is loaded with nutrients like nitrate, phosphoric acid, and potassium sulphate which play a key role in high yielding of crops. Energy and nutrients can be recovered in this way more maturely, integrating into the circular economy concept and extending the life of nutrients. Still, the need for extensive agricultural acreage to produce digestate continuously poses a problem for counties with limited land. As a result, an excess of digestate is available, and digestate must be transported over large distances. Meanwhile, it has also been suggested that bio-based materials and chemicals can generate significantly better economic returns than biofuels themselves (biogas in this case). On the other hand, when combined with appropriate bulking materials, microbial agents, and nutrient additives, some of the processes like composting, co-composting, and vermicomposting are the processes that can produce organic fertilizer with excellent agronomic characteristics. Nowadays, anaerobic co-digestion is widely used to minimize the issues that are linked with mono-digestion of food waste. Co-digestion also contributes an increase in biogas generation as compared to mono-digestion. Various wastewater management systems currently use AD to handle sewage sludge and food waste and can also be utilized as a co-substrate with sludge in the future. This condition can help with energy recovery and the high capital costs associated with new biogas facilities. As a result, additional research should be conducted to ascertain the best operating characteristics for creating a large-scale co-digestion system.Biogas produced through AD contains CH_4, CO_2, various gases (e.g. H_2S), contaminants (e.g. siloxanes), and a small amount of water vapour. Biogas is a clean and environmentally friendly renewable energy source that may generate electricity in place of fossil fuels via combined heat and power. Biogas technology has to update not just to use as a vehicle fuel or in a fuel cell but also into natural gas systems as an inoculants. Pressure swing adsorption, water scrubbing, amine scrubbing, cryogenic, and membrane separation are commercially available biogas-upgrading technologies. Technologies based on biological and hybrid approaches are still in the development. Additional research should be conducted to examine the economics and technological advancements associated with biogas-upgrading techniques.

As mentioned earlier, combining farm animal dung with agricultural plant wastes optimizes the C:N ratio of the reaction while minimizing ammonia inhibition.

Carbon produced in this process plays a key role for plant growth and development, and as a carbon-rich material, biochar can be used as a carbon fertilizer in the soil. It can increase crop growth and development by extending the time between fertilizer applications, altering the structure and quantity of soil flora, and increasing crop disease resistance.

12.11.5 Environmental Benefits of Anaerobic Digestion

AD provides a slew of ecological benefits like odour regulate, improves air and water quality, enhances nutrient management, increases flexibility, and reduces greenhouse gas emissions. Furthermore, AD is an efficient method of lowering GHG emissions, particularly NH_3 and CH_4. AD produces a stable substance, and this can be divided into liquid and solid portions. The liquid portion is an excellent source of plant nutrients, while the particles make an excellent soil conditioner or bedding for animals. Carbon and nitrogen are the two most significant components that transform the anaerobic digester. It is expected that the conversion of carbon to biogas results in the reduction in biological oxygen consumption which is approximately 85%.

AD also helps water quality in numerous ways. The digestive process transforms some of the nitrogen from its comparatively stable organic form to the more volatile ammoniacal form, enhancing its effectiveness. For example, when digestate is applied to land, phosphorus and metals such as copper and zinc settle in the process cells, minimizing the release of these compounds into surface waters.

The AD process residue exhibited increased SSA, pH, and CEC values and increased anion exchange capacity (AEC), hydrophobicity, and adverse surface charge, all of which might be employed to improve soil quality and absorb water pollutants. In addition to these facts, the AD digester is a highly stable material that also works as a soil conditioner. It maintains soil health balance, regulates soil microbial composition, and enhances soil enzyme activity, and residues of biogas have a high adsorption capacity and are highly effective at treating contaminated water. Biogas residues were found to be capable of removing Cd and Ni ions from wastewater.

12.12 Concluding Remarks

For environmental protection and sustainable development, the anaerobic treatment process is considered as the core method. In recent years, the high-density agro-based systems have led to excess production of wastes which ultimately have a severe adverse effect on ecosystems, if these wastes are not disposed properly. Nevertheless, these are valuable resource for biogas production with less emission of greenhouse gases. As we have mentioned in this chapter, AD process plays a key role as an advanced technology for protecting the environment and preserving resources. Application of appropriate and successful method of treating wastes (including solid, wastewater, and agro-based wastes) using AD process is valuable in the biogas production. Application of high-rate reactors to degrade solid agro-based wastes will also play a key role in forthcoming days.

References

Ahring, B.K., Sandberg, M., and Angelidaki, I. (1995). Volatile fatty acids as indicators of process imbalance in anaerobic digesters. *Applied Microbiology and Biotechnology* 43: 559–565.

Alagöz, B.A., Yenigün, O., and Erdinçle, A. (2015). Enhancement of anaerobic digestion efficiency of wastewater sludge and olive waste: synergistic effect of co-digestion and ultrasonic/microwave sludge pre-treatment. *Waste Management* 46: 182–188.

Archer, D.B., Hilton, M.G., Adams, P., and Wiecko, H. (1986). Hydrogen as a process control index in pilot scale anaerobic digester. *Biotechnology Letters* 8: 197–202.

Ayalon, O., Avnimelech, Y., and Shechter, M. (2001). Solid waste treatment as a high-priority and low cost alternative for greenhouse gas mitigation. *Environmental Management* 27: 697–704.

de Baere, L. (2000). Anaerobic digestion of solid waste: state of the art. *Water Science and Technology* 41: 283–290.

Bjornsson, L., Mattiasson, B., and Henrysson, T. (1997). Effects of support material on the pattern of volatile fatty acid accumulation at overload in anaerobic digestion of semi-solid waste. *Applied Microbiology and Biotechnology* 47: 640–644.

Bjornsson, L., Murto, M., and Mattiasson, B. (2000). Evaluation of parameters for monitoring ananaerobic co-digestion process. *Applied Microbiology and Biotechnology* 54: 844–849.

Callander, I.J. and Barford, J.P. (1983). Recent advances in anaerobic digestion technology. *Process Biochemistry* 18: 24–23.

Choi, J.D.R., Chang, H.N., and Han, J.I. (2011). Performance of microbial fuel cell with volatile fatty acids from food wastes. *Biotechnology Letters* 33: 705–714. https://doi.org/10.1007/s10529-010-0507-2.

Chynoweth, D.P., Owens, J.M., and Legrand, R. (2001). Renewable methane from anaerobic digestion of biomass. *Renewable Energy* 22: 1–8.

Dinsdale, R.M., Hawkes, F.R., and Hawkes, D.L. (1997). Mesophilic and thermophilic anaerobicdigestion with thermophilic pre-acidification of instant-coffee-production wastewater. *Water Research* 31: 1931–1938.

Elefsiniotis, P. and Oldham, W.K. (1994). Influence of pH on the acid-phase anaerobic digestion of primary sludge. *Journal of Chemical Technology and Biotechnology* 60: 89–96.

Elias, A., Barona, A., Ormazaba, J. et al. (1999). Anaerobic treatment of acidified and non-acidified substrata in UASB reactors. *Journal of Chemical Technology and Biotechnology* 74: 949–956.

Fox, P. and Pohland, F.G. (1994). Anaerobic treatment applications and fundamentals: substrate specificity during phase separation. *Water Environment Research* 66: 716–724.

Garcia, J.L., Patel, B.K.C., and Ollivier, B. (2000). Taxonomic, phylogenetic, and ecological diversity of methanogenic *Archaea*. *Anaerobe* 6: 205–226.

Gijzen, H.J. (2001). Anaerobes, aerobes and phototrophs, a winning team for wastewater management. *Water Science and Technology* 44: 123–132.

Hwang, S., Lee, Y., and Yang, K. (2001). Maximisation of acetic acid production in partial acidogenesis of swine wastewater. *Biotechnology and Bioengineering* 75: 521–529.

Kaseng, K., Ibrahim, K., Paneerselvam, S.V., and Hassan, R.S. (1992). Extracellular enzymes and acidogen profiles of a laboratory-scale two-phase anaerobic digestion system. *Process Biochemistry* 27: 43–47.

Kashyap, D.R., Dadhich, K.S., and Sharma, S.K. (2003). Biomethanation under psychrophilic conditions: a review. *Bioresource Technology* 87: 147–153.

Lema, J.M. and Omil, F. (2001). Anaerobic treatment: a key technology for sustainable management of wastes in Europe. *Water Science and Technology* 44: 133–140.

Lepisto, S.S. and Rintala, J.A. (1997). Start-up and operation of laboratory-scale thermophilic upflow anaerobic sludge blanket reactors treating vegetable processing wastewaters. *Journal of Chemical Technology and Biotechnology* 68: 331–339.

Lettinga, G. (1995). Anaerobic digestion and wastewater treatment systems. *Antonie Van Leeuwenhoek* 67: 3–28.

Leu, S., Cao, L., Xu, F. et al. (2021). Chapter Five – Integration of algae cultivation to anaerobic digestion for biofuel and bioenergy production. In: *Advances in Bioenergy* (ed. Y. Li and W. Zhou) Volume 6, Issue 1, 199–300.

Lin, C.Y. (1992). Effect of heavy metals on volatile fatty acid degradation in anaerobic digestion. *Water Research* 26: 177–183.

Lin, C.Y. (1993). Effect of heavy metals on acidogenesis in anaerobic digestion. *Water Research* 27: 147–152.

Llamas, M., Magdalena, J.A., González-Fernández, C., and Tomás-Pejó, E. (2020). Volatile fatty acids as novel building blocks for oil-based chemistry via oleaginous yeast fermentation. *Biotechnology and Bioengineering* 117: 238–250. https://doi.org/10.1002/bit.27180.

Moosbrugger, R.E., Wentezel, M.C., Ekama, G.A., and Marais, G.V. (1993). Weak acid/bases and pH control in anaerobic systems – a review. *Water SA* 19: 1–10.

Mosey, F.E. and Fernandes, X.A. (1989). Patterns of hydrogen in biogas from the anaerobic digestion of milk sugars. *Water Science and Technology* 21: 187–196.

Nordberg, A. and Edstrom, M. (2003). Treatment of animal waste in co-digestion biogas plants in Sweden. In: *The Future of Biogas in Europe II* (ed. T. Al Seidi and J.B. Holm-Nielsen) European Biogas Workshop. 2-4 October. Esbjerg, Denmark: University of southern Denmark.

Noykova, N., Muller, T.G., Gyllenberg, M., and Timmer, J. (2002). Quantitative analysis of anaerobic wastewater treatment processes: identifiability and parameter estimation. *Biotechnology and Bioengineering* 78: 89–103.

Nyns, E.J. (1986). Biomethanation processes. In: *Microbial Degradations*, vol. 18 (ed. W. Schonborn), 207–267.

Rajeshwari, K.V., Balakrishnan, M., Kansal, A. et al. (2000). State of- the-art of anaerobic digestion technology for industrial wastewater treatment. *Renewable and Sustainable Energy Reviews* 4: 135–156.

Reddy, A.K.N. (2004). Lessons from the pura community biogas project. *Energy Sustainable Development* 8: 68–73.

Schink, B. (1997). Energetics of syntrophic cooperation in methanogenic degradation. *Microbiology and Molecular Biology Reviews* 61: 262–280.

Solera, R., Romero, L.I., and Sales, D. (2002). The evolution of biomass in a two-phase anaerobic treatment process during start-up. *Chemical and Biochemical Engineering Quarterly* 16: 25–29.

Sonesson, U., Jörklund, A.B., Carlsson, M., and Dalemo, M. (2000). Environmental and economic analysis of management systems for biodegradable waste. *Resources, Conservation and Recycling* 28 (1–2): 29–53. https://doi.org/10.1016/S0921-3449(99)00029-4.

Zhang, Z., Gonzalez, A.M., Davies, E.G.R., and Liu, Y. (2012). Utilization of byproducts and waste materials from meat, poultry and fish processing industries: a review. *Water Environment Research* 84: 1386–1406.

Zinder, S.H. (1984). Microbiology of anaerobic conversion of organic wastes to methane: recent developments. *American Society for Microbiology* 50: 294–298.

13

Expansion of Agricultural Residues to Biofuel Processing and Production

Varucha Misra, A.K. Mall, Santeshwari Srivastava, and A.D. Pathak

Division of Crop Improvement, ICAR-Indian Institute of Sugarcane Research, Lucknow, Uttar Pradesh, India

CHAPTER OUTLINE

13.1 Introduction

Biofuel is defined as the products produced from renewable biological products. These products include liquid fuels and blending components that are used in transportation (Wu et al. 2010). The products so obtained are a supplement to fossil fuels and possess several benefits like use in transportation, as a mitigator of carbon dioxide emissions, etc (Reijinders 2006; Yan and Lin 2009). Plant biomass is considered a major nonfossil fuel material for the production of biofuel (Demirbas 2007). This is due to its high potential for many years. Zhao et al. (2009) had shown that plant biomass as a biofuel producer contributes to 14% of the world's energy demand. The transportation fuels required today for various purposes emit polluted air that could be minimized by the use of biofuels produced from agricultural biomass. Ethanol production is from the biomass of crops like sugarcane and corn, while biodiesel production is from oil palm, rapeseed, etc., which are mostly known for biofuel production (Allakhverdiev and Ramakrishna 2015). The utilization of

Agriculture Waste Management and Bioresource: The Circular Economy Perspective, First Edition.
Edited by Suruchi Singh, Pardeep Singh, Anu Sharma, and Moharana Choudhury.

biomass (whether a by-product or residue) as a source for the production of energy is not new to the world. Dried wood, leaves, and twigs of trees were used in bonfires for heating purposes (Renewables 2015). Similarly, the use of biofuel in engines was also been started much before. But at that time the cheap production of fuel from fossil products dominated biofuel production. However, the production of biofuel from agricultural biomass came into existence due to difficulties associated with the use of fossil fuel towards the environment (Voloshin et al. 2016). The eco-friendly nature of biofuel made it much more important (Allakhverdiev and Ramakrishna 2015). The production of energy through biomass has increased recently due to the low carbon content and a major constituent in mixing a significant part in petroleum (Reid et al. 2019).

13.2 Agricultural Residues and Importance in Biofuel Production

Agricultural residues are defined as materials (having carbon as the main component) that are produced as a by-product (implying left out after harvesting or during processing) from agricultural crops (Mohammed et al. 2018). These are of two kinds, *viz*., primary and secondary residues (Kyryzyuk et al. 2020). Primary residues are also referred to as field-based residues, implying that they are generally obtained once the crop is harvested while secondary residues are meant to be those residues that are left out either during or after processing of the crop. These residues are heterogeneous compounds; frequently fibrous; differ in their bulk densities, geographic site, particle size, amount of moisture, etc. (Smith 1989). These are often produced as waste material from agricultural crops like sugarcane, maize, and rice. These residues are scarcely used for fodder; however, a major portion has been either burned off or leftover in the soil so as to get rid of the residues of the previous crop before the new crop is planted (Mohammed et al. 2018). Panthapulakkal and Swain (2015) showed that agricultural residues are easily available and ready to use as, besides being a good resource of ligno-cellulosic fibre. Ruane et al. (2010) showed that a significant portion of these agricultural residues is being used for biofuel production. The reason behind is the diverse distinctiveness of the ligno-cellulosic materials. Three types of biopolymers are involved in lignocelluloses, *viz*., cellulose, lignin, and hemicellulose. These lignocelluloses components are associated with covalent linkages and the forces occurring between noncovalent bonds having a lesser amount of oils, starch, monosaccharides, and proteins (Sanchez 2009). Studies further illustrated that many processes are involved in converting residues obtained from agricultural fields into biofuel (Demirbas 2010; Lim et al. 2012).

In general, there are four ways by which crop residues are able to be used up (Figure 13.1). However, using processed residues obtained from agricultural crops for making biofuel has a high prospective. There are many benefits of using agricultural residues in the production of biofuel (Pattanaik et al. 2019). A major one is the utilization of agricultural residues that does not interfere with the production of food (Mohammed et al. 2018). However, its continued use as biofuel raw material will also help in serving a role towards reduced food prices. Virmond et al. (2013) had shown that for every one tonne of harvested grains, there is roughly one-tonne production of residue. The most common and important residues are obtained from crops like corn, rice, and wheat (Adhikari et al. 2018).

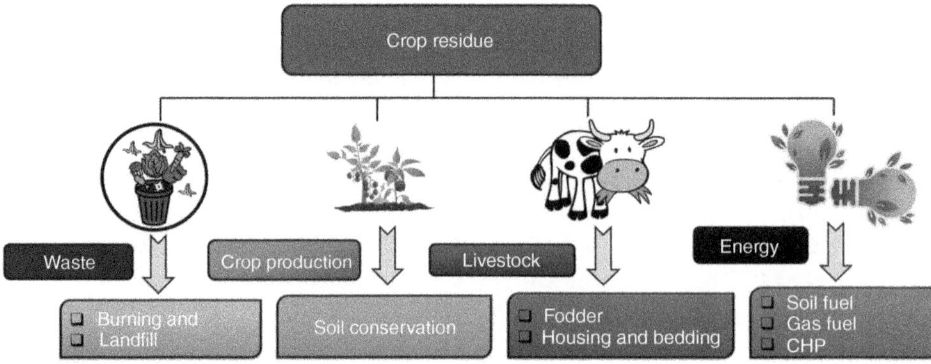

Figure 13.1 Different ways of utilization of crop residues.

13.3 Process of Conversion of Biomass Residue to Biofuels

There are many ways by which agricultural biomass residue is known to be transformed to biofuel; however, physical and chemical processes are broadly divided into two categories. Of these two processes, chemical processes are the most crucial ones. The chemical processes are further classified into two divisions, *viz.*, thermochemical and biochemical processes (Ahmad et al. 2018). Pyrolysis, gasification, and liquefaction are the three subprocesses involved in the thermochemical process (Neto et al. 2019).

13.3.1 Pyrolysis

An anaerobic atmosphere is necessary for heating of biomass at a temperature of 300–600 °C to obtain pyrolysis oil, gas, and charcoal (Antal 1983). The pyrolysis oil is crude-like liquid material. The product obtained is influenced by temperature, the sort of material used, and the process of treatment. This process can be further classified into slow, intermediate, and fast pyrolysis. The process was classified on the basis of the tenure of reaction time (DOE/EERE 2010; Jones et al. 2009; Wright et al. 2010). Gasification is a part of pyrolysis where combustible gas is the end product. The lengthy chains of C, H, and oxygen compounds present as the main constituent in biomass are degraded into smaller molecules under pyrolysis conditions (Zaman et al. 2017).

13.3.2 Gasification

In this technology, oxygen is allowed to pass through biomass for the production of a blend of carbon monoxide and hydrogen, besides carbon dioxide (Wu et al. 2010). This technology occurs through two processes, *viz.*, catalytic and noncatalytic methods. The two methods differ in temperature usage where the catalytic method involves high temperatures (1300 °C), while the noncatalytic method involves low temperature (900 °C) (Lee et al. 2007). Producer and synthesis gas (Syngas) produced in this process also differs from each other where producer gas is involved for stationary power generation, but the syngas is involved for fuels used in transportation and chemical intermediates (Drift and Boerrigter 2006). In the process of gasification, the lignocellulosic biomass resources pour into a gasifier. Obtained raw syngas is full of many contaminants, *viz.*, acid gas, ammonia, alkali metals, tars, etc. To remove contaminants from syngas, a cleanup process is essential to get pure syngas (Abdoulmoumine et al. 2015). Talmadge et al. (2013) illustrated that the pure syngas so obtained may be able to utilize by Fischer–Tropsch synthesis. By the use of methanol synthesis and water-gas-shift process, syngas, methanol, and hydrogen can also be produced (Bukhtiyarova et al. 2017).

13.3.3 Liquefaction

This process involves the use of a mixture of alkalis, glycerin, butanol, and propanol (Demirbas 2005). In this process, high viscosity oil and despise property produce, which is difficult to handle hence some organic solvents should be added to the reaction process. The organic solvents may be propanol, butanol, methyl ethyl ketone, etc. In this process, high viscosity oil and despise property are produced, which is difficult to handle hence some organic solvents like acetone, butanol, propanol, ethyl acetate, and methyl ethyl ketone should be added to the reaction system. The most favourable thing is that all these solvents are retained back after the process; however, ethyl acetate is completely consumed. The high viscosity oil can be used as a solvent to decrease the gases (Rowlands et al. 2008). By using appropriate catalysts, the transformation of lignocellulosic component to heavy viscosity fuel oil can be done. Demirbas (2004) reported that approximately 63% oil yield is obtained from liquefaction through catalyst, but the noncatalytic produces only 31% (Demirbas 2004). Aqueous liquefaction contains severance of wood ultra-structure and consequent depolymerization occurs in structural components. By the process of decarboxylation, the hydroxyl group produces ester in alkali liquefaction and a derivative of carbonate is produced. Alkali salts serve as a catalyst and liberate minor products during hydrolysis of cellulose and hemicelluloses.

13.3.4 High-Pressure Liquefaction (HPL) Process and Thermochemical Process

High-pressure liquefaction process is the way by which bio-oil is manufactured. The manufacturing involves the mixing of many volatile and nonvolatile organic compounds. The use of catalyst in this process helps in the higher production of oil (Demirbas 2004).

Thermochemical processes, a common process, are engaged in chemical control (direct combustion) and in the production of products at the industry level. The production of methanol belongs to this category. Plant biomass is used as the main source material for combustion instead of fossil fuel (Abdeshahian et al. 2010).

13.3.5 Biochemical Processes

Biochemical processes involved three processes: aerobic digestion, anaerobic digestion, and alcoholic fermentation. Aerobic digestion occurs in the presence of oxygen, where microbial aerobic metabolism causes biomass to produce heat and carbon dioxide emission. The aerobic digestion process is of utmost importance to the biological carbon cycle but is of no significance for bioenergy production (Silvestre et al. 2015). Anaerobic digestion occurs in the absence of oxygen as some microbes attain their own energy by reducing the carbon compounds for the production of carbon dioxide. Alcoholic fermentation involves the use of ethanol in place of refined petroleum and is produced by the action of microbes. This process is often termed fermentation.

13.4 Agricultural Residue Sources for Biofuel Production

In the twenty-first century, biotechnological approaches have been focused on identifying the sources for the production of biofuel from either industrial feedstock or from green processes (Stevens and Verhe 2004). The first generation of biofuel defines the biofuels produced from agricultural feedstock mainly edible ones like sugarcane, sugar beet, and soybean oil (Moore 2008). The carbohydrate feedstocks undergo a fermentation process to produce ethanol. Agricultural crops consisting of sugar, starch, and cellulosic biomass are the three groups in particular for ethanol production. These three groups are classified on the basis of carbohydrate sources. Sugarcane, wheat, and sugar beet are some of the agricultural crops consisting of sugar, while starch sources, grain, corn, and sweet sorghum are its feedstocks. Wood, agricultural residues, pine, etc., belongs to cellulosic biomass (Abdeshahian et al. 2010). The alcohol obtained from lignocellulosic biomass is known as biomass ethanol or in other words as bioethanol (Minteer 2006). The second generation defines the ones that are obtained from nonedible sources and in particular from lignocellulosic biomass (Figure 13.2). Goh et al. (2010) illustrated that the combustion of these biofuels helps in the emission of low carbon dioxide gas. This will help in the mitigation of greenhouse gases (Abdeshahian et al. 2010).

The biofuel production from agricultural residues obtained from crops is well known. Tahir et al. (2019) had reported the use of canola residue for biofuels. The biofuels so produced required two variable temperatures in the pyrolysis process. The higher temperature gave conversion of residue to biofuel (65–70% of weight), while low temperatures gave conversion of residue to biofuel (70% of weight). Nitsos et al. (2018) described the production of bioethanol from vineyard pruning, olive pruning, and almond shells. Antwi et al. (2019) revealed another crop residue for biofuel production. The cocoa pod residues

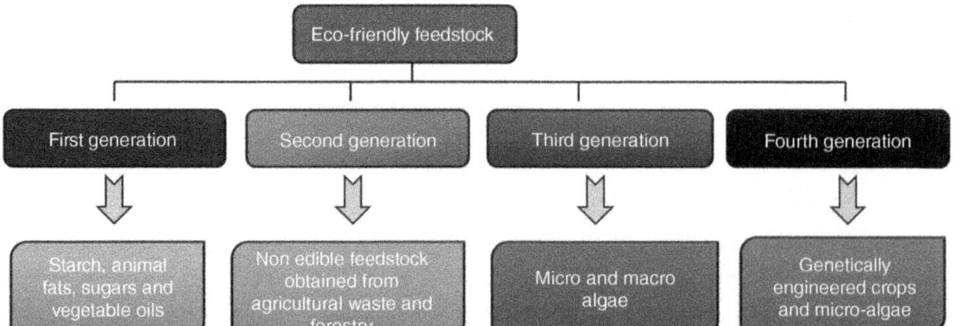

Figure 13.2 Feedstocks belonging to different generation for biofuel production.

undergo pretreatment *via* hydrothermal process (150–220 °C for 5–15 minutes, respectively) to produce biogas (5 261(N)/kgVS at 150 °C for 15 minutes). The prospective of drained pomace for biomass is also known through thermochemical process in which biochar is maximally produced at the end of reaction at 500 °C. Liu and Rajagopal (2019) described energy generation from different types of residues like agricultural, forestry, and municipal waste. The energy produced by these sources amounts to 2.4–3.2 EJ; however, there was the liberation of 103–178 million tons of carbon dioxide.

Sugar crops such as sugarcane, sugar beet are mostly used as sources for biofuel production are described below:

13.4.1 Sugarcane

Bagasse and molasses are the two surpluses of sugarcane produced in sugar industries. These residues are being known to be utilized in the number of other processes like in the production of electricity, paper, bioethanol, and also as an animal feed (Mohlala et al. 2016). Bagasse is defined as the fibrous residue remaining once the cane is crushed (Mbohwa 2013). Cellulose, lignin, and hemicelluloses make the cane bagasse a material of choice for biofuel production of second generation (Nsaful 2012). Due to the high calorific value it attains, sugarcane bagasse has a high potential for biofuel production (Mohlala et al. 2016). Ogwo et al. (2012) had revealed that one tonne of sugarcane stalk has the capability of generating 270 kg of bagasse. The difficulty while utilizing this residue for biofuel production is its low bulk density that causes problems during handling, storing, transporting, and processing. For easy transporting, bagasse in the form of pellet is easier for transportation and processing. Bioethanol produced from this feedstock is amongst the ones that are considered as an appropriate alternative for hydrocarbon fuel. This is so as it offers a renewable source for the production of biofuel and also contains fewer amounts of carbons than in gasoline (Awoyale and Lokhat 2018). Biofuel application in vehicles and other purposes helps not only in the reduction of pollution in the environment but also helps in mitigation of climate change by reducing the amount of greenhouse gases (Canilha et al. 2012).

During sugarcane processing, molasses is another product that even possesses the potential for biofuel production (Elbehri et al. 2013). The molasses constituent of 50–60% of sugar content (Ghorbani et al. 2011). The cane molasses consisting of 15% fermentable sugars can produce ethanol by 6% v/v in a continuous system (Raharja et al. 2019). This yield may be enhanced by 7.0% with an overall rise of 11% in fermentation efficiency by using acid clarified cane molasses medium (Sheoran et al. 1998). Hassib et al. (2013) described that the application of molasses for ethanol production is becoming eye catching for researchers due to their low cost, high yield, rich sucrose content, and even does not require many pretreatment facilities as in the case of other materials.

13.4.1.1 Processing of Sugarcane to Biofuel Production

Quality sugarcane goes through standardized analysis for processing. The possibility of ethanol production from cane bagasse has increased after its utilization to produce electricity. Keeping this point in view, some of the new varieties had been produced by BioVertix® that can produce more fibre than sugar (Castro et al. 2018). Sugarcane transport and mechanization add dirt to it, so the cleaning process is very necessary. If sugarcane is harvested manually, washing it is possible before sending it to shredder or cutting process in the industry. In present days, mechanization processes are in progress and because of this pieces of sugarcane come in factories. Loss of sugar has been recorded from cutting edge pieces of sugarcane, hence it cannot be washed. To avoid such losses, many industries apply dry cleaning technology to cast aside dirt particles.

Ahmed and Alam-Eldin (2015) reported that the increase of 1 kg dirt/100 kg of sugarcane is expected 0.1% loss of sugar recovery by the processes of industry. This technology consumes about 0.5–1.0 kW per tonne of sugarcane (Castro et al. 2018). After dry cleaning, next process is extraction, which separates fibre and solid stream from the liquid stream. In the cane-conveying system, rotating knives and swing hammers are applied that is why size reduction takes place.

As mechanically harvested billeted cane is already in pieces so there is no requirement to cut it again and directly fed it into the shredder. Juice extraction is done by milling and diffuser (Bartens and Rein 2007). The diffuser is an additionally improved technique that extracts greater sugar, but more saturated water and steam are only two main drawbacks observed in milling. Hydrodynamic extraction, also known as 'Rivière Juice Extractor', is a new technology introduced to improve the extraction process to decrease the use of imbibition water and lower the cast (Walsh 1998).

13.4.1.1.1 Juice Treatment

The juice extracted after milling contains so many impurities. It is necessary to separate these impurities before the concentration and fermentation process. The separation process is done through maintaining pH, heating, setting the precipitate of juice, and filtration process (Castro et al. 2018). The heating process is done by vertical shell-and tube steam heaters in most industries. The pH of obtained juice after treatment reaches up to 5.3 that requires balance. To maintain pH at 7, lime $Ca(OH)_2$ is added to the juice, and the refineries produce sugar with ethanol. Sulfitation, phosphatization, and carbonation processes are used to lower turbidity and colour (Bartens and Rein 2007). As the pH adjusts at 7, the juice is transferred into a clarifier to settle down the insolubilities. Multitray clarifier's installation units are mostly used to separate the impurities without using chemicals. Only the disadvantage of this tray is a long duration of three hours, although single-tray clarifiers also known as rapid clarifiers take a single hour, however, in this process, the use of chemicals (flocculants and polymer) is necessary that promotes mud coagulation and settling. The advantage of single-tray clarifiers is their lower installation cost and less withholding time that prevents degradation of sugar.

13.4.1.1.2 Juice Concentration

For juice concentration, attached and autonomous distilleries are used. From the attached distilleries, we get sugar and ethanol both but autonomously produce only ethanol. After production of sugar from juice, it goes through a concentration process by using multiple-effect-evaporators (MEVs) to obtain suitable Brix, and crystallization process takes place which is having 60% sucrose by weight. MEVs produce lower pressure steam and reduce the steam requirement so that juice becomes concentrated (Castro et al. 2018).

Ethanol can be extracted from molasses (a residue of sugar production), the amount depends on the quality of molasses mixture and juice. The rest evaporated steam is used from industrial condensation turbines to save bagasse for second-generation ethanol extraction or for only selling bagasse as a product. For crystallization of sugar, water withdrawal is necessary, total consumption of steam is also required, hence, the steam is transferred into MEV and used in the production of ethanol. The advantage of this process is the exhaust steam requirement of MEV. The combination of sugar and ethanol production together results in an energy-efficient refinery.

13.4.1.1.3 Pretreatment of Cane Biomass (Sugarcane Bagasse/Straw)

The lignocelluloses of the cane bagasse constitute many polymers, of which the three chief ones are (i) lignins, enclosed in the cellulosic part of cell walls (ii) hemicelluloses, present on cell wall; and (iii) microfibrils that enclose cellulose. Lignins provide rigidity to the cell wall. Hemicellulose and microfibrils both confer strength to the cell wall. The strength given by hemicelluloses is by the lignin cellulose bonding while microfibrils give it in the tensile manner (Sadhukan et al. 2014). Polymers of guaiacyl, para-hydroxyphenyl, and syringyl alcohol are the chief lignin compounds (Hernández et al. 2017). The lignocellulosic content in the bagasse is the raw material for ethanol production (Sabiha-Hanim and Halim 2018). Hemicelluloses were the first component released from the pretreatment process of sugarcane bagasse as hydrolysis of hemicelluloses occurs at a quicker pace. This causes the separation of two other components from bagasse, *viz*., lignin and cellulose. The pretreatment of celluloses releases glucose by breaking down the β-(1,4)-glycosidic bond (Martinez-Hernandez et al. 2017).

The sugarcane bagasse or straw first undergoes pretreatment following different methods as revealed by several studies. Chandel et al. (2007) stated that the specificity of the method varies from method to method depending on the mechanical treatment of the cell wall components and the conditions required for doing so. For the pretreatment of lignocellulosic biomass, the following condition is important and ideal for the process to occur properly (Holtzapple and Humphrey 1984; Margeot et al. 2009):

1) There should be a rise in the available surface area.
2) Decrystallization and partial depolymerization of cellulose should occur.
3) Hemicellulose and lignin solubilization should take place.
4) Modification in the structure of lignin is essential.
5) The pretreated material should undergo maximum enzymatic digestibility.
6) There should be a minimum loss of sugars at a minimum rate.
7) The cost involved in this process should be minimum.

Majorly four types of pretreatment, in the case of sugarcane processing for bioethanol production, are commonly being in practice. These are mechanical, chemical, physicochemical, and biological.

13.4.1.1.3.1 Mechanical Pretreatment In this method, to improve the surface area of the material, the size of the cane bagasse or straw (the raw material for ethanol production) is reduced (Mtui 2009). There are various methods by which mechanical/physical pretreatment can be done. Some of these are mentioned briefly later.

i) ***Milling:*** Sun and Cheng (2002) had illustrated that during the process of milling, the lignocellulosic component of bagasse and straw were broken down and the cellulose crystallinity was reduced. The ball milling method is a frequently used eco-friendly milling method (Inoue et al. 2008). As the names suggest, balls are involved in this method and no chemical is used in the process. The contact of biomass with the balls in a cyclic device causes a reduction in the size of the particles (da Silva et al. 2010). The advantage of this method is no generation of inhibitors (da Silva et al. 2010) while the drawback is it requires high power energy that results in high costs. Hideno et al. (2009) had illustrated that in the case of sugarcane bagasse processing for ethanol production, the number of cycles involved is more resulting in a longer time duration for the process to complete.

ii) ***Pyrolysis:*** In this method, degradation of cellulosic content in the bagasse occurs when it is heated at high temperatures (>300 °C) releasing hydrogen carbon monoxide and char as products (Kumar et al. 2009). When char is separated, the solution obtained contains glucose as a major component. This glucose then undergoes fermentation for ethanol production (Sarkar et al. 2012). Acikalin et al. (2012) showed that two types of reactions occur in this process, *viz.*, primary pyrolysis reaction and secondary pyrolysis reaction. High temperatures are the main requirement for primary pyrolysis reaction to begin that causes the release of volatiles which later on undergoes condensation process. After condensation, secondary pyrolysis reactions were initiated (Acikalin et al. 2012). Agblevor et al. (1995) showed that the yield and quality of the products obtained so depend on many factors, one being feedstock properties and the other being the parameters associated with processing (Acikalin et al. 2012).

13.4.1.1.3.2 Chemical Pretreatment The hydolysate obtained after pretreatment (using dilute acid) contains high content of hemicelluloses (Hernåndez et al. 2017). These hemicelluloses were degraded by chemicals, and lignin was also removed. This causes the breakdown in the bonding of lignin and hemicelluloses in the structure of the bagasse and straw (Canilha et al. 2012). The acid used in the process was hydrochloric acid (Hernåndez-Salas et al. 2009), sulphuric acid (Gomez et al. 2010; Dussan et al. 2014; Timung et al. 2016; Soares et al. 2017), nitric acid (Rodríguez-Chong et al. 2004), phosphoric acid (Gámez et al. 2006; Gomez et al. 2010), but sometimes alkalies like NaOH was also used for pretreatment (Gámez et al. 2006; Hu et al. 2017). The use of alkali for pretreatment is termed alkali pretreatment while in acid use, it is termed acid pretreatment (Sabiha-Hanim and Halim 2018).

1) ***Alkali pretreatment:*** The alkali treatment on cane bagasse was revealed to be efficient for the production of ethanol as it helps in breaking the linkages present in the structure of fibre between various molecules (Sabiha-Hanim and Halim 2018). Kaur et al. (2020) illustrated that the use of NaOH as alkali had the highest yield obtained for ethanol, while calcium hydroxide usage in alkali treatment also does not lack much behind in the race of high ethanol yield. Carvalho et al. (2019) showed that pretreatments with alkalies (NaOH) also favour glucose production as NaOH help in the swelling of the fibres and also in rising their hydrophilicity. This in turn causes fibres to be more susceptible to reagents involved in hydrolysis. The result of which is more production of total reducing sugars (Carvalho et al. 2019).
2) ***Acid treatment:*** The acid treatment is of two types, *viz.*, dilute and concentrated acid treatment. Of which dilute acid treatment is the most frequently used (Barrera

et al. 2016). This is so due to certain drawbacks of acid treatment like the formation of by-products (like furan (Jiang et al. 2013)) that inhibit the production of glucose (Sabiha-Hanim and Halim 2018) and loss of such sugars that can undergo fermentation increased the costs of acid and the extra cost for its neutralization (Kaur et al. 2020). The degradation of hemicellulosic content in the substrate was seen to be easily degraded in dilute acid pretreatment. This results in rising in biomass porosity and augmenting enzymatic hydrolysis of cellulose. This process causes the glycosidic bond present either in two hemicelluloses or two lignins or between lignin-hemicellulose to break down. In doing so, sugar gets dissolved in hemicelluloses. Furthermore, a rise in cell wall porosity was also seen that will later help in the enzyme digestibility (Jiang et al. 2013). The dilute acid pretreatment with sulphuric acid on lignocellulosic biomass gave high yields after hydrolysis in comparison with other acids (nitric, phosphoric, and hydrochloric acid) (Alvira et al. 2010). The concentration of acid used in dilute acid pretreatment was 0.5–0.6% and the temperature ranges from 120 to 170 °C (Sabiha-Hanim and Halim 2018).

Concentrated acid treatment requires a low temperature (30–60 °C) with an acid concentration of 40–80%. The process yields high sugar yields, but the drawbacks of this process were so severe that generally this process is not being carried out. The use of a large amount of acid, its high toxicity, and its corrosive nature are some of the main drawbacks. Due to its corrosive property, the reactors used in the process should be resistant to corrosion. Even the concentration of acid used in the process should recover back so that the process becomes economical (Sun and Cheng 2002).

13.4.1.1.3.3 Physiochemical Pretreatment In this method, an essential requirement is the controlling of the conditions required for the proper working of this method. This is so as the reactions occurring in this method involve high temperature and pressure (Taherzsadeh and Karimi 2007). Steam explosion and liquid hot water treatment are two of its method. The temperature requirement in the steam explosion and liquid hot water treatment is approximately 180–195 °C and 170–200 °C, respectively. The physicochemical treatment by any means (steam explosion (You et al. 2016) and liquid hot water) enhanced the cellulosic content in the sugarcane bagasse. The increase in cellulose content is associated with lignin removal in both methods (Sabiha-Hanim and Halim 2018). Gurgel et al. (2016) reported that by the application of liquid hot water treatment the glucan content in the material was properly maintained.

13.4.1.1.3.4 Biological Pretreatment This pretreatment is the most effective, ecofriendly, and cheap of all the pretreatments (Wan and Li 2012). The role of cellulolytic and hemicellulolytic microbes was revealed in the method where lignocellulosic biomass was treated with them. Chandel et al. (2007) revealed that lignocellulosic biomass is dignified in this method. Vats et al. (2013) had shown that the most frequently being used microbe in this method is the filamentous fungi. These fungi are isolated from the soil. Kumar et al. (2009) illustrated that the highest efficient microbe for this method is white-rot fungi. These fungi help in the breakdown of lignin by the enzyme secreted by them. The enzymes are peroxidases and lactases that belong to lignin-degrading enzyme category (Kumar et al. 2009). Brown, white, and soft rot fungi were also known to be used in this method. The brown fungi degrade cellulose while the other two degrade cellulose and lignin, both. Cost-effectiveness, low needs of energy, no use of chemicals, and mild conditions of the environment are some of its benefits. Substantial carbohydrate loss, longer time periods for incubation, low efficiency, and restrain in space are some of the chief drawbacks of this method (Zheng et al. 2009). Jiraprasertwong et al. (2014) showed that pretreatment of bagasse with white-rot fungus had a maximum concentration of glucose. Hernández et al. (2017) revealed that pretreatment of bagasse with *Pycnoporus sanguineus* had an enhanced rate of breakdown of lignin content, higher liberation of glucose.

13.4.1.1.4 Fermentation Process

Cleavage of the glycolic bonding present in between the molecules of glucose and fructose is the primary step of alcoholic fermentation (Castro et al. 2018). The fermentation requires the use of yeast for conversion of the product obtained after pretreatment to bioethanol. *Saccharomyces cerevisiae* is the most common yeast used in this process followed by several others like *Kluyueromyces* sp., *Schizosaccharomyces pombe,* and *Saccharomyces uvarum.* The anaerobic condition required in this process stimulates yeast to metabolize the glucose into ethanol. The process used in doing so is termed as Embden-Meyerhof pathway. A certain quantity of oxygen is necessary to the yeast for causing the reaction to occur. The oxygen content present in the broth is 0.05–0.10 mm Hg oxygen tension. Besides, very small amounts are required of minerals and organic factors in this process (Badger 2002).

When this process gets completed, the fermented solution undergoes centrifugation so that the removal of yeast may take place. The yeast so removed is recycled back for further fermentation of the products obtained from pretreatment. The product obtained after the fermentation process is thereafter sent for purification process (Castro et al. 2018).

13.4.1.1.5 Purification

Purification processes take place via distillation and dehydration methods.

13.4.1.1.5.1 Distillation Method The distillation method is further divided into two methods, i.e. batch or continuous method. Both the methods are almost similar with heating as the main principle. The use of which method is dependent on the plant type.

i) ***Batch system:*** In this, the complete product obtained after fermentation was sent to a vessel of large size known as still. The mash present in the still was heated, and vapours were made to pass into the distillation column. The operation or processing of a batch system is far much easier than a continuous feed distillation system, but this method is less effective (Ayhan 2005).
ii) ***Continuous feed distillation system:*** In this, the complete product obtained after fermentation was heated in the distillation column at a controlled rate. The unprocessed mash was again returned back to the distillation column through the aperture present on the top of the column. This method is effective that surpasses the drawback of complexity in the system (Ayhan 2005).

13.4.1.1.5.2 Dehydration This method is required for the removal of water present in the ethanol obtained after the distillation process. The amount of water present is 5%. This step is basically necessary when ethanol needs to be blended with gasoline. The phase separation in storage tanks is the resultant of water in the ethanol-gasoline blend. If ethanol is used in places as an alternative to gasoline then, in that case, this step is not required. When the ethanol (95%) is heated for further purification, the quantity of water present in the vapour phase is the same as in the liquid phase. But the difference is that in the liquid phase the mixture becomes azeotropic. Due to this, the process of simple distillation is not so effective in purifying ethanol. The role of a small quantity of benzene does wonders in distilling the complete amount of ethanol. The benzene so added may also be recovered. The production cost of ethanol is known to be high due to the large amounts of energy involved in producing ethanol (Badger 2002; Ayhan 2005).

13.4.2 Sugar Beet

Sugar beet is a rich source of ethanol production (Misra et al. 2020). The sugar beet pulp contains arabinose, rhamnose, galactose, and ferulic acid (Kühnel et al. 2011), which is dietary fibre, but it can also be used as an ethanol ingredient or for lactic acid fermentation coupled with various catalytic processes for obtaining valuable bioproducts and production of biogas and single-cell protein. Alcohol that is produced from sugar beet can serve as superior gasohol (Bowen et al. 2010). Sugar beet is considered highly capable of converting sunlight into stored energy. The interesting part of this crop is particularly is its high productivity and high carbohydrate content (61%). The highest yield of ethanol (alcohol) from sugar beet is up to 4400 l/ha/year (Rodriguez et al. 2010). The process of pulp to ethanol production has been reported to be much cheaper and less complicated. Methanol and ethanol have advantages over liquid hydrocarbon fuels. These are clear fuels, and their use will reduce the air pollution caused by gasoline-run transport and industry. This will help in maintaining a better and clean ecosystem and solve habitat pollution. The use of raw materials as biofuels gives double credit to increase farmer's income.

The various impurities contained in the raw juice after the extraction of sugar are used for bioethanol production. In general, fermentable sugars (hydrolysis); fermentation of simple sugars to ethanol; and separation and purification are the steps in ethanol processing (Joanna et al. 2018).

13.4.2.1 Processing of Sugar Beet to Biofuel Production

During the sugar processing from sugar beet, pulp and molasses are the two waste materials that act as fermentation substrates for biofuel production (van der Poel et al. 1998; Mall et al. 2021). The sugar beet pulp, molasses, and all by-products from sugar production can be further processed. By the process of hydrolysis, sugar beet pulp produces yeast biomass and other valuable products. Furthermore, by using some sophisticated and innovative technologies, propylene glycol production is the main constituent of by-product.

The description of the ethanol production process from sugar beet pulp and molasses has been briefly mentioned later:

13.4.2.1.1 Sugar Beet Pulp

The sugar beet pulp is made up of hemicelluloses, cellulose, lignin, and pectin and is insoluble in water. The constituents present in the sugar beet roots are the reason behind the role of feedstock in ethanol fermentation. Many by-products formed by the processing of sugar beet can be used to produce biofuels, i.e. bioethanol (Anasontzis et al. 2016), biobutanol (Zheng et al. 2015), biomethane (De Vriezea et al. 2014), and biohydrogen (Abbasi and Abbasi 2011). The production of biofuel occurs by fermentation either in aerobic conditions or in an anaerobic condition wherein the use of microbes was involved. The production of biogases occurs during the fermentation process; however, the production of alcohols occurs after fermented broth distillation. The obtained sugary juices can be used to produce many essential products like organic acids (butyric) described by Zigová et al. (1999), succinic (McKinlay et al. 2007), itaconic (Klement and Bachs 2013) lactic acid (Hofvendahl and Hahn-Hagerdal 2000), fumaric acid (Roa Engel et al. 2008) malic acid (Battat et al. 1991), muconic acid (Curran et al. 2013), triacylglycerols (Kosa and Ragauskas 2011), and polyhydroxyalkanoates (Du et al. 2001). The cost involved in the production of sugar beet can be made up by the revenue obtained by selling the by-products obtained.

After the sugar is extracted from beet roots, fibrous waste is left out that is termed as sugar beet pulp. For the production of biofuel, the pulp undergoes pretreatment either by chemical or enzymatic means. For rapid and complete enzymatic hydrolysis, biological, thermal, and chemical pretreatments are the various compulsory steps along with high cellulolytic activities of enzymes used (Kracher et al. 2014). Enzymatic hydrolysis is reported as a common process for the treatment of agricultural biomass. In sugarbeet pulp, hemicelluloses, cellulose, and pectin are the three main carbohydrates. Zheng et al. (2012) illustrated that when cellulase and pectinase were used for the treatment of the pulp, the saccharification process was much highly effective. The application of the pectinase enzyme has a strong impact on the galactose content of the pulp that causes the liberation of 91.3% of its entire content present in the pulp. Furthermore, the degradation of hemicelluloses to arabinose and cellulose to glucose has also been reported as being increased by the usage of pectinase on the sugar beet pulp (Rezic et al. 2013). The use of pectinase on beet pulp is known for the production of ethanol. The deadly combination of pectinase with cellulose or β-glucosidase had shown a strong impact on the degradation of the polymeric structure of sugar beet pulp into its smaller constituents (Zheng et al. 2012). Several studies reported the use of many other enzymes for enzymatic hydrolysis of sugar beet pulp; however, significant and effective results were seen when pectinase and cellulases were added (Foster et al. 2001; Zheng et al. 2012; Zieminski and Kowalska 2015).

The process of enzymatic hydrolysis is the process where the sugar liberated was further used as ethanol raw material that utilizes micro-organisms. When lignocelluloses were used as a substrate for the process of fermentation, many difficulties were seen to occur. The polymeric compounds were dehydrated and the production of fermentative sugars occurs. The involvement of microbes in the fermentation process helps in the effective conversion of hexoses and pentoses into by-products. But certain bacterial and yeast strains available commercially cause difficulties in the assimilation of pentose (Joanna et al. 2018). Zheng et al. (2012) showed that arabinose or galacturanose was not able to convert into ethanol by the use of *Saccharomyces cerevisiae* available commercially. Xylose and hemicelluloses may also be constituted of the sugars produced from the lignocellulosic hydrolysis that causes the requirement of such strains that can help in improving the effectiveness of the fermentation process (Joanna et al. 2018). Yücel and Aksu (2015) reported the use of *Pichia stipitis* for ethanol production from beet pulp.

13.4.2.1.1.1 Pretreatment Process

13.4.2.1.1.1.1 Biological Hydrolysis The decomposition of biomass for the use of biorefining industries takes place by the process of biological ensilage. The raw sugar beet pulp enzymatic digestibility is up to 71%, whereas the ensiled pulp has 74–95% enzymatic digestibility (Zhang et al. 2011). The cost of enzymes required is somewhat high priced.

13.4.2.1.1.1.2 Thermal Hydrolysis Singh et al. (2015) illustrated that two favourable environments, viz., acid or alkaline condition, in combination with other factors like temperature, time, and dry mass amount, were required for thermal hydrolysis, a common process of pretreatment. Many studies illustrated that when the pulp undergoes the hydrothermal pretreatment process, lignin content in the pulp was partially removed. This causes an apparent rise in pore measurements (Mosier et al. 2005; Singh et al. 2014). Mosier (2013) showed that even hemicellulosic content has deteriorated after the hydrothermal pretreatment process. Studies also reported alteration in the structure of sugar beet pulp after hydrolysis (Zheng et al. 2013; Yücel and Aksu 2015). In the pretreatment process of sugarbeet pulp, lignin removal is essential. However, this removal needs care as hemicelluloses should not be broken down during the process. It is known that when the pulp undergoes $Ca(OH)_2$ treatment, the pulp becomes amorphous that causes difficulty in identifying the fibres (Dredge et al. 2011). However, Foster et al. (2001) revealed that using the ammonia pressurization depressurization method during the pretreatment process causes improvement in the pulp structure that helps in enhancing the effectiveness of hydrolysis (particularly of cellulosic content). Contrastingly by using enzymes degrading hemicelluloses and pectin, there was no difference in the hydrolysis efficiency that may be due to alteration in the structure of pectin and hemicelluloses (Foster et al. 2001). Zheng et al. (2013) revealed that a combination of acidic concentration with temperature plays an important role in the removal of hemicellulosic content in the pulp but in the case of cellulose solubilization, the essential part is played by solid loading.

13.4.2.1.1.1.3 Chemical Pretreatments The use of dilute acid (such as dilute nitric acid, dilute sulphuric acid, and dilute hydrochloric acid) during the pretreatment process is well known (Ninju et al. 2020). The pretreatment with these chemicals helps in the breakage of the pulp structure (especially the polymeric bonding) resulting in the improvement of the hydrolysis process and even in the yield obtained after the fermentation process (Zheng et al. 2013). Chemical pretreatments are applied to reduce the number of enzymes required to achieve significantly fermentable sugar yields (Zheng et al. 2012). The use of sulfurous acid as a pretreatment for beet pulp helps in obtaining high effectiveness in both essential processes, namely, enzymatic hydrolysis and carbohydrate extraction. The regeneration capability of this chemical was the main reason for using it for pretreatment purposes. The use of sulfurous acid helps in obtaining 33.6% of carbohydrate (9.4 g/l arabinose and 7.3 g/l glucose) from sugar beet pulp pretreatment (Kharina et al. 2016). The best and commonly used chemical hydrolysis method is the acidic hydrolysis where the use of acids causes partial hydrolysis of cellulose and completion of hemicelluloses. The concentrated solution of acids causes severe issues like corrosion, production of toxic and harmful compounds that inhibit the process to occur (Joanna et al. 2018).

13.4.2.1.1.2 Other Pretreatment Methods

13.4.2.1.1.2.1 Hydrothermal Pretreatment Hydrothermal pretreatment is another way where a combination of high temperatures and pressure plays a role in proton release by autoionization of water. The process also causes a reduction in pH of the solution. Treatment with hot water is an efficient way in sugar beet pulp that involves about 40–60% of the entire biomass to be solubilized. Reduced amount of reagent requirement, lower cost in comparison to other pretreatments, and lesser production of fermentation inhibitory compounds or degrading products are some of the benefits of this treatment (Zieminski et al. 2014).

13.4.2.1.1.2.2 Mild Hydrothermal Pretreatment The mild hydrothermal pretreatment method involves a temperature of 140 °C with a blend of enzymes responsible for the digestion process. This helps in increasing the efficiency of hydrolysis of beet pulp. By this method, almost the entire cellulosic content in the pulp was known to be hydrolyzed within a day. This method has several benefits with respect to the chemical pretreatment method (Kühnel et al. 2011).

13.4.2.1.1.2.3 Steam Explosion Method Singh et al. (2014) showed that during this process, heating of particles of biomass was required with the help of high pressure saturated steam (160–260 °C; 0.7–4.83 MPa) for a small duration (seconds or a few minutes) after which the pressure was released to stop the reactions occurring inside. This results in the decompression of the biomass in a sudden manner. The hemicellulosic content of sugar beet pulp undergoes hydrolysis and produces acids. These acids might help in further hydrolysis of the cellulosic content present in the pulp by increasing the availability of cellulosic content (Joanna et al. 2018).

13.4.2.1.1.3 Fermentation Process for Pulp

The fermentation process needs the use of ethanol fermenting yeasts as required in the sugarcane fermentation process. Grohmann et al. (1994) reported that *Saccharomyces cerevisiae* is not able to metabolize two components, i.e. arabinose and galacturonic acid, of beet pulp into ethanol. Doran et al. (2000) showed that those bacteria that were genetically engineered were able to ferment the carbohydrates (arabinose and galacturonic acid) that *Saccharomyces cerevisiae* could not. These bacteria are *Escherichia coli* KO 11 (with the highest ethanol yield of 25.5 g/l), *Klebsiella oxytoca* P2, and *Erwinia chrysanthemi* EC 16. A comparison of fermentation processes was made where *E. coli* KO11 and *S. cerevisiae* (Type II -YSC2) were used for the conversion of beet pulp to ethanol. Results revealed that *E. coli* KO11 had maximum ethanol content (0.144 g ethanol/g) and maintained its superiority over *S. cerevisiae* (0.092 g ethanol/g) (Rorick et al. 2009).

13.4.2.1.2 Molasses

By the process of trans-esterification process, we can get biodiesel from molasses. In this process, utilization of ethanol and large triglycerides with small straight chain molecules react in the presence of catalyst.

13.4.2.1.2.1 Pretreatments For increasing the concentration of monomeric sugars during the process of hydrolysis, pretreatment is required to degrade the cellulosic content, in such a way that there should be lesser losses in sugar content and reduced production of products that can be further degraded (Bellido et al. 2015). Furthermore, it is important that the pretreatments done by the enzymes should be effective to such an extent that the bonding of cellulose with hemicelluloses in sugar beet molasses and cellulose with lignin may be damaged or gets broken down (Kühnel et al. 2011). As a concept, Kühnel et al. (2011) revealed that the amount of production of nonfermentable or inhibitory components is dependent on the severity of the pretreatment process. The beet molasses are alkaline in nature and a huge quantity of inorganic salts is present in it. The presence of these salts causes difficulties in pH adjustment.

13.4.2.1.2.2 Fermentation Process The fermentation process of the rest of the uncrystallized syrup residue (termed as molasses) is a conventional and extensively considered method of waste streams treatment, as well as for ethanol synthesis. Primarily, the molasses are firstly diluted to a particular sugar concentration and pH is also adjusted (Dodić et al. 2009). The difference in the application of fermentation methods was seen to have a strong impact on associated parameters required for the production of ethanol from beet molasses (Zayed and Foley 1987; El-Refai et al. 1992). El-Refai et al. (1992) found that with the increase in sugar concentration from 200 g/l to 250 or 350 g/l, the ethanol yield obtained was reduced. This may be due to the effect of a high concentration of sugar on yeast that caused plasmolysis. Zayed and Foley (1987) reported that with the changes in the usage of yeast strains in the fermentation process of beet molasses, the production of ethanol content was seen. This is due to the variation in sugar concentration. Dodić et al. (2009) and El-Refai et al. (1992) revealed that higher sucrose concentration in beet molasses, more will be the ethanol yield. Studies illustrated the usage of immobilized yeast cells for ethanol production from beet molasses (Roukas 1996; Goksungur and Zorlu 2001). Roukas (1996) revealed that when immobilized yeast cells were used, the highest ethanol yield was achieved when the concentration of sugar was 250 g/l. Goksungur and Zorlu (2001) stated that a reduction in ethanol yield was observed when sugar concentration was enhanced. The ethanol yield was further increased by the addition of urea (1.2 g/l), phosphorus (0.4 g/l), and magnesium (0.3 g/l). The presence of nitrogen in the sugar beet molasses seems not to be available to yeast for the fermentation process (Zayed and Foley 1987). It does not imply that the increase in nitrogen concentration is responsible to increase the yield of thanol. Yield of ethanol increase is a result of increase in pH value of solution because concentrated nitrogen increases its value. This rise does not allow the yeast to grow at high pH (Zayed and Foley 1987). Furthermore, the addition of selenium during the fermentation process of molasses does not have significant results on the ethanol yield obtained (Faramarzi et al. 2019). Mikulski and Klosowski (2021) proposed that the application of hydrolysate with dilute sulphuric acid after the distillery stillage treatment may help in achieving benefits from the concentration of fermentable sugars (available in mash form). This will further help in obtaining a good amount of alcohol through the fermentation process.

It is well known that many strains and forms of *Saccharomyces cerevisiae* were involved in ethanol production during the process of fermentation. Ranković et al. (2009) showed that five different strains of yeast were effective in utilizing the sugar content present in the beet molasses to ethanol production. Dodić et al. (2009) illustrated that using different strains of yeast (yeast for beverage production; yeast for wine; and bakery yeast), the ethanol yield ranges from as low as 0.485 to as high as 0494 g/l. Usage of wine yeasts reported having the maximum ethanol yield. Furthermore, no association of different yeast forms (dried and pressed blocks) on ethanol content was also observed (Dodić et al. 2009).

13.4.2.1.3 Purification Step for Ethanol Production

The fermentation products obtained after the pretreatment step of beet pulp and molasses were made to undergo purification. The purification step requires the removal of other unwanted by-products. The volatile by-products are much more difficult to remove from ethanol. For doing so, there are three methods, distillation, adsorption, and ozonation. Of all, the most common process is distillation.

13.4.2.1.3.1 Distillation Distillation is the separation method wherein the volatile components present in the mixture are easily differentiated by heating the mixture. As per its principle, the components in the mixture having a low boiling point were concentrated in the vapor. This vapor, later on, was allowed to condense in the liquid phase to obtain less volatile components. Though this method is commonly used yet its high cost (due to recurring steps of vaporization and condensation) and lodging of impurities having a similar boiling point as that of ethanol causes difficulties in ethanol extraction (Onuki et al. 2008).

13.4.2.1.3.2 Adsorption This separation method involves the use of adsorbent and its characteristics. The products obtained after pretreatments were made to be adsorbed into the adsorbent. The large particles having low diffusivities and alike polarity were adsorbed on the adsorbent. For ethanol purification, nonpolar surface and extensive pore distribution range were recommended. This is so as ethanol belongs to the polar compounds category. The mixture obtained for ethanol purification may at times constitute impurities. Activated carbon and alumina are the most commonly used adsorbents (Demirbas et al. 2008; Tripathy and Raichur 2008).

13.4.2.1.3.3 Ozonation The ozonation method involves the use of ozone (which contains three atoms of oxygen) for a breakdown of many compounds due to its oxidation potential. This breakdown causes alteration in the physiochemical properties of the compounds. Bailey (1982) revealed that oxidation of ethanol may also be performed, but pure ethanol was not obtained under atmospheric conditions. For this involvement of ozone came into play where removal of impurities occurs without any damage to ethanol. Though this method is used for ethanol purification yet it has disadvantages. The production of new by-products was required to be removed after the ozonation process gets completed (Onuki et al. 2008).

13.5 Conclusion

Global warming is the result of no restriction on the use of fossil fuels as carbon dioxide is produced from its combustion, which increases greenhouse gas, particularly CO_2. This has caused interest in renewable energy sources like bioethanol and biofuel. Agricultural residues as an essential ingredient for biofuel production have been paid attention in recent past years due to environmental issues. The high content of lignocellulosic fibre, easy availability, and ready-to-use property made agricultural residues some of the finest feedstock for biofuel production. A large number of processes are involved in the conversion of agricultural residues to biomass-based energy. Sugarcane bagasse, molasses, beet pulp, cocoa pod residues, canola residue, etc., are some of the agricultural residues from which ethanol is being produced by many processes like fermentation and hydrothermal process.

Lignocellulosic biomass as an agricultural biomass residue is gaining importance and recognition as a source of fermentable sugars. Biotechnological studies are being focused on searching potential feedstocks for the production of bioethanol with eco-friendly properties. The process of enzymatic treatment subsequent to lignocellulosic biomass is much more costly and therefore there is a need for wide-ranging research for finding a solution to this important problem. Lignocellulosic biomass as a raw material for the production of bioethanol and biofuel is cost-effective and environmentally safe. With respect to the high needs of second-generation biofuel production, this material reveals to have great potential and promising raw material.

References

Abbasi, T. and Abbasi, S.A. (2011). Renewable hydrogen: prospects and challenges. *Renewable and Sustainable Energy Reviews* 15 (6): 3034–3040.

Abdeshahian, P., DAshti, M.G., Khalil, M.S., and Yusoff, W.Y.M. (2010). Production of biofuel using biomass as a sustainable biological resource. *Biotechnology* 9 (3): 274–282.

Abdoulmoumine, N., Adhkari, S., Kulkarni, A., and Chattanathan, S.A. (2015). A review on biomass gasification syngas clean up. *Applied Energy* 155: 294–307.

Acikalin, K., Karaca, F., and Bolat, E. (2012). Pyrolysis of pistachio shell: effects of pyrolysis conditions and analysis of products. *Fuel* 95: 169–177.

Adhikari, S., Nam, H., and Chakraborty, J.P. (2018). Conversion of solid wastes to fuels and chemicals through pyrolysis. In: *Waste Biorefinery Potential and Perspectives* (ed. T. Bhaskar, A. Pandey, S.V. Mohan, et al.), 239–263. Elsevier.

Agblevor, F.A., Besler, S., and Wiselogel, A.E. (1995). Fast pyrolysis of stored biomass feedstocks. *Energy and Fuels* 9 (4): 635–640.

Ahmed, A.E. and Alam-Eldin, A.O.M. (2015). An assessment of mechanical vs manual harvesting of the sugarcane in Sudan – the case of Sennar sugar factory. *Journal of the Saudi Society of Agricultural Sciences* 14 (2): 160–166.

Ali Ahmad, S., Karbassi, A.R., Ibrahim, G. et al. (2018). A review of the production and upgrading of biofuel; raw materials, processes and products. *Journal of Energy Resources and Conversion* 1 (1): 102–116.

Allakhverdiev, S.I. and Ramakrishna, S. (2015). A random walk to and through the photoelectrochemical cells based on photosynthetic systems. *Biofuel Research Journal* 6: 222.

Alvira, P., Tomas-Pejo, E., Ballesteros, M., and Negro, M.J. (2010). Pretreatment technologies for an efficient bioethanol productionprocess based on enzymatic hydrolysis: a review. *Bioresource Technology* 101: 4851–4861.

Anasontzis, G.E., Kourtoglou, E., Villas-Boas, S.G. et al. (2016). Metabolic engineering of *Fusarium oxysporum* to improve its thanol-producing capability. *Frontier in Microbiology* 7: 1–10.

Antal, M.J. (1983). Biomass pyrolysis: a review of literature part 1-carbohydrate pyrolysis. In: *Advances in Solar Energy*, vol. 1 (ed. K.L. Boer and J.A. Duffie), 61–111. Springer.

Antwi, E., Engler, N., Nelles, M., and Schüch, A. (2019). Anaerobic digestion and the effect of hydrothermal pretreatment on the biogas yield of cocoa pods residues. *Waste Management* 88: 131–140.

Awoyale, A.A. and Lokhat, D. (2018). *Harnessing the Potential of Bio-Ethanol Production from Lignocellulosic Biomass in Nigeria – A Review*, 1–16. Society of Chemical Industry and John Wiley & Sons, Ltd.

Ayhan, D. (2005). Bioethanol from cellulosic materials: a renewable motor fuel from biomass. *Energy Sources* 27: 327–337.

Badger, P.C. (2002). Ethanol from cellulose: a general review. In: *Trends in New Crops and New Uses* (ed. J. Janick and A. Whipkey), 17–21. Alexandria, VA: ASHS Press.

Bailey, P.S. (1982). *Ozonation in Organic Chemistry*, 288. New York, N.Y.: Academic press, Inc.

Barrera, I., Amezcua-Allieri, M.A., Estupiñan, L. et al. (2016). Technical and economical evaluation of bioethanol production from lignocellulosic residues in Mexico: case of sugarcane and blue agave bagasses. *Chemical Engineering Research and Design* 107: 91–101.

Bartens, A.K.G. and Rein, P. (2007). *Cane Sugar Engineering*, 943. Berlin: Verlag.

Battat, E., Peleg, Y., Bercovitz, A. et al. (1991). Optimization of L-malic acid production by *Aspergillus flavus* in a stirred fermenter. *Biotechnology and Bioengineering* 27: 1108–1116.

Bellido, C., Infante, C., Coca, M. et al. (2015). Efficient acetone–butanol–ethanol production by *Clostridium beijerinckii* from sugar beet pulp. *Bioresource Technology* 190: 332–338.

Bowen, E., Kennedy, S.C., and Miranda, K. (2010). *Ethanol from sugar beets: A process and economic analysis. Project Report*, 1–143.

Bukhtiyarova, M., Lunkenbein, T., Kahler, K., and Scholgl, R. (2017). Methanol synthesis from industrial CO_2 sources: a contribution to chemical energy conversion. *Catalysis Letters* 147: 416–427.

Canilha, L., Chandel, A.K., dos Santos Milessi, T.S. et al. (2012). Bioconversion of sugarcane biomass into ethanol: an overview about composition, pretreatment methods, detoxification of hydrolysates, enzymatic saccharification, and ethanol fermentation. *Journal of Biomedicine and Biotechnology* 1–15. (Article Id 989572).

Carvalho, J.A., Pego, M.F.F., Hein, P.R.G. et al. (2019). Physical and chemical pretreatment of sugarcane bagasse for enhanced acid hydrolysis. *Cellulose Chemistry and Technology* 54 (7-8): 699–704.

Castro, R.E.N.D., Alves, R.M.D.B., Nascimento, C.A.O.D., and Giudici, R. (2018). Assessment of sugarcane-based ethanol production. In: *Fuel Ethanol Production from Sugarcane* (ed. T.P. Basso and L.C. Basso). Intech Open https://doi.org/10.5772/intechopen.78301.

Chandel, A.K., Chan, E.C., Rudravaram, R. et al. (2007). Economics and environmental impact of bioethanol production technologies: an appraisal. *Biotechnology and Molecular Biology Reviews* 2: 14–32.

Curran, K.A., Leavitt, J.M., Karim, A.S., and Alper, H.S. (2013). Metabolic engineering of muconic acid production in *Saccharomyces cerevisiae*. *Metabolic Engineering* 15 (1): 55–66.

De Vriezea, J., Hennebela, T., Van den Brandea, J. et al. (2014). Anaerobic digestion of molasses by means of a vibrating and non-vibrating submerged anaerobic membrane bioreactor. *Biomass Bioenergy* 68: 95–105.

Demirbas, A. (2005). Potential applications of renewable energy sources, biomass combustion problems in boiler power systems and combustion related environmental issues. *Progress in Energy and Combustion Science* 31: 171–192. http://dx.doi.org/10.1016/j.pecs.2005.02.002.

Demirbas, A. (2007). Progress and recent trends in biofuels. *Progress in Energy and Combustion Science* 33: 1–18. http://dx.doi.org/10.1016/j.pecs.2006.06.001.

Demirbas, B. (2010). Biomass business and operating. *Energy Education Science and Technology Part A* 26: 37–47.

Demirbas, E., Kobya, M., and Sulak, M.T. (2008). Adsorption kinetics of a basic dye from aqueous solutions onto apricot stone activated carbon. *Bioresource Technology* 99 (13): 5368–5373.

Dodić, S., Popov, S., Dodić, J. et al. (2009). Bioethanol production from thick juice as intermediate of sugar beet processing. *Biomass and Bioenergy* 33: 822–827.

DOE/EERE (2010). Biopower technical strategy workshop summary. US Department of Energy, Energy Efficiency & Renewable Energy, p 96. https://www.energy.gov/sites/prod/files/2014/04/f14/biopower_workshop_report_december_2010.pdf

Doran, J.B., Caripe, J., Sutton, M., and Foster, B. (2000). Fermentations of pectin-rich biomass with recombinant bacteria to produce fuel ethanol. *Applied Biochemistry and Biotechnology* 84–86: 141–152.

Dredge, R., Radloff, S.E., Susan van Dyk, J., and Pletschke, B.I. (2011). Lime pretreatment of sugar beet pulp and evaluation of synergy between ArfA, ManA and XynA from *Clostridium cellulovorans* on the pretreated substrate. *3 Biotechology* 1: 151–159.

Drift, A.V.D. and Boerrigter, H. (2006). Synthesis gas from biomass for fuels and chemicals. *Environmental Science* 1–29.

Du, G.C., Chen, J., Yu, J., and Lun, S. (2001). Continuous production of poly-3-hydrobutyrate by *Ralstonia eutropha* in a two-stage culture system. *Journal of Biotechnology* 88: 59–65.

Dussan, K.J., Silva, D.D.V., Moraes, E.J.C. et al. (2014). Dilute-acid hydrolysis of cellulose to glucose from sugarcane bagasse. *Chemical Engineering Transactions* 38: 433–438.

Elbehri, A., Segerstedt, A., and Liu, P. (2013). *Biofuels and the Sustainability Challenge: A Global Assessment of Sustainability Issues, Trends and Policies for Biofuels and Related Feedstock. Trade and Markets Division*, 188. Rome: Food and agriculture Organization of the United Nations.

El-Refai, A.H., El-Abyad, M.S., El-Diwany, A.I. et al. (1992). Some physiological parameters for ethanol production from beet molasses by Saccharomyces cerevisiae Y-7. *Bioresource Technology* 42: 183–189.

Faramarzi, S., Anzabi, Y., and Jafarizadeh-Malmiri, H. (2019). Selenium supplementation during fermentation with sugar beet molasses and *Saccharomyces cerevisiae* to increase bioethanol production. *Green Process Synthesis* 8: 622–628.

Foster, B.L., Dale, B.E., and Doran-Peterson, J.B. (2001). Enzymatic hydrolysis of ammonia treated sugar beet pulp. *Applied Biochemistry and Biotechnology* 91-93: 269–282.

Gámez, S., Cabrialesa, J.J.G., Ramírez, J.A. et al. (2006). Study of the hydrolysis of sugar cane bagasse using phosphoric acid. *Journal of Food Engineering* 74: 78. https://doi.org/10.1016/j.jfoodeng.2005.02.005.

Ghorbani, F., Younesi, H., Sari, A.E., and Najafpour, G. (2011). Cane molasses fermentation for continuous ethanol production in an immobilised cell reactor by *Saccharomyces cerevisiae*. *Renewable Energy* 39: 503–509.

Goh, C.S., Tan, K.T., Lee, K.T., and Bhatia, S. (2010). Bioethanol from lignocelllulose: status, perspectives and challenges in Malaysia. *Bioresource Technology* 101: 4834–4841.

Goksungur, Y. and Zorlu, N. (2001). Production of ethanol from beet molasses by Ca-Alginate immobilized yeast cells in a packed-bed bioreactor. *Turkish Journal of Biology* 25: 265–275.

Gomez, M.R.S., Andrade, R.R., Santander, C.G. et al. (2010). Pretreatment of sugar cane bagasse with phosphoric and sulfuric diluted acid for fermentable sugars production by enzymatic hydrolysis. *Chemical Engineering Transactions* 20: 321–326.

Grohmann, K., Baldwin, E.A., Buslig, B.S., and Ingram, L.O. (1994). Fermentation of galacturonic acid and other sugars in orange peel hydrolysates by the ethanologenic strain of *Escherichia coli*. *Biotechnology Letter* 16: 281–286.

Gurgel, L.V.A., Pimenta, M.T.B., and Curvelo, A.A.S. (2016). Ethanol–water organosolv delignification of liquid hot water (LHW) pretreated sugarcane bagasse enhanced by high–pressure carbon dioxide (HP–CO_2). *Industrial Crops and Products* 94: 942–950.

Hassib, B., Youssef, T., Nedia, H. et al. (2013). Performances comparison between three technologies for continuous ethanol production from molasses. *Biomass and Bioenergy* 48: 25–32.

Hernández, C.A., Ziarelli, F., Gaime, I. et al. (2017). Chemical and biological pretreatments on sugarcane bagasse to enhance its enzymatic hydrolysis. *ChemistrySelect* 2 (15): 4213–4218.

Hernández-Salas, J.M., Villa-Ramírez, M.S., Veloz-Rendón, J.S. et al. (2009). Comparative hydrolysis and fermentation of sugarcane and agave bagasse. *Bioresource Technology* 100 (3): 1238–1245.

Hideno, A., Inoue, H., Tsukahara, K. et al. (2009). Wet disk milling pretreatment without sulfuric acid for enzymatic hydrolysis of rice straw. *Bioresource Technology* 100 (10): 2706–2711.

Hofvendahl, K. and Hahn-Hagerdal, B. (2000). Factors affecting the fermentative lactic acid production from renewable resources. *Enzyme and Microbial Technology* 26: 87–107.

Holtzapple, M.T. and Humphrey, A.E. (1984). The effect of organosolve pretreatment on the enzymatic hydrolysis of poplar. *Biotechnology and Bioengineering* 26 (7): 670–676.

Hu, Y., Feng, S., Xu, C.C., and Bassi, A. (2017). Production of low-nitrogen bio-crude oils from microalgae pre-treated with pre-cooled NaOH/urea solution. *Fuel* 206: 300–306. https://doi.org/10.1016/j.fuel.2017.06.021.

Inoue, H., Yano, S., Endo, T. et al. (2008). Combining hot-compressed water and ballmilling pretreatments to improve the efficiency of the enzymatic hydrolysis of eucalyptus. *Biotechnology for Biofuels* 1, article 2: 1–9.

Jiang, L., Fang, Z., Li, X. et al. (2013). Combination of dilute acid and ionic liquid pretreatments of sugarcane bagasse for glucose by enzymatic hydrolysis. *Process Biochemistry* 48: 1942–1946.

Jiraprasertwong, A., Gulari, E., and Chavadej, S. (2014). Effect of different microbial strains on biological pretreatment of sugarcane bagasse for enzymatic hydrolysis. *World Academy of*

Science, Engineering and Technology International Journal of Bioengineering and Life Sciences 8 (9): 968–972.

Joanna, B., Michal, B., Piotr, D. et al. (2018). Sugar beet pulp as a source of valuable biotechnological products. In: *Advances in Biotechnology for Food Industry: Handbook of Food Bioengineering*, vol. 14 (ed. A.M. Holban and A.M. Grumezescu), 359–390. Academic Press.

Jones F.S.B., Holladay, J.E., Valkenburg, C. et al. (2009). Production of gasoline and diesel from biomass via fast pyrolysis, hydrotreating and hydrocracking: a design case. Pacific Northwest National Laboratory. PNNL-18284.

Kaur, M., Neetu, Y.P.V., and Chauhan, S. (2020). Effect of chemical pretreatment of sugarcane bagasse on biogas production. *Materials Today: Proceedings* 21: 1937–1942.

Kharina, M., Emelyanov, V., Mokshina, N. et al. (2016). Pretreatment of sugar beet pulp with dilute sulphurous acid is effective for multipurpose usage of carbohydrates. *Applied Biochem Biotechnology* 179 (2): 307–320.

Klement, T. and Bachs, J. (2013). Itaconic acid: a biotechnological process in change. *Bioresource Technology* 135: 422–431.

Kosa, M. and Ragauskas, A. (2011). Lipids from heterotrophic microbes: advances in metabolism research. *Trends in Biotechnology* 29: 53–61.

Kracher, D., Oros, D., Yao, W. et al. (2014). Fungal secretomes enhance sugar beet pulp hydrolysis. *Biotechnology Journal* 9: 483–492.

Kühnel, S., Schols, H.A., and Gruppen, H. (2011). Aiming for the complete utilization of sugar beet pulp: examination of the effects of mild acid and hydrothermal pretreatment followed by enzymatic digestion. *Biotechnology for Biofuels* 4: Article number 14.

Kumar, P., Barrett, D.M., Delwiche, M.J., and Stroeve, P. (2009). Methods for pretreatment of lignocellulosic biomass for efficient hydrolysis and biofuel production. *Industrial and Engineering Chemistry Research* 48 (8): 3713–3729.

Kyryzyuk, S., Krupin, V., Borodina, O., and Was, A. (2020). Crop residue removal: assessment of future bioenergy generation potential and agro environmental limitations based on a case study of Ukraine. *Energies* 13: 5343–5366.

Lee, S., Speight, J.G., and Loyalka, S.K. (2007). *HandbooK of Alternative Fueld Technologies*. Boca Raton, USA: CRC Taylor and Francis Group. ISBN: 9780824740696.

Lim, J.S., Manan, Z.A., Wan Alwi, A.R., and Hashim, H. (2012). A review of utilization of biomass from rice industry as a source of renewable energy. *Renewable Sustainable Energy Review* 16: 3084–3094.

Liu, B. and Rajagopal, D. (2019). Life-cycle energy and climate benefits of energy recovery from wastes and biomass residues in the United States. *Nature Energy* 4: 700–708.

Mall, A.K., Misra, V., Santeshwari et al. (2021). Sugar beet cultivation in India: prospects for bioethanol production and value-added co-products. *Sugar Tech* 23: 1218–1234.

Margeot, A., Hahn-Hagerdal, B., Edlund, M. et al. (2009). New improvements for lignocellulosic ethanol. *Current Opinion in Biotechnology* 20 (3): 372–380.

Mbohwa, C. (2013). Energy management in the South African sugar industry. *Proceedings of the World Congress on Engineering 2013*, Vol I, WCE 2013, London, UK (3–5 July 2013).

McKinlay, J.B., Vieille, C., and Zeikus, J. (2007). Prospects for a bio-based succinate industry. *Applied Microbiology and Biotechnology* 76: 727–740.

Mikulski, D. and Klosowski, G. (2021). Integration of first and second generation bioethanol production from beet molasses and distillery stillage after dilute sulfuric acid pretreatment. *Bioenergy Research*, https://doi.org/10.1007/s12155-021-10260-w.

Minteer, S. (2006). *Alcholic Fuels*, 1e, 293. Boca raton, USA: CRC Press. ISBN: 9780849339448.

Misra, V., Mall, A.K., and Pathak, A.D. (2020). Sugarbeet: a sustainable crop for salt stress conditions. In: *Agronomic Crops* (ed. M. Hasaanzuman), 40–62. Springer Nature Singapore Pte Ltd https://doi.org/10.1007/978-981-15-0025-1_10.

Mohammed, N.I., Kabbashi, N., and Alade, A. (2018). Significance of agricultural residues in sustainable biofuel development. In: *Agricultural Waste and Residues* (ed. A. Aladjadjiyan), 71–88. *InTech Open.*

Mohlala, L.M., Bodunrin, M.O., Awosusi, A.A. et al. (2016). Beneficiation of corncob and sugarcane bagasse for energy generation and materials development in Nigeria and South Africa: a short overview. *Alexandria Engineering Journal* 55 (3): 3025–3036.

Moore, A. (2008). Biofuels are dead: long live biofuels? Part One. *New Biotechnology* 25: 6–12.

Mosier, N.S. (2013). Fundamentals of aqueous pretreatment of biomass. In: *Aqueous Pretreatment of Plant Biomass for Biological and Chemical Conversion to Fuels and Chemicals* (ed. C.E. Wyman), 129–143. Hoboken, New Jersey: Wiley.

Mosier, N., Wyman, C., Dale, B. et al. (2005). Features of promising technologies for pretreatment of lignocellulosic biomass. *Bioresource Technology* 96: 673–686.

Mtui, G.Y.S. (2009). Recent advances in pretreatment of lignocellulosic wastes and production of value added products. *African Journal of Biotechnology* 8 (8): 1398–1415.

Neto, J.M., Komesu, A., Martins, L.H.D.S. et al. (2019). Third generation biofules: an overview. In: *Sustainable Bioenergy: Advances and Impacts* (ed. M. Rai and A.P. Ingle), 283–298. Elsevier.

Ninju, S., Swathika, M., and Balaji, M. (2020). Pretreatment of lignocellulosic sugarcane leaves and tops for bioethanol treatment. In: *Lignocellulosic Biomass to Liquid Biofuels* (ed. A. Yousuf, D. Pirozzi and F. Sannino), 301–324. Academic Press.

Nitsos, C., Matsakas, L., Triantafyllidis, K. et al. (2018). Investigation of different pretreatment methods of Mediterranean-type ecosystem agricultural residues: characterization of pretreatment products, high-solids enzymatic hydrolysis and bioethanol production. *Biofuels* 9 (5): 545–558. https://doi.org/10.1080/17597269.2017.1378988.

Nsaful, F. (2012). Process Modelling of Sugar Mill Biomass to Energy Conversion Processes and Energy Integration of Pyrolysis. M.Sc. Thesis, Stellenbosch University, Stellenbosch. p. 171.

Ogwo, J.N., Dike, O.C., Mathew, S.O., and Akabuogu, E.U. (2012). Overview of biomass energy production in Nigeria: implications and challenges. *Asian Journal of Natural and Applied Sciences* 1 (4): 46–51.

Onuki S., Koziel, J.A., Leeuwen, J.V. et al. (2008). Ethanol production, purification, and analysis techniques: a review. Proceeding of SAABE, Paper No. 8, Agricultural and Biosystems Engineering, Iowa State University, pp. 1–14.

Panthapulakkal, S. and Swain, M. (2015). The use of wheat straw fibers as reinforcements in composites. In: *Biofibre Reinforcements in Composite Materials* (ed. O. Frauk and M. Sain), 423–453. Woodhead Publishing.

Pattanaik, L., Pattnaik, F., Saxena, D.K., and Naik, S.N. (2019). Biofuels from agricultural wastes. In: *Second and Third Generation of Feedstocks* (ed. A. Basile and F. Dalena), 103–142. Elseveir Inc.

van der Poel, P.W., H. Schiweck, T. Schwartz. 1998. Sugar technology. Beet and cane sugar manufacture. Verlag Dr. Albert Martens KG, Berlin, Pp. 1005.

Raharja, R., Murdiyatmo, U., Sutrisno, A., and Wardani, A.K. (2019). Bioethanol production from sugarcane molasses by instant dry yeast. IOP Conference Series: Earth and Environmental Science 230, 012076, pp. 1–9.

Ranković, J., Dodić, J., Dodić, S., and Popov, S. (2009). Bioethanol production from intermediate products of sugar beet processing with different types of Saccharomyces cerevisiae. *Chemical Industry and Chemical Engineering Quarterly* 15: 13–16.

Reid, W.V., Ali, M.K., and Field, C.B. (2019). The future of bioenergy. *Global Change Biology* 26: 274–286.

Reijinders, L. (2006). Conditions for the sustainability of biomass based fuel use. *Energy Policy* 34: 863–876.

Renewables (2015). Global status report. REN21. Paris: REN21 Secretariat; 2015. 978-3-9815934-6-4.

Rezic, I., Oros, D., Markovi, I. et al. (2013). Integrated hydrolyzation and fermentation of sugar beet pulp to bioethanol. *Journal of Microbiology and Biotechnology* 23 (9): 1244–1252.

Roa Engel, C.A., Straathof, A.J.J., Zijlmans, T.W. et al. (2008). Fumaric acid production by fermentation. *Applied Microbiology and Biotechnology* 78: 379–389.

Rodriguez, L.A., Toro, M.E., Vazquez, F. et al. (2010). Bioethanol production from grape and sugar beet pomaces by solid state fermentation. *International Journal of Hydrogen Energy* 35 (11): 5914–5917.

Rodríguez-Chong, A., Ramírez, J.A., Garrote, G., and Vázquez, M. (2004). Hydrolysis of sugarcane bagasse using nitric acid: a kinetic assessment. *Journal of Food Engineering* 61: 143–152.

Rorick, R.E., Nabar, N., and Pryor, S. (2009). Enzymatic hydrolysis and fermentation of sugar beet pulp. Reno, Nevada, paper No. 096193.

Roukas, T. (1996). Ethanol production from non-sterilized beet molasses by free and immobilized *Saccharomyces cerevisiae* cells using fed-batch culture. *Journal of Food Engineering* 27: 87–96.

Rowlands, W.N., Masters, A., and Maschmeyer, T. (2008). The biorefinery challenges, opportunities and an Australian perspective. *Bulletin of Science Technology and Society* 28: 149–158.

Ruane, J., Sonnino, A., and Agostini, A. (2010). Bioenergy and the potential contribution of agricultural biotechnologies in developing countries. *Biomass Bioenergy* 34: 1427–1439.

Sabiha-Hanim, S. and Halim, N.A.A. (2018). Sugarcane bagasse pretreatment methods for ethanol production. In: *Fuel Ethanol Production from Sugarcane* (ed. T.P. Basso and L.C. Basso), 63–79. Intech Open http://dx.doi.org/10.5772/intechopen.81656.

Sadhukhan, J., Ng, K.S., and Martinez-Hernandez, E. (2014) Novel integrated mechanical biological chemical treatment systems for the production of levulinic acid from fraction of municipal solid waste: a comprehensive techno-economic analysis. *Bioresource Technology* 215: 131–143.

Sanchez, C. (2009). Lignocellulosic residues: biodegradation and bioconversion by fungi. *Biotechnology Advances* 27: 185–194.

Sarkar, N., Ghosh, S.K., Bannerjee, S., and Aikat, K. (2012). Bioethanol production from agricultural wastes: an overview. *Renewable Energy* 37 (1): 19–27.

Sheoran, A., Yadav, B.S., Nigam, P., and Singh, D. (1998). Continuous ethanol production from sugarcane molasses using a column reactor of immobilized *Saccharomyces cerevisiae* HAU-1. *Journal of Basic Microbiology* 38 (2): 123–128.

da Silva, A.S., Inoue, H., Endo, T. et al. (2010). Milling pretreatment of sugarcane bagasse and straw for enzymatic hydrolysis and ethanol fermentation. *Bioresource Technology* 101 (19): 7402–7409.

Silvestre, G., Fernandez, B., and Bonmati, A. (2015). Signficance of anaerobic digestion as a source of clean energy in waste water treatment plants. *Energy Conversion and Management* 101: 255–262.

Singh, R., Shukla, A., Tiwari, S., and Srivastava, M. (2014). A review on delignification of lignocellulosic biomass for enhancement of ethanol production potential. *Renewable and Sustainable Energy Reviews* 32: 713–728.

Singh, J., Suhag, M., and Dhaka, A. (2015). Augmented digestion of lignocellulose by steam explosion, acid and alkaline pretreatment methods: a review. *Carbohydrate Polymers* 117: 624–631.

Smith O.B. (1989). Utilization of crops residues in the nutrition of sheep and goats in the humid tropics of West Africa. In: Atta Krah, A.N. and Reyholds, L. (eds.) Sheep and Goat Meat Production in the Humid Tropics of West Africa; Yamoussoukro; 21–25 September 1987. FAO Animal Production and Health Paper.

Soares, K.C., Mendes, S., Benachour, M., and Abreu, C.A.M. (2017). Evaluation of the effects of operational parameters in the pretreatment of sugarcane bagasse with diluted sulfuric acid using analysis of variance. *Journal Chemical Engineering Communications* 204 (12): 1369–1390.

Stevens, C. and Verhe, R. (2004). *Renewable bioresources scope and modification for non food application*, 328. England: Wiley. ISBN: 978-0-470-85446-4.

Sun, Y. and Cheng, J. (2002). Hydrolysis of lignocellulosic materials for ethanol production: a review. *Bioresource Technology* 83 (1): 1–11.

Taherzsadeh, M.J. and Karimi, K. (2007). Enzymatic-based hydrolysis processes for ethanol from lignocellulosic materials: a review. *Bioresources* 2: 707–738.

Tahir, M.H., Çakman, G., Goldfard, J.L. et al. (2019). Demonstrating the suitability of canola residue biomass to biofuel conversion via pyrolysis through reaction kinetics, thermodynamics and evolved gas analyses. *Bioresource Technology* 279: 67–73.

Talmadge, M., Biddy, M., Dutta, A. et al. (2013). Syngas upgrading to hydrocarbon fuels technology pathway. In NREL/TP-5100-58052, PNNL-22323. National Renewable Energy Laboratory, Pacific Northwest National Laboratory.

Timung, R., Deshavath, N.N., Goud, V.V., and Dasu, V.V. (2016). Effect of subsequent dilute acid and enzymatic hydrolysis on reducing sugar production from sugarcane bagasse and spent citronella biomass. *Journal of Energy* 2016: 1–12.

Tripathy, S.S. and Raichur, A.M. (2008). Enhanced adsorption capacity of activated alumina by impregnation with alum for removal of As(V) from water. *Chemical Engineering Journal* 138: 179–186.

Vats, S., Maurya, D.P., Shaimoon, M. et al. (2013). Development of a microbial consortium for the production of blend enzymes for the hydrolysis of agricultural waste into sugars. *Journal of Scientific and Industrial Research* 72 (9-10): 585–590.

Virmond, E., Rocha, J.D., Moreira, R.F.P.M., and José, H.J. (2013). Valorization of agro-industrial solid residues and residues from biofuel production chains by thermochemical conversion: a review, citing Brazil as a case study. *Brazilian Journal of Chemical Engineering* 30 (2): 197–229.

Voloshin, R.A., Rodionova, M.V., Zharmukhamedov, S.K. et al. (2016). Review: biofuel production from plant and algal biomass. *International Journal of Hydrogen Energy* 30: 1–17.

Walsh, G.H. (1998). The Riviere Association. Juice extractor: a new approach to the extraction of juice from cane. In: *Proceedings of the South African Sugar Technologists*, pp. 173–180.

Wan, C. and Li, Y. (2012). Fungal pretreatment of lignocellulosic biomass. *Biotechnology Advances* 30: 1447–1457.

Wright, M.M., Satrio, J.A., Brown, R.C. et al. (2010). Technoeconomic analysis of biomass fast pyrolysis to transportation fuels. In NREL/TP6A20-46586. National Renewable Energy Laboratory.

Wu, X., McLaren, J., Madl, R., and Wang, D. (2010). Biofuel from lignocellulosic biomass. In: *Sustainable Biotechnology* (ed. O.V. Singh and S.P. Harvey), 19–41. The Netherlands: Elsevier.

Yan, J. and Lin, T. (2009). Biofuels in Asia. *Applied Energy* 86: S1–S10.

You, Y., Zhou, Z., Zhou, P. et al. (2016). Comparison of pretreatment methods for production of ethanol from sugarcane bagasse. *BioResource* 11 (1): 2297–2307.

Yücel, H.G. and Aksu, Z. (2015). Ethanol fermentation characteristics of *Pichia stipitis* yeast from sugar beet pulp hydrolysate: use of new detoxification methods. *Fuel* 158: 793–799.

Zaman, C.Z., Pal, K., Yehye, W.A. et al. (2017). Pyrolysis: a sustainable way to generate energy from waste. In: *Pyrolysis* (ed. M. Samer), 1–35. InTech Open https://doi.org/10.5772/intechopen.69036.

Zayed, G.Z.A. and Foley, J. (1987). The influence of fermentation conditions on ethanol yields from sugar beet molasses and fodder beet juice using *Saccharomyces cerevisiae* strains. *Irish Journal of Food Science and Technology* 11: 119–133.

Zhang, B., Shahbazi, A., Wang, L.J. et al. (2011). Hot-water pretreatment of cattails for extraction of cellulose. *Journal of Industrial Microbiology and Biotechnology* 38: 819–824.

Zhao, Y.L., Dolat, A., Steinberger, Y. et al. (2009). Biomass yield and changes in chemical composition of sweet sorghum cultivars grown for biofuel. *Field Crop Research* 111: 55–64.

Zheng, Y., Pan, Z., and Zhang, R.N. (2009). Overview of biomass pretreatment for cellulosic production. *International Journal of Agricultural and Biological Engineering* 2: 51–68.

Zheng, Y., Cheng, Y., Chaowei, Y. et al. (2012). Improving the efficiency of enzyme utilization for sugar beet pulp hydrolysis. *Bioprocess and Biosystems Engineering* 35 (9): 1531–1539.

Zheng, Y., Lee, C., Yu, C. et al. (2013). Dilute acid pretreatment and fermentation of sugar beet pulp to ethanol. *Applied Energy* 105: 1–7.

Zheng, J., Tashiro, Y., Wang, Q., and Sonomoto, K. (2015). Recent advances to improve fermentative butanol production: genetic engineering and fermentation technology. *Journal of Bioscience and Bioengineering* 119 (1): 1–9.

Zieminski, K. and Kowalska, M. (2015). Effect of enzymatic pretreatment on anaerobic co-digestion of sugar beet pulp silage and vinasse. *Bioresource Technology* 180: 274–280.

Zieminski, K., Romanowska, I., Kowalska, M., and Cyran, M. (2014). Effects of hydrothermal pretreatment of sugar beet pulp for methane production. *Bioresource Technology* 166: 187–193.

Zigová, J., Šturdk, E., Vandak, D., and Schlosser, A. (1999). Butyric acid production by *Clostridium butyricum* with integrated extraction and pertraction. *Process Biochemistry* 34: 835–843.

14

Creating Wealth from Agrowaste: Success Stories from India

Joystu Dutta[1], Tirthankar Sen[2], Payal Sen[3], Pradeep Kumar Tiwari[4], Amrita Kumari Panda[5], and Joysurya Dutta[6]

[1] Department of Environmental Science, Sant Gahira Guru Vishwavidyalaya, Sarguja, Ambikapur, Chhattisgarh, India
[2] Department of Biosciences and Bioengineering, Indian Institute of Technology Guwahati, Guwahati, Assam, India
[3] Independent Researcher, Chembur East, Mumbai, Maharashtra, India
[4] New Housing Board Colony, Shahdol, Madhya Pradesh, India
[5] Department of Biotechnology, Sant Gahira Guru Vishwavidyalaya, Sarguja, Ambikapur, Chhattisgarh, India
[6] Symbiosis Institute of Operations Management (SIOM), Nashik, Maharashtra, Kolkata, West Bengal, India

CHAPTER OUTLINE

14.1 Global Scenario of Agriculture

Since 10 000 years, cereals such as wheat, maize, and rice have constituted the staple food supply for humans and animals alike throughout the globe (FAO 2001a, 2015b). The United States of America, China, India, Brazil, and the European Union are the major cereal producing countries of the world (FAO 1999, 2003). The 1970s saw the green revolution transform farming techniques and processes in developing nations such as India and Mexico. These improvements resulted in the bumper yield of a variety of cereal crops, thus improving the livelihoods of millions of marginal farmers and

Agriculture Waste Management and Bioresource: The Circular Economy Perspective, First Edition.
Edited by Suruchi Singh, Pardeep Singh, Anu Sharma, and Moharana Choudhury.

ameliorating national food insecurity. Agricultural advancements have undoubtedly impacted the socio-economic graphs of several nations. Agricultural production has experienced a growth of over 300% over the last 50 years, primarily due to agricultural soil expansion and the technological advancements of the green revolution. The global average of agricultural production amounts to approximately 23.7 million tons of food per day with livelihood generation for approximately 2.5 billion people worldwide as per the reports published by the CBD (Convention on Biological Diversity). This humongous and unbalanced growth in agricultural productivity unnecessarily pressurizes natural resources such as soil, water, air and forests, further negatively impacting the environment. This unbalanced development is a negative externality of nation's role in sustainable development and global food security issues. Agriculture is largely influenced by climate vagaries and increases greenhouse gas production, as proposed by the Kyoto Protocol. Rampant use of fertilizers, pesticides, herbicides, insecticides, and the generation of farm-based animal wastes contributes approximately 30% of total greenhouse gas emissions. Organizations under the leadership of FAO are now focusing on more sustainable models such as cyclical agricultural production systems. For example, traditional aquaculture in the rice-based farming systems of Southeast Asia boosts rice productivity by increasing nutrient availability to the plants. Rice-fish farmers enjoy higher productivity and better economic returns as compared to their traditional counterparts. The fish further provides a rich source of proteins, fatty acids, and micronutrients especially needed by children and pregnant women perennially malnourished in developing economies. Practices like these can also soothe mosquito growth in stagnant water as fishes feed on mosquito larvae, thus heavily reducing the risk of vector-borne disease outbreaks like malaria, dengue, and chikungunya. The biological control of mosquitoes and other insects also reduces insecticides and pesticides in rice crops, thus reducing the risk of environmental pollution. Water quality and its potable value are also maintained as a direct consequence. The above is an ideal example of integrated and sustainable farming practices proposed by FAO. Increased agricultural productivity has significantly contributed to agricultural waste generation in significant Indian as well as overseas farmlands.

Agricultural wastes can be classified into two main categories: organic (natural) and inorganic (non-natural) wastes, both of which can exist in a wide variety of physical and chemical forms. They can be liquids, slurries, or solids. They can also be water-soluble or water-insoluble, combustible or incombustible, toxic or nontoxic. This heterogeneity, however, opens more exciting possibilities for the exploration and exploitation of agricultural waste resources. Farming is one sector that contributes significantly to biomass production. Policymakers are increasingly promoting the concept of a bio-economy for which agricultural waste is an essential feedstock. One of the main motivations for transitioning into a bio-economy is that it is aimed to reduce mankind's dependence on nonrenewable, non-sustainable, environmentally detrimental natural resources such as petroleum and coal. Bioeconomy can be conceptualized as the knowledge driven use of renewable biological resources to fulfil mankind's ever-increasing needs for food, materials, and energy in an environmentally sustainable manner. A functional bioeconomy promotes the conversion of agricultural waste into economically valuable products, which not only facilitate positive ecological footprint but also results in employment generation and the socio-economic upliftment of marginal farmer families. Therefore, agricultural waste management and the concept of a functional bioeconomy go hand in hand. Society's transition towards a sustainable circular economy/bioeconomy (the terms are often used interchangeably) promises intelligent, sustainable, and inclusive growth. Global economies are increasingly developing and adopting national policies and strategies in sync with this new economic model to establish an efficient resource management system. Therefore, agricultural waste management remains a global priority. Such a system would attempt to maximize the lifespan of various used materials and products by incorporating them into other production processes, thereby massively promoting the use–reuse--recycle paradigm instead of the traditional manufacture–use–dispose paradigm.

14.2 National Scenario

Agriculture is the most predominant occupation in the rural hinterlands of India. Despite this, a substantial fraction of farmers, especially the small and marginal farmers, spend most of the year in financial distress. They often migrate to cities and towns far from their homes for a significant part of the year except the monsoon months. In the towns and cities, they toil endlessly, often in deplorable living and working conditions, just to make ends meet. This tragic disparity is propagated by a variety of factors including unsustainable agricultural practices, fragmented land holdings, market and weather information inaccessibility, outdated technology, lack of quality input supplies and storage facilities, unaffordable institutional credit, and the absence of proper irrigation facilities. These factors, combined with the overall lack of farmers' awareness regarding government schemes and welfare programs, prevent a significant portion of our nation's rural farmers from enjoying a state of financial solvency.

Various government and private organizations are intervening at multiple points of the agricultural value chain in a two-pronged effort to support small and marginal farmers by saving costs and optimizing income. These efforts are crucial steps to creating food and financial security for these resources deprived communities. Some such approaches entail organizing individual farmers into self-help groups (SHGs) and producer groups (more formally referred to as producer companies and cooperatives). These organized communities provide farmers with better access to markets and institutional credit, extension services, training on better farming practices, productivity improvement, technology support, and most importantly, the bargaining power to protect themselves from the exploitative practices employed by many brokers. These linkages are crucial steps towards increasing their incomes. Labour, machinery, resources such as seeds, insecticides, pesticides, fertilizers, electricity, water, and lease and loan interest payments are some of the predominating costs farmers have to bear. Farmers are trained and encouraged by extension services to modify their practices to reduce cultivation costs. The use and production of organic or bio-inputs is one such approach. There are various ways of producing bio-inputs. They are mainly produced from plant and animal remains or excreta. This way, the farmers recycle their farm's wastes, save on cultivation costs by processing locally available materials, produce chemical-free products for themselves and their customers, and replenish essential nutrients. These practices improve the physical and biological properties of the soil in the long run, which are otherwise depleted and miss balanced by the excessive use of chemical inputs.

14.3 Types of Organic Inputs for Farms

A variety of techniques are available with farmers to incorporate organic materials into their farms, some of which are briefly described later.

- **Composting**: Composting is the microbiological decomposition of organic wastes, primarily made of animal excreta stored in a pit or a shed for a few weeks.
- **Vermicomposting**: Vermicomposting turns organic wastes such as crop residues, weed biomass, vegetable waste, leaf litter, hotel refuse, agricultural waste, and even rural waste such as animal excreta into compost by storing it in a moist area to promote microbiological growth and decomposition. Earthworms are used to convert the waste into solid additives on a small scale.
- **Coir Composting**: Coir composting finds use in horticulture. It is made from coconut husk, which is stored in a moist environment and turned into compost.
- **Sugarcane Trash Composting**: Waste from sugarcane crops coupled with animal excreta are treated with fungi such as *Aspergillus, Penicillium, Trichurus*, and *Trichoderma* compost. Residues from one batch of sugarcane compost are recycled into the subsequent batch of compost.
- **Composting Poultry Wastes**: A combination of poultry droppings and coir pith is treated with a fungal consortium for compost preparation.
- **Crop Residue Composting**: Green biodegradable wastes such as fresh leaves and weeds and brown biodegradable wastes such as coir dust, straw, dried grasses, and leaves wastes are combined to prepare compost.
- **Green Manure**: Leaves from *Crotalaria juncea* (sun hemp), *Sesbania bispinosa* (dhaincha), and *Cyamopsis tetragonoloba* (cluster beans), along with other shrubs, herbs, leaves, and twigs, are used as green manure in agriculture.
- ***Panchagavya* and *Dasagavya***: *Panchagavya* comprises nine products: cow dung, cow urine, curd, jaggery, ghee, milk, banana, tender coconut, and water. When suitably mixed and used, it improves the quality and quantity of horticultural products. *Dasagavya* comprises cow dung and urine, mixed with extracts from plants that grow on wastelands or roadsides such as *Jatropha curcas* (Jatropha), *Tephrosia purpurea* (kolingi), *Azadirachta indica* (neem), *Calotrophis* (erukam), and *Vitex negundo* (notchi). Its usage helps in seed germination, root development, and controls pest and disease infestations.
- **Biofertilizers**: These are preparations of microbial consortia that catalyse the growth of plants and efficiently support microbial processes in soil. Azolla is one such biofertilizer, which is also used as animal fodder. (TNAU 2016)

14.4 Production by Self-Help Groups (SHGs): A Microunit of Circular Economy

Since marginal and small farming households have a limited production capacity, their waste generation is also limited. They generally own one cow or a few poultry birds. Since they sell off goats to resolve their emergency financial requirements, the predominating source of animal excreta is often in short supply. They understand the apparent advantages of using organic inputs and hence are open to using them. However, due to the minimal livestock resources in their possession, they are often unable to domestically produce organic at capacities that satisfy their annual crop production requirements. Thus, the use of organic inputs is limited to kitchen gardens.

However, this current scenario presents an entrepreneurial opportunity for SHGs that collect wastes from all households in the village, produce composts, manures, biofertilizer, and sell to the farmers at affordable rates leading to an overall upliftment of the village economy. Using rural and farm wastes to produce organic inputs improves soil quality and farm productivity and brings economic benefits to communities by helping them save and earn. It extends landfill longevity, thereby delaying the construction of newer landfills. It also regulates the growth of weeds and confers protection against damaging pests and pathogens.

14.5 Linkages to United Nations Sustainable Development Goals (UNSDGs) and Government Schemes

Goal 12 of the SDGs, 'Sustainable Consumption and Production', has been globally recognized as an important goal to sustain life on earth (Figure 14.1). Achieving this goal calls for concerted holistic efforts from both the industrial as well as agricultural sectors. Farm waste recycling is a positive step towards this sustainable development goal. It improves the quality of natural resources, slows down degradation, ensures human well-being, and promises economic growth of the poor people heavily dependent on natural resources for their survival and livelihoods.

Farm waste recycling is a part of subgoal 12.5, which calls for action such that 'By 2030, sustainably reduce waste generation through prevention, reduction, recycling and reuse'. Both central and state governments in India have several sectoral schemes to support natural (chemical-free) farming, thus promoting the recycling of wastes, thus

Figure 14.1 UN SDG goal on responsible consumption and production.

ensuring long-term soil fertility to produce healthy food options. Many of these schemes outdate SDGs' adoption on 1 January 2016, demonstrating India's zeal and foresight in sustainability and sustainable practices. Examples of some welfare schemes for Indian farmers as proposed by Ministry of Agriculture, Government of India are delineated in Table 14.1.

According to some conservative estimates, around 800–1000 Mt of inappropriately managed agricultural waste are available annually in India that contributes significantly to

Table 14.1 Some welfare schemes for Indian farmers as proposed by Ministry of Agriculture, Government of India (Agri Farming 2021).

Scheme names	Brief description
Paramparagat Krishi Vikas Yojana (PKVY)	62% of the scheme's incentive, which amounts roughly to Rs. 31 000, is aimed at encouraging individual farmers to use organic inputs.
Rashtriya Krishi Vikas Yojana (RKVY)	This scheme encourages convergence with other schemes and promotes among other things, soil health and productivity, production and usage of biofertilizers and organic inputs.
MOVCDNER (Mission Organic Value Chain Development for North-Eastern Region) Scheme for Organic Farming	This scheme establishes standards and certifications for organic farm produce.
National Project on Organic Farming	This project seeks to build capacities of farmers on organic farming, document traditional practices and techniques in this regard and has also established a National Centre on Organic Farming and 6 regional centres to promoting organic farming.
Capital Investment Subsidy Scheme under SHM (Soil Health Management Scheme)	This scheme seeks to prevent pollution by converting and using organic waste to reduce farmers' dependence on chemical inputs and increase availability of biofertilizers and composts.
National Mission on Oilseeds and Oil Palm (NMOOP) for Organic Farming	Under this scheme, 50% subsidy is provided, amounting to Rs. 300 per hectare for some components used in producing biofertilizers and vermicomposting.
National Horticulture Mission	The mission provides financial assistance to the tune of Rs. 5 lakhs for organic certification. A group of farmers cultivating 50 hectares of land can apply for this assistance.
Godhyan Nyay Scheme	The Government of Chhattisgarh scheme targets the promotion of organic compost to reduce farmers' overwhelming reliance on chemical fertilizers. Organic compost promotes the improvement of soil health that in turn helps boost farmers' incomes. The scheme also proposes to sell vermicompost to farmers at the nominal rate of Rs. 8/kg.
Odisha Integrated Irrigation Project for Climate Resilient Agriculture	The Odisha government encourages farmers to produce millets organically.

(*Continued*)

Table 14.1 (Continued)

Scheme names	Brief description
Organic Agriculture Act (OAA)	Government of Uttarakhand has declared 10 blocks (Dunda, Pratapnagar, Jaiharikhal, Jakholi, Augustmuni, Ukhimath, Dewal, Salt, Betalghat, and Munsyari) as fully organic and makes it a criminal offense under this act to sell or purchase chemical inputs.
Savayava Bhagya Yojane	Under this scheme, the Government of Karnataka, through various non-profit organizations, facilitates the formation and registration of organic farmers' associations.
Raitha Siri Scheme	The Government of Karnataka, through this scheme, provides cash incentive worth Rs. 10 000/hectare to organic millet producers.
NARWA-GARWA-GHURWA-AUR-BADI-Current Scheme of Chhattisgarh Government for Farmers. *Gothan* Scheme	Small river streams and seasonal canals are referred to as *narwa*, which literally translates to canal. Both the male and female cattle are referred to as *garwa*, which literally translates to cow. *Ghurwa* refers to a place where cow dung is stored for the purpose of processing into farmyard manure. It is a big pit filled with cow dung, cow urine, as well as other agricultural wastes. *Badi* refers to a kitchen garden located in front of or behind a farmer's house. This Chhattisgarh state government scheme talks about the holistic development of farm and farm-based livelihood options while ensuring zero waste generation and maximum recycling. This scheme is an eco-friendly and economically feasible initiative for marginal farmers or land-based labourers.

Source: Agri Farming (2021).

environmental pollution (Bhuvaneshwari et al. 2019). In the recent years, the national media has publicized how the practice of stubble burning carried out by the farmers residing in the states neighbouring the National Capital Region and Delhi have severely degraded the quality of air in the city. Stubble burning has been recognized as a major contributor of air pollution in India and major parts of South Asia (Abdurrahman et al. 2020). This has brought to the forefront the government's need to prioritize the productive and environmentally sustainable management as well as recycling of agricultural waste. Emphasis has been placed on converting waste biomass into valuable animal feed, biofuel, organic manure, and ethyl alcohol. Selling the waste biomass to processing units in exchange for financial compensation is also a viable option. Many low-cost models are available in the deep rural pockets of India that demonstrate approaches to utilize these agricultural wastes into valuable materials productively. Although composting has been traditionally followed in rural India, the proliferation of chemical fertilizers has led to the diminishing utilization of farm waste to produce good-quality manure in recent times. A significant fraction of farm waste can be converted into manure, boosting soil fertility, and consequently, farmers' income.

14.6 Statistics of Stubble Burning in India

The estimates of food grains production in India by the Directorate of Economics and Statistics (DES) in (2019–2020) proposed 370 Mt of sugarcane, 118 Mt of rice, and 107 Mt of wheat (Table 14.2). The Indian farmers burn the bulk crop residues of sugarcane, rice, and wheat. As these crops fetch high revenues, farmers do not prefer cultivating alternative crops producing a lower bulk of residues. According to FAO, 679 Mt of crops were produced in 2014, which generated 916 Mt of residues. Out of these 916 Mt only 47% was used as domestic fuel (Kainthola et al. 2019). As per the Intergovernmental Panel on Climate Change (IPCC) report, on-farm stubble burning is most prominently practised in Uttar Pradesh, followed by Punjab and Haryana (Jain et al. 2014). They also reported that rice accounts for up to 43% of all crop residues burnt, whereas wheat accounts for only 21%. Furthermore, sugarcane contributes 19%, whereas oilseed crops contribute only 5% to stubble burning. India produces 500 Mt of stubble each year, out of which 92 Mt is charred every year (NPMCR report 2014).

According to FAO, farmers from various states of India, i.e. Uttar Pradesh, Punjab, Rajasthan, and Haryana, burn around 35 Mt of rice and wheat stubble during the month of October to November.

This kind of stubble burning approach increased in our country in current years due to use of advance machines that shed straw on field and only gather grains. In the recent years, FAO sponsored few start-ups to collect, store, and process the rice straws into economical and valuable products. Rice straw is rich in C6 and C5 sugars, which are the most favoured energy source used by many fermentative micro-organisms to produce ethanol (Bhattacharyya et al. 2021). There are few reports of addition of binders (cassava starch, sawdust, sodium silicate, and molasses) to improve the strength and durability of rice straw briquettes (Olugbade and Ojo 2021). These alternative commercial uses of rice and wheat straw such as bioethanol production, pellets, and briquettes generation and paper, pulp production offer more options to farmers to double their income. There is an urgent need of strategies by the policy makers and government to create awareness and to convince various farmers to adopt these alternative successful nonburning stubble management practices. A study was reported among the farmers of Thailand to test the psychological factors of farmers to adopt stubble management approaches for socio-demographic developments and climate smart agriculture (Sereenonchai and Arunrat 2022).

Table 14.2 Estimated production of major Indian crops.

Crop	Estimate of production (Mt)
Rice	118
Wheat	107
Sugarcane	370
Oil seeds	33
Cotton	36
Jute	9
Pulses	23

Source: DES (2021).

14.7 The Adverse Impacts of Stubble Burning on the Environment

Stubble burning creates several environmental complications, including greenhouse gas emissions, increased smog and particulate matter (PM), and even soil fertility loss (Lohan et al. 2018). Stubble burning also leads to the loss of organic C, N, and other nutrients generally retained in the soil (Jain et al. 2014). The stubble burning of 98.4 Mt during 2008–2009 resulted in the release of 141.15 Mt CO_2 (91.6% of the total emissions), 8.57 Mt CO, 0.23 Mt NOx, and 0.037 Mt SOx into the atmosphere. Twenty-five lakh farmers charred twenty-three million metric tons of rice stubble in North-West India during the harvest season of 2017 (NAAS 2017). Many stubbles burning from Uttar Pradesh, Punjab, and Haryana during October-December causes excessive PM concentrations and a dense fog throughout the Delhi NCR region (Greenpeace 2019). Particles with aerodynamic diameter between 2.5 and 10 μm are reported to effect adversely on health of a larger population in many Indian cities (Saini and Sharma 2020). $PM_{2.5}$ was found to be one of the most abundant lethal pollutants in NCR region (Pant et al. 2017). In the year 2018, the concentration of $PM_{2.5}$ significantly during 20 October to November last week due to huge stubble burning (Beig et al. 2020). Open stubble burning releases delicate particulate matter and aerosol particles that cause chronic asthma, pneumoconiosis, and bronchitis. Chakrabarti et al. (2019) reported that approximately 14.4% of the severe respiratory infections were ascribed directly to stubble burning.

14.8 Alternative Strategies to Stubble Burning in India

Various sustainable stubble management strategies are practised in Punjab, Haryana, Uttar Pradesh, and West Bengal, as summarized in Figure 14.2. In situ management strategies such as incorporating agricultural stubble into the soil for three weeks before new cropping, enhances soil fertility and organic content. Every year in Punjab, around 14% of rice stubble is incorporated into the soil (Kumar et al. 2015). There are reports that stubble straw contains many essential nutrients for plant growth. Microbial consortia have been inoculated with rice and wheat stubble straw to effect the enrichment of its protein content (Zayed 2018). A May 2019 report by the Ministry of Agriculture and Farmers, Govt. of India stated that 59.79% of paddy stubble straw was managed through soil incorporation in Punjab and Haryana (https://farmech.dac.gov.in/). Borah et al. (2016) reported a simple

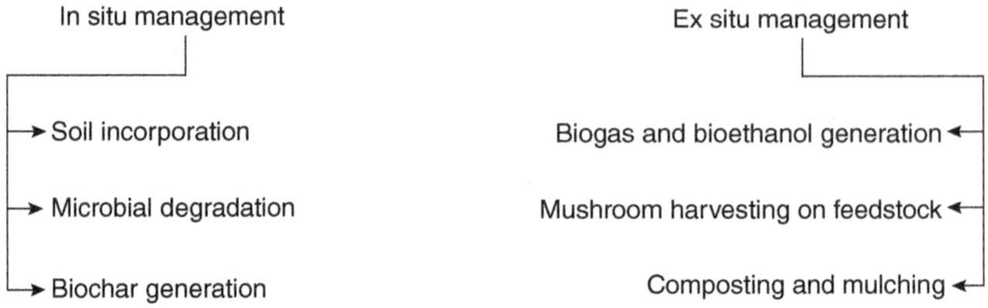

Figure 14.2 Sustainable stubble management strategies.

Table 14.3 Stubble straw based 2G ethanol commercial projects in India (up to August 2020).

Location	2G ethanol capacity	Client
Panipat, Haryana	100 KL/day	Indian Oil
Bargarh, Odisha	100 KL/day	Bharat Petroleum
Bhatinda, Punjab	100 KL/day	Hindustan Petroleum
Davangere, Karnataka	60 KL/day	Oil and Natural Gas Corporation

stubble straw management approach involving in situ decomposition by spraying cellulose-degrading microbial consortia along with *Bacillus* sp. and *Azospirillium* sp. Crop stubble can also be converted to carbon-rich biochar by means of anaerobic pyrolysis (Jyothsna 2019). Biochar increases the soil carbon sequestration capacity. It was reported that a 38–49% decline in CO_2 emissions was attained by biochar production in India (Mohammadi et al. 2016).

Agricultural stubble has also been utilized for biofuel production in various states of India. For example, Sukhbir Agro Energy Ltd. (SAEL) began biomass plants of 30 MW in Faridkot & Ferozepur districts and installed 15 MW plants at Ghazipur, UP, and Muktsar (Channu) Punjab. Furthermore, various bioethanol production plants have been developed by Praj's second generation (2G) cellulosic ethanol technology 'Enfinity', as delineated in Table 14.3.

More than 500 Power plants with a 9075.5 MW capacity, which are partially or entirely fuelled by biomass, have been installed across India (MNRE 2016). For example, Sukhbir Agro Energy Limited established two 18 MW biomass power plants in Ferozepur and Faridkot districts, Punjab, in December 2019. Eleven thermal power plants (97.5 MW) are functional in Punjab that utilizes around 8.8 Lac Mt of paddy straw per annum.

Alternate stubble management strategies such as nano-silica and nano-zeolite production, pulp, and paper production are also practised (Zhang et al. 2017). The use of waste stubble in the construction sector has also been reported (Bories et al. 2015). The Govt. of India supports many biomass-based biogas plants and has launched a new initiative named Sustainable Alternative Towards Affordable Transportation (SATAT). The SATAT initiative tackles stubble burning and converts agricultural stubble to compressed biogas for vehicular fuel (MNRE 2018).

14.9 Riches from Rags' Studies from Rural India: Case Studies of Recycling Farm Wastes into Valuable Products

Case Study 1 Small Steps toward Sustainability

Puran Pao, 48 years, son of Mangal Pao from *Badi Deori*, belongs to a remote tribal village Shahdol district of Madhya Pradesh, India. He lives with a big family of eleven members, farming being their primary source of income. His four children who live with him help him in farming operations. They have a farm area of seven acres where they mainly grow Kharif crops. In the vicinity Rabi and summer crops cannot be taken up due to paucity of water. Single cropping and dependency on farming make them economically vulnerable. Occasionally, they also work as daily wage labourers to fulfil their day-to-day financial requirements.

Puran got to know of importance of manuring in improving soil health and fertility. But due to lack of knowledge, he used to dump cow dung in the open field under the misconception that he would benefit from using natural organic fertilizers. However, he saw the area soon became infested with termites due to the usage of raw cow dung. Moreover, weed infestations also impacted his farm's agricultural productivity. During 2018, some local development agencies made him aware of the modern farming methods including of using good quality far yard manure. One crucial learning for Puran was the utility of a biocompost pit, as it presented a vastly improved and effective composting process, that too at a very low cost with major investment being his own manual labour of an hour or two. He started using good manure from his compost pit, when it became ripe in a span on three months. He kept composting the cow dung and farm waste and utilized the farm yard manure in his field on regular basis. Next cropping season he observed good microbial growth in his farm soil. Seeing the benefit, he gradually constructed 3–4 compost pits and subsequently replaced a significant chunk of DAP and Urea with his compost, thereby reducing his fertilizer consumption by more than 75%. These interventions saved him over INR 15 000 annually in expenses. Puran Pao, who keenly got interested in knowing other natural means like bio-pesticides made of natural ingredients such as neem leaves, kernels, various local leaves, and cow urine. He also got informed about the various advantages biopesticides had over chemical pesticides and how using biopesticides would be advantageous in the long term. Puran says, 'Within two years of shifting to bio-fertilisers and bio-pesticides, I am astonished to observe much lower incidence of weed and termites. I have also seen improvements in agricultural productivity'. Puran does his own composting and maximally relies on well-digested compost on his farm. His farming income has also improved, which has boosted his confidence to continue with farming and fellow farmers are also following his path (Figure 14.3-right)

Figure 14.3 Ramlakhan selling his manure to another farmer (left) and Puran Rao with his newly constructed bio-compost pit (right).

Case Study 2 Ramlakhan Earns from Waste Made Manure

Shahdol is a tribal-dominated district in the Eastern Madhya Pradesh state of India. Farming is the predominant occupation here, with communities relying on traditional farming practices.

With the advent of modern techniques, farmers adopted them, albeit on a limited scale. Ramlakhan Mishra, 45 years, son of Ramsewak Mishra, a Jarwahi village resident grows paddy as a Kharif season crop, which is primarily rain dependent. Harvesting a single crop was not enough to fulfil his daily needs, so Ramlakhan's family kept exploring alternative income routes. However, apart from daily wage labour, they seldom got to explore any other viable income route. Unfortunately, even this additional earning was paused during the 2020 SARS-CoV-2 pandemic induced lockdown. Under these circumstances, he was unsure as to how he could sustain his five-member family. However, even in these testing times, things took a positive turn for Ramlakhan and his family. Some local development agencies skilled him in using farm waste and cattle dung to produce good quality farm yard manure. After being upskilled, Ramlakhan has been establishing compost units for his farm. He observed that many farmers were unaware of making high-quality compost or did not have cattle and thus struggled for manure during the farming seasons. Seeing a business opportunity, he started mass production of compost by constructing more compost pits. He has now become a known seller of compost in his village as well as the nearby villages. He sells good quality compost at Rs. 15/kg. He even offers bulk discounts (Rs. 5/kg) to farmers ordering over 10 kg of manure. He has already sold compost worth Rs. 10 000/ till the current year, which is a handsome amount given the minimal investments he had to make (Figure 14.3-left).

Ramlakhan says, 'farming is like a cyclical endeavour because every product in the entire agriculture supply chain is of use to him. It feeds his family and earns him a substantial income'. He plans to enhance further his earning from farm waste by vermicomposting, for which he has also been trained by the local government agency. Farmers like Ramlakhan show the path to fellow farmers on ways to utilize farm waste and improve earnings. The manure made by Ramlakhan is shown in Figure 14.4.

Over the past century, aggressive industrialization, exponential population growth, and technological developments have resulted in environmental pollution, global warming, and ecosystem destruction. We as a species are standing at a highly critical juncture in the history of human civilization, where we need to carefully consider the detrimental environmental consequences of our actions to ensure that our existence as a species is not threatened for the decades and centuries to come.

Agriculture is one of the pillars of human civilization. The development of agriculture resulted in the shift of ancient man from a hunter-gatherer society to an agrarian society. This transition was critical to catalysing the formation of increasingly complex societal structures. It is not surprising that much of the complexity and efficiency underlying modern societal systems exist due to agriculture. Agriculture, has therefore, since time

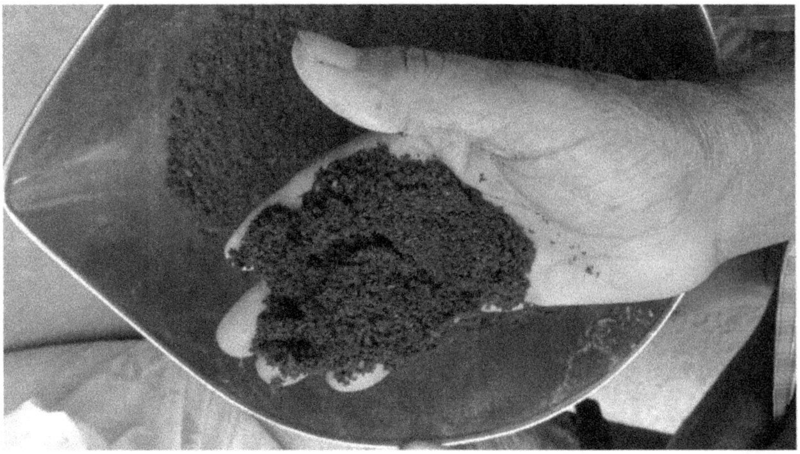

Figure 14.4 Manure made by Ramlakhan.

immemorial, played a critical and central role in shaping the course of human civilization.

We require agriculture to ensure food security, maintain global economies, and feed supply chains requiring the input of natural substrates. Agriculture is also instrumental in ensuring the future generations of this world's fast-growing population do not face malnutrition and starvation. There have been steady developments in farming processes since the Green Revolution in our country. Scientific advancements in several disciplines such as plant biotechnology, bioprocess engineering, microbiology, metagenomics, and biochemistry have allowed us to increase the productivity as well as the nutrient richness of our produce.

However, all this fast-paced agricultural development has a flipside. Agricultural waste and the lack of effective agricultural waste management is a burning problem in our country. Positive steps towards a resolution are, however, being taken at an increasingly accelerated pace. The deeply entrenched perception that agricultural waste is a valueless product is changing fast as more perspectives with regards to agricultural waste management are being designed, implemented, and even monetized. Several exciting, novel, and innovative approaches to agricultural waste management have surfaced in recent years.

The wide range of current options for effective agricultural waste management range from the development of low-value high-volume products like handmade paper, organic manure, and biochar to the industrial-scale production of high-value low-volume products such as cellulases. Some effective agricultural waste management approaches even involve the production of edible substances such as microbial proteins, oyster mushrooms, and soy-based food products. More and more innovative approaches are expected to surface with the progression of time.

The Sustainable Development Goals (SDGs) are a constellation of seventeen interconnected targets for humanity, as illustrated in Figure 14.5. The SDGs present a shared blueprint for peace, prosperity, and a more sustainable future. Agricultural waste

Figure 14.5 The sustainable development goals.

management and shifting to a bio-based sustainable economy positively impact our progress towards each goal and pushes humanity substantially closer to a brighter future.

The conversion of agricultural waste into economically valuable products reinforces the positive economic impact of recycling agricultural waste. This is particularly beneficial for marginal farmers (farmers having less than one hectare of land) and other financially downtrodden communities who may have access to agricultural waste materials but lack the knowledge and expertise to convert the waste to wealth. The two case studies discussed above reinforce this assertion and provide real-life context into how robust a tool the proper knowledge of agricultural waste management can be in the fight against poverty and hunger. Ramlakhan's case study is interesting as it illustrates how economically viable biofertilizer and biopesticide production can be carried out in a domestic setting. Biofertilizer and biopesticides are two very prominent examples of economically valuable substances produced from recycling agricultural waste products. Wide scale adoption of these will incentivize innovation, entrepreneurship, and promote the upward economic mobility of society's economically downtrodden strata. Moreover, it will also ameliorate the financial pressure farmer's face due to the recurring expenses of costly chemical fertilizers. Furthermore, wide scale adoption of biofertilizers and biopesticides will significantly benefit the environment as the deleterious environmental impacts of chemical fertilizers and insecticides can be avoided to a significant extent.

The traditional approach to dealing with agricultural waste was its incineration or disposal. Although effective in the short term, these unsustainable approaches to dealing with a recurring problem eventually resulted in notable environmental pollution and significantly increased public health risks. The damaging aspects of this practice gained particular attention over the past few years after the gravity of the air pollution situation in Delhi and the NCR grabbed the national media's attention. Even though the government and private bodies have taken a variety of steps to facilitate this massive problem, it is still far from a resolution. Valorization of this agricultural waste would prevent the release of all the biomass into the atmosphere as harmful carbon dioxide and instead be redirected into other valuable supply chains, several examples of which have been delineated in Table 14.4. One of the most exciting aspects underlying this proposition is that agricultural waste valorization is a pivotal step in producing biofuels and other chemicals directly or indirectly fuelling our industrial growth and expansion.

Finally, looking at the bigger picture, the social, economic, and cultural benefits of agricultural waste management carried out at scale promises an unprecedented level of promise and hope. The upward economic mobility of the socially downtrodden will result in a notable improvement in their quality of life. This will incentivize education, personal development, sustainable city and community development, and more environmentally responsible production-consumption patterns. These positive changes occurring at the grassroot level of society can, in the long run, potentially pave the way for peace, justice, and strong institutions promoting equality, growth, and socio-cultural inclusivity. We can be optimistic that national prosperity will increase as a consequence, which may, in turn, promote the development of stronger international collaborations, ties, technology transfer agreements, and hopefully a brighter tomorrow.

Humanity is just starting to realize the transformative potential agricultural waste management can have not just in the short term but also in the medium and long term. With the recent focus and attention agricultural waste management has had in the past several years, we can be optimistic about the future of humanity and the planet.

Table 14.4 Some innovative technologies undertaken by the agricultural scientists of India to convert waste into economically valuable products.

S.N.	Technology developed	Innovator(s)	Institute affiliation of innovator(s)	Description and salient feature(s)
1.	Handmade Paper Production from Jute Waste	1) S.N. Chattopadhyay	ICAR-NINFET, Kolkata	The development of technologies that allow the production of handmade paper from residual jute fibre will discourage its disposal as waste. Instead, it can be utilized to make handmade paper of substantial commercial value.
2.	Lac Mud based Organic Manure	1) A.K. Singh 2) S. Ghosal	ICAR-IINRG, Ranchi	The raw material, sticklac, is processed by the lac processing industry to generate the final product. Lac mud is a waste product of this process that has several properties that make it excellent organic manure.
3.	Biochar from Agricultural Waste Material	NA	Division of Agricultural Engineering	Biochar is a form of charcoal produced by biomass pyrolysis in anaerobic conditions and is often used to enrich the soil. Agricultural biomass can be processed to biochar using a modified portable metallic kiln within 2 h at a sustainable conversion efficiency (25–35%).
4.	Broken Rice and Dal based Fortified Rice Analogues	1) Udaykumar Nidoni 2) Ambrish G 3) P F. Mathad 4) Sudha Devi G	University of Agricultural Science (UAS), Raichur (AICRP on PHET)	Rice kernels broken in the milling process are not accepted by consumers and hence contributes to food wastage. Using extrusion technology, milling by-products such as broken dal and rice are mixed to produce a composite flour-based fortified rice analogue.
5.	Microbial Protein using Corn Cob	1) Surya Tushir 2) S.K. Tyagi 3) V. Chanasekar, 4) A.K. Jaiswal	ICAR-CIPHET, Ludhiana	Microbial protein is the dried cells of bacteria, fungi, yeast, and algae, used as a dietary protein supplement. Scalable and inexpensive production of microbial protein requires cheap yet readily available substrates. Lignocellulosic waste like discarded corn cobs constitutes an excellent carbon source required for microbial protein production.
6.	Industrial Scale Cellulase Production by Microbial Fermentation of Groundnut Shells	NA	ICAR-Directorate of Groundnut Research, Junagadh	Fungal strains such as *Trichoderma* sp. and *Phanerochaete* sp. are commonly used for industrial scale cellulase production using solid-state and submerged fermentation. Since the purified cellulase enzyme is a low volume-high value product of high economic significance, easily accessible low-cost substrates are desirable to make the bioprocess as financially lucrative as possible. Using discarded groundnut shells as the feedstock for an industrial scale cellulase production bioprocess is an excellent example of waste to wealth generation.

(*Continued*)

Table 14.4 (Continued)

S.N.	Technology developed	Innovator(s)	Institute affiliation of innovator(s)	Description and salient feature(s)
7.	Oyster Mushroom Cultivation using Cotton Stalks	NA	ICAR-CIRCOT Mumbai	Oyster mushrooms are a highly nutritious type of mushroom having significant culinary and economic value. Growing oyster mushrooms on lignocellulosic biomass such as cotton stalks, the primary leftover waste product of cotton harvesting, can appreciably curb agricultural waste production and present a financially lucrative opportunity to various farming communities.
8.	Bioconversion of Organic Waste into Low-cost Phospho-Sulpho-Nitro (PSN) Compost	1) M. C. Manna, 2) J. K. Thakur, 3) B. Singh, 4) Sahu, 5) S. Bhattacharjya, 6) Asit Mandal, 7) K. Tripathi, 8) Ashok K. Patra	ICAR-IISS, Bhopal	Nutrient poor organic waste such as crop residues and horticultural wastes can be effectively bioconverted into nutrient-enriched phospho-sulpho-nitro (PSN) compost capable of enriching the soil and enhancing its productivity, thus promoting sustainable agricultural practices.
9.	Kulhad Making using Corn Cob Powder	1) Surya Tushir 2) S.K. Tyagi 3) V. Chanasekar, 4) A.K. Jaiswal	ICAR-CIPHET, Ludhiana	Corn cobs, a waste product, are powdered and mixed with mud in varying proportions. The mixture is subsequently used to make earthen cups, thus converting agricultural waste into a value-added product.
10.	Soy-based food Product Development Using Okara	1) L.K. Sinha 2) Sumedha Deshpande	ICAR-CIAE, Bhopal	Okara is the by-product of soy milk and tofu production, consisting of insoluble soybean parts. Instead of disposing of the okara, which has considerable nutritional value, using it to prepare soy-based food products like biscuits, cakes, muffins, idli, dosa can significantly help reduce soy processing waste.
11.	Livestock Enteric Methane Emission Reduction using Tamarind Seed Husks	1) Raghavendra Bhatta 2) P.K. Malik 3) A.P. Kolte	ICAR-NIANP, Bengaluru	Methane is a greenhouse gas, and ruminant livestock methane emissions is a known contributor to global warming. A livestock diet constituting 5% tamarind seed husks has been documented to reduce enteric methane emissions by up to 20%. Tamarin Plus, a product developed using tamarind seed husk, can be added ad libitum to the feed of lactating, growing, and adult animals to reduce the greenhouse gas footprint of livestock rearing.

12.	Pineapple Fruit Residue (PFR) Silage-based Livestock Fodder	1) N.K.S. Gowda 2) D.T. Pal 3) S. Anandan 4) C.S. Prasad	ICAR-NIANP, Bangalore	Only 30% of the pineapple fruit is edible, while the remaining 70% is the pineapple fruit residue (PFR) consisting of the inedible crown, peels, and pulp. Although PFR can be used as cattle fodder, it has a high moisture and sugar content, making it very susceptible to fungal growth and thus challenging to store. Any crop preserved by partial fermentation while maintaining its succulence in airtight storage such as a silo is referred to as silage. Technology to prepare silage from PFR in drums and bags to market as livestock fodder has been licenced by an organization named M/s. Fresh Fruits Processing Pvt. Ltd. to ameliorate the problem of rapid PFR spoilage and ultimately reduce edible waste generation.
13.	Palm Oil Mill Waste-based Alternative Animal and Fish Feeds	1) M. Kochu Babu 2) M. Vijayalakshmi, ANU, Guntur 3) P. Kumaraiah, 4) P. V. Rangacharyulu 5) E. Raghava Rao 6) Shinoj Subramannian	1) ICAR-IIOPR, Pedavegi, Andhra Pradesh 2) ANU, Guntur 3) ICAR- CIFA, Odisha 4) ICAR- CIFA, Odisha 5) NTR College of Veterinary Science, Gannavaram 6) ICAR-IIOPR, Pedavegi	Palm oil mill leftovers can serve as a viable yet cheaper alternative to traditional animal husbandry and pisciculture feeds without having any detrimental effects such as depressed growth rate or reduced milk production.
14.	Banana Peel-based Food Products	1) P. Suresh Kumar 2) K.N. Shiva 3) Amelia Keran D 4) K. Kamaraju 5) S. Uma	ICAR-NRCB, Tamil Nadu	The banana peel accounts for nearly 25–35% of the banana fruit. Although it is deemed inedible by humans, it is rich in nutrients and antioxidants. Therefore, banana peels can serve as a raw ingredient to prepare various food products like pickles, muffins, and even pizza bases.
15	Cotton stalk Briquetting and Pelleting Technology	NA	ICAR-CIRCOT	Briquettes and pellets made from cotton stalks are suitable for applications requiring renewable and energy-rich fuel sources. For example, briquettes can be used as a viable substitute fuel in appliances meant for burning wood or coal.
16.	Decorative Wall Panels made from Banana Pseudostem Fibres and Natural Binders	NA	ICAR-CIRCOT	Banana pseudostem fibres are easy to cut into any shape or size. Moreover, they can be coloured using a novel colouring technology to make attractive decorative items such as wall panels using eco-friendly binders.

(*Continued*)

Table 14.4 (Continued)

S.N.	Technology developed	Innovator(s)	Institute affiliation of innovator(s)	Description and salient feature(s)
17.	Eco-friendly Natural Fibre Biomass-based Paper Pates	NA	ICAR-CIRCOT	Technologies have been developed to produce disposable and biodegradable paper plates from discarded agricultural feedstock-like banana pseudostem fibres and sugarcane bagasse. The plastic plate coating has also been replaced by a novel bee or sugarcane wax-based material that is functionally comparable to plastic coats but is devoid of detrimental environmental impacts.
18.	Accelerated Bio-enriched Compost Preparation Process	NA	ICAR-CIRCOT	The institute has developed an accelerated process of cotton stalk composting using microbial consortium. Due to this process, composting of cotton stalk happens 15–30 d faster than regular composting. It also has a higher NPK value compared to regular composting, resulting in better soil health. This manure further reduces chemical fertilizer use, prevents the burning of cotton stalks.
19.	Mushroom Production Using Palm Oil Factory Waste as Substrate	M. Kochu Babu	ICAR-IIOPR, Pedavegi, Andhra Pradesh	The mesocarp waste produced from palm oil factories provides a suitable substrate that promotes the growth of edible mushrooms such as paddy straw mushrooms, oyster mushrooms, summer white milky mushrooms, summer white button mushrooms and many more. Furthermore, the mesocarp waste can be used directly as a growth substrate or developed into compost before mushroom cultivation.
20.	Cheap Pomegranate Peel Waste Extract-based Poultry Feed Supplement	1) Sarbaswarup Ghosh 2) Paresh Nath Chatterjee	1) Sasya Shyamala Krishi Vigyan Kendra (SSKVK), W.B 2) West Bengal University of Animal & Fishery Sciences (WBUAFS), Kolkata	Pomegranate peels are rich in polyphenols having antimicrobial, antioxidant, and hepatoprotective properties. However, they are mostly discarded by humans in favour of the juicy edible seeds. The valuable phytochemicals can be extracted from the discarded pomegranate peels using a hot infusion method-based aqueous extraction technique. Furthermore, daily supplementation of pomegranate peel infusions in poultry feed is beneficial for both the broiler and layer birds.

References

Abdurrahman, M.I., Chaki, S., and Saini, G. (2020). Stubble burning: effects on health and environment, regulations and management practices. *Environmental Advances* 2: 100011.

Beig, G., Sahu, S.K., Singh, V. et al. (2020). Objective evaluation of stubble emission of North India and quantifying its impact on air quality of Delhi. *Science of the Total Environment* 709: 136126.

Bhattacharyya, P., Bisen, J., Bhaduri, D. et al. (2021). Turn the wheel from waste to wealth: economic and environmental gain of sustainable rice straw management practices over field burning in reference to India. *Science of the Total Environment* 775: 145896.

Bhuvaneshwari, S., Hettiarachchi, H., and Meegoda, J.N. (2019). Crop residue burning in India: policy challenges and potential solutions. *International Journal of Environmental Research and Public Health* 16: 832.

Borah, N., Barua, R., Nath, D. et al. (2016). Low energy rice stubble management through in situ decomposition. *Procedia Environmental Sciences* 35: 771–780.

Bories, C., Aouba, L., Vedrenne, E., and Vilarem, G. (2015). Fired clay bricks using agricultural biomass wastes: study and characterization. *Construction and Building Materials* 91: 158–163. https://doi.org/10.1016/j.conbuildmat.2015.05.006.

Chakrabarti, S., Khan, M.T., Kishore, A. et al. (2019). Riskof acute respiratory infection fromcrop burning in India: estimatingdisease burden and economic welfare from satellite and nationalhealth survey data for 250 000 persons. *International Journal of Epidemiology* 48: 1113–1124. https://doi.org/10.1093/ije/dyz022.

Directorate of Economics and Statistics (2021). https://eands.dacnet.nic.in/PDF/Monthly%20bulletin%20April%202021%20final.pdf (accessed 25 June 2021).

FAO (2001a). Food and Agriculture Organization of the United Nations (FAO) Wheat: Post-harvest Operations (2001). http://www.fao.org/3/a-ax448e.pdf (accessed 5 July 2019).

FAO (2015b). Food and Agriculture Organization of the United Nations (FAO) Prospects by Major Sector Crop Production. http://www.fao.org/3/y3557e/y3557e04.pdf (accessed 8 July 2019).

FAO (1999). Food and Agriculture Organization of the United Nations (FAO) RICE Post-harvest Operations (1999). http://www.fao.org/3/a-ax442e.pdf (accessed 5 July 2019).

FAO (2003). Food and Agriculture Organization of the United Nations (FAO) Maize: Post-harvest Operations. http://www.fao.org/3/a-av007e.pdf (accessed 6 July 2019).

Greenpeace (2019. https://www.greenpeace.org/india/en/press-center/ (accessed 4 August 2019).

Jain, N., Bhatia, A., and Pathak, H. (2014). Emission of air pollutants from crop residue burning in India. *Aerosol and Air Quality Research* 14: 422–430.

Jyothsna, J. 2019. Biochar: An Ingredient to Redress Stubble Burning and Boost Crop Production. *International Journal of Current Microbiology Applied Science* 8(12): 20-27. doi: https://doi.org/10.20546/ijcmas.2019.812.004

Kainthola, J., Kalamdhad, A.S., and Goud, V.V. (2019). A review on enhanced biogas production from anaerobic digestion of lignocellulosic biomass by different enhancement techniques. *Process Biochemistry* 84: 81–90.

Kumar, P., Kumar, S., and Joshi, L. (2015). The extent and management of crop stubble. In: *Socioeconomic and Environmental Implications of Agricultural Residue Burning, Springer Briefs in Environmental Sciences*, 13–34. New Delhi, India: Springer.

Lohan, S.K., Jat, H.S., Yadav, A.K. et al. (2018). Burning issues of paddy residue management in north-west states of India. *Renewable and Sustainable Energy Reviews* 81: 693–706.

Ministry of New and Renewable Energy Resources (2016). Govt. of India, New Delhi. www.mnre.gov.in/biomass resources (accessed 29 May 2020).
Mohammadi, A., Cowie, A., Mai, T.L.A. et al. (2016). Biochar use for climate-change mitigation in rice cropping systems. *Journal of Cleaner Production 116*: 61–70.
NAAS (2017). Innovative viable solution to rice residue burning in rice-wheat cropping system through concurrent use of super straw management system-fitted combines and turbo Happy Seeder. Policy Brief No. 2. National Academy of Agricultural Sciences. https://naasindia.org/documents/CropBurning.pdf (accessed 10 September 2019).
NPMCR (2014). NPMCR_1.pdf (agricoop.nic.in). http://agricoop.nic.in/sites/default/files/NPMCR_1.pdf (accessed 6 March 2019).
Olugbade, T.O. and Ojo, O.T. (2021). Binderless briquetting technology for lignite briquettes: a review. *Energy, Ecology and Environment* 6 (1): 69–79.
Pant, P., Habib, G., Marshall, J.D., and Peltier, R.E. (2017). $PM_{2.5}$ exposure in highly polluted cities: a case study from New Delhi, India. *Environmental Research* 156: 167–174.
Saini, P. and Sharma, M. (2020). Cause and age-specific premature mortality attributable to $PM_{2.5}$. Exposure: an analysis for Million-Plus Indian cities. *Science of the Total Environment* 710: 135230.
Sereenonchai, S. and Arunrat, N. (2022). Farmers' perceptions, insight behavior and communication strategies for rice straw and stubble management in Thailand. *Agronomy* 12 (1): 200.
Zayed, M.S. (2018). Enhancement the feeding value of rice straw as animal fodder through microbial inoculants and physical treatments. *International Journal of Recycling of Organic Waste in Agriculture* 7: 117–124. https://doi.org/10.1007/s40093-018-0197-7.
Zhang, H., Hu, J., Qi, Y. et al. (2017). Emission characterization, environmental impact, and control measure of $PM_{2.5}$ emitted from agricultural crop residue burning in China. *Journal of Cleaner Production* 149 (1): 629–635. https://doi.org/10.1016/j.jclepro.2017.02.092.

Web Links

Agri Farming (2021). Government schemes for organic farming in India. Last updated 19 March 2021. https://www.agrifarming.in/government-schemes-for-organic-farming-in-india.
MNRE (2018). https://pib.gov.in/newsite/PrintRelease.aspx?relid=183842.
TNAU Agritech Portal (2016). *Organic Farming: Organic Inputs and Techniques*. https://agritech.tnau.ac.in/org_farm/orgfarm_inputs_tech.html (accessed 5 September 2021).

Index

Agriculture Waste Management and Bioresource: The Circular Economy Perspective, First Edition.
Edited by Suruchi Singh, Pardeep Singh, Anu Sharma, and Moharana Choudhury.